Soils of the World

Soils of the World

by J. PAPADAKIS

Soil Geographer, Instituto de Suelos y Agrotecnia,
Buenos Aires, Argentina

ELSEVIER PUBLISHING COMPANY

Amsterdam   London   New York

1969

Elsevier Publishing Company
335 Jan van Galenstraat, P.O.Box 211, Amsterdam, The Netherlands

Elsevier Publishing Company, Ltd.
Barking, Essex, England

American Elsevier Publishing Company, Inc.
52 Vanderbilt Avenue, New York, N. Y. 10017

Library of Congress Card Number: 67-12778

Standard Book Number 444-40709-x

With 46 illustrations and 9 tables

Printed in The Netherlands

Preface

The thin mantle of fine earth that covers land surfaces, suffers the influence of weather, maintains vegetation, and provides mankind with food and other goods, has a tremendous scientific and practical importance. We cannot imagine a physical or economic geography that ignores soils. Geology cannot ignore the processes that take place on the earth's surface. Agriculture, animal husbandry and forestry are often defined as the art of using soil.

The systematic study of soil, as a natural body, is a young science. Modern pedology is hardly 100 years old; and it is only 50 years ago that it spread outside Russia. Nevertheless all countries have their pedologic services, more or less detailed maps of their soils, and it is taught in almost all universities.

These are certainly great achievements. But they should not prevent us from seeing the many gaps that exist in our understanding of soils.

Although much work has been done recently in experimental pedology, the results of this research have not yet sufficiently influenced our understanding of soil genesis. In spite of much work done in humus chemistry, humus fractionation has not yet entered routine pedologic analysis in many countries, and the quality of humus is not taken in consideration in many soil classifications. There is a wide gap between soil morphology and classification on the one hand and soil genesis on the other; many classifications are not sufficiently genetic. An analogous gap exists between the pedologic study of soils and that of their agricultural potentialities; they are often determined empirically with little reference to soil genesis and classification. Finally, soil horizons and groups are so loosely defined that the same horizon, or soil, are often classified very differently by different persons, applying the same definition.

In this respect the situation has greatly improved in recent years, due to the preparation of the 7th Approximation by the Soil Survey Service of the United States. Diagnostic horizons have been introduced. Both diagnostic horizons and soil taxa are precisely defined. We may disagree with the nomenclature of this approximation, and we may prefer a different system of classification, but the introduction of diagnostic horizons, and of the principle that all classification taxa and all terms should be defined precisely, marked an era in the evolution of pedology.

After its spreading outside Russia pedology developed separately in different countries and various schools have been formed. The concept of each soil group, even of soil processes, horizons, etc., differs substantially from country to country. Differences are so profound, that until recently understanding between pedologists was very difficult.

As a consequence of this situation, and of the fact that soil units and other pedologic terms are loosely defined, the preparation of international maps, including various countries, a continent or the whole world, was an almost impossible task. And almost all world maps that appeared hitherto are very deficient.

Fortunately this situation has greatly improved in recent years. Contacts between pedologists of different countries, meetings, etc., became more frequent. International technical assistance has given the opportunity to pedologists of different schools to work together, with the same soils, during many months, often years. And what has been still more decisive, FAO and UNESCO decided to prepare a world soil map and founded the World Soils Resources Office of FAO.

Naturally the work of an institution depends on the man who directs it, and this office has encountered in Doctor L. Bramão the right man. Panels including the most outstanding pedologists of the world have been formed and charged to prepare regional, continental, or world maps. Working together and obliged to prepare a map acceptable to all, pedologists began to understand one another and to unify their concepts. The result is that the various pedologic schools approach one another now, dropping what is wrong in each of them, and accepting what is right in others. Possibly the maps prepared are not perfect; in order to obtain a general agreement, a certain compromise is necessary, and compromise is not always the best way to solve scientific problems. But the progress achieved in mutual understanding, unification of concepts, and improving each school, by taking advantage of the experience and ideas of the others, is far more important than the maps themselves.

This book is a modest contribution to fill the gaps previously mentioned. And its preparation has been possible due to the fore-mentioned progress.

It begins with soil genesis and more especially with weathering, which, by determining the type of clay, determines the type of soil; recent research has considerably elucidated this problem. We proceed with humification and types of humus, leaching and biological accumulation of bases in soil surface, clay eluviation, braunification, rubification, formation of concretionary horizons and laterite, gleization, solonization, podsolization, etc. And on the basis of all that, the influence of climate, drainage, parent material and time on soil type is examined. As the reader will observe, we can now have a coherent theory of soil genesis that may be used as basis of soil classification, etc.

Following the example of the 7th Approximation we give great emphasis to diagnostic horizons. Special attention is given to the type of clay; "adjusted cation exchange capacity/ clay" ratio is used as a criterium; the influence of organic matter is discarded by applying the criterium to a lower horizon and extending the characterization to overlying ones. As the reader will observe, the author's diagnostics are diagnostics of genetic processes; that is why their definition may be both precise and simple. And since soil type is determined by soil formation, they also serve as diagnostics of soil taxa.

In soil classification the author tries to depart as little as possible from the concepts and nomenclature consecrated by time. All taxa are precisely defined on the basis of characters that are easy to measure, chiefly chemical, and which are determined in routine analyses. But since the classification is genetic the author's definitions are both precise and simple.

Great emphasis is given to the type of clay: 1 : 1 clays that characterize the kaolisols (tropical soils formed under heavy rainfall); 2 : 1 clays that characterize the soils of temperate climates; and amorphous clays that characterize ando; cation exchange / clay ratio serves as a criterium. In the subdivision of kaolisols (tropical soils) emphasis is given to cation exchange capacity and base status, both in relation to clay content, presence of concretionary horizons or laterite; palaeo-kaolisols and palaeo-lateritic form special subdivisions. The rubified soils with 2 : 1 clays of dry climates are separated from those of grassland (chernozemic) and form the group of cinnamonic soils (korichnevie). Clay illuviation is considered of great importance in soil classification, when it results in distinctly higher cation exchange capacity. Planosols, in which clay illuviation is due to natrium, are separated from podsolic. Following the example of the 7th Approximation, we separate podsolic from podsols. Gleisolic soils form a special group. All groups are further and further subdivided, but since soils form a continuum, some subgroups belong simultaneously to two or more higher groups. Soils are always classified on the basis of their properties. For each soil group, major or minor, numerous examples of profiles are given that have been described in the world literature; and the correlation with various classification systems are discussed[1]. That gives the reader a concrete idea of what each soil is and facilitates correlation. It also shows the adequacy of the diagnostics and definitions used.

The problem of a rational nomenclature is solved by using symbols formed with letters, each of which shows a definite soil character; being brief these symbols are printed in the maps within each region; this fact greatly facilitates both the reading and the printing of maps. Since many types of soils are encountered in a small area and individual mapping

---

[1] Correlation with the soil units proposed for the soil map of the world is not discussed because the document that has been circulated is not for reference.

of each taxum is often impossible, the concept of soil region is emphasized and the fundamental patterns of soil distribution, according to climate, vegetation, parent material, etc., are described.

A special chapter deals with the soils of each country and their agricultural potentialities. Regional or country maps always accompany such descriptions. The same legend is used in all maps. This is, of course, the longest chapter of the book.

The question of agricultural potentialities is approached from an ecologic point of view[1], taking account of the ways soil affects plant growth, and considering soil as a part of the whole environment. Special emphasis is given to parent material and soil genesis. The agricultural potentialities of each soil group are discussed separately. And since the environment forms an indivisible whole, and the ecologic influence of the same soil varies according to climate, the question is also discussed region by region.

Finally the question of soil survey is treated, giving emphasis to the concept of land type, and inviting surveyors to take better advantage of the correlation between soil characteristics and crop behaviour, as shown by land use and agricultural experience.

The subject index helps to find the definition of each soil group, higher or lower, its synonyms, examples, profile descriptions in world literature, correlation, etc.; it helps also to find the definition of soils, diagnostics, and other terms, description of processes, etc.; it also serves as a vocabulary. A geographic index shows the page numbers on which the soils of each country or region are dealt with. A bibliography and author index complete this part of the book.

The idea to publish this book is due to the outstanding geographer W. van Royen, ex-Professor of the University of Maryland, now Head of the Division of Environmental Sciences of the United States Army, author of well-known treatises of geography and of *Agricultural Resources of the World*; he also edited a great part of the manuscript. In 1964 I published a small book (141 pp.) *Soils of the World*. I sent this book to Dr. Van Royen, who appreciated its clarity and conciseness, in spite of its comprehensiveness, and considered it very useful not only to pedologists, agronomists and geographers, but also as a textbook for teaching pedology in the universities; he recommended me to find a publisher for a commercial edition. This is the origin of the book and I am greatly indebted to Dr. W. van Royen.

I am also indebted to Dr. J. Vallega of FAO, who gave me the opportunity to study the soils of West Africa and check my concepts on tropical soils; to the Instituto de Technología Agropecuaria and Instituto de Suelos y Agrotecnia de Buenos Aires, Argentina, who gave me the opportunity to study the very interesting soils of Argentina; and to the World Soil Resources of FAO for very valuable information.

J. Papadakis

---

[1] The author is also crop-ecologist.

Contents

PREFACE.   V

CHAPTER 1. SOIL FORMATION   1
  Clay synthesis   1
    Types of weathering   1
  Humification   3
    Types of humification.   3
  Leaching: eluviation–illuviation   5
    General considerations   5
    Leaching rainfall   5
    Decalcification and calcification   5
    Acidification.   7
    Biologic migration of substances from the lower horizons to the surface . .   7
    Clay illuviation   7
    Iron eluviation.   9
    Humus eluviation   9
    Formation of pans   9
  Processes adding mineral substances to soil profile   10
    Capillarity rise of water from a water table.   10
    Flooding and irrigation   10
    Moisture regime   10
    Lateral drainage   10
    Aeolian additions: dust   10
  Other processes involving iron   11
    Formation of iron silicates or free iron: ferrugination, rubification, braunification   11
    Formation of concretionary horizons   11
    Gleization.   12
    Iron and clay eluviation by gleization.   12
    Formation of laterite   12
  Podsolization   12
    General considerations   12
    Conditions favouring the accumulation of raw humus   13
    Podsolic weathering   13
    Podsolic eluviation of clay.   13
    Bleaching (iron eluviation).   13
    Humus eluviation   13
    Gleization.   14
    Formation of pans (by podsolization)   14
    Depth and rapidity of podsolization   14
    Modification of podsols and podsolic soils by cropping   14
  Influence of climate on soil formation.   14
  Influence of drainage and moisture regime on soil formation.   15
  Influence of parent material on soil formation   15
  Influence of time on soil formation: polygenesis   15

CHAPTER 2. SOIL DIAGNOSTICS.   17
  Introduction.   17
  Degree of leaching   17

Degree of weathering: mineralogical composition of the non-clay fraction . . . 18

Mineralogical composition of the clay fraction: adjusted Tca/clay ratio 18

Base status: adjusted S/clay ratio. 19

Diagnostic horizons 19

    Horizons according to clay mineralogy 19

    Horizons produced by eluviation–illuviation 21

    Horizons according to base status and salt content 23

    Horizons according to colour and iron form 23

    Horizons according to organic matter content and composition 24

    Concretionary horizons and pans. 24

    Miscellaneous horizons 24

    Horizons according to texture 24

    Horizons nomenclature: meaning of some terms 25

Other methods of soil diagnosis 25

CHAPTER 3. SOIL CLASSIFICATION: INTRODUCTION. SOILS WITH DEGREE OF LEACHING 1–2

(DOMINATED BY 2:1 CLAYS AND NOT STRONGLY ELUVIATED) 27

Introduction. 27

    Natural and artificial classifications. 27

    Characteristics of natural classifications 27

    Soils form a continuum, soil classification cannot be rigid 27

    The names and other terms used should be precisely defined 28

Nomenclature 29

Soil description 29

Highest groups of soil classification and their diagnostics 29

Soil symbols. 32

Rendzina group (soils with degree of leaching 1) 32

    General considerations 32

    General diagnostics. 33

    Subgroups and their diagnostics 33

    Correlation and examples 33

Ando group (soils rich in amorphous clays) 34

    General considerations 34

    General diagnostics 34

    Subgroups and their diagnostics 34

    Correlation and examples 34

Dark clays (soils rich in 2:1 clays, well-saturated with bases) 35

    General considerations 35

    General diagnostics. 35

    Subgroups and their diagnostics 35

    Correlation and examples 35

Raw (undifferentiated) soils (undifferentiated soils with degree of leaching 2). . 36

    General considerations 36

    Subgroups and their diagnostics 36

    Correlation and examples 37

Rankers (soils with degree of leaching 2 consisting of a humic horizon resting on

bed-rock or permafrost). 37

    General considerations and diagnostics 37

    Subgroups and their diagnostics 37

    Correlation and examples 38

Brunisolic group (soils with degree of leaching 2 and "braunified" or "acid"

horizons) ·. 38

    General considerations 38

    General diagnostics. 38

    Subgroups and their diagnostics 38

    Correlation and examples 39

Cinnamonic group (rubified soils with degree of leaching 2) 39

    General considerations 39

General diagnostics. 39
Subgroups and their diagnostics 40
Correlation and examples 40
Chernozemic group (soils with degree of leaching 2 and deep neutral dark humic
horizons) 41
General considerations 41
General diagnostics. 41
Subgroups and their diagnostics 41
Correlation and examples 42

CHAPTER 4. SOIL CLASSIFICATION: SOILS WITH DEGREE OF LEACHING 3–6 (DOMINATED BY
1:1 CLAYS AND/OR PODSOLIZED). 45
Kaolisols (tropical group) (soils with 1:1 clays) 45
General considerations 45
Classification 45
General diagnostics. 46
Subgroups and their diagnostics 46
Correlation and examples 47
Podsolic (lessivé) soils (soils with moderate podsolization) 50
General considerations 50
General diagnostics. 50
Subgroups and their diagnostics 50
Illitic podsolic, 50 — Kaolinitic podsolic, 51
Correlation and examples 51
Podsols (soils with advanced podsolization) 52
General considerations 52
General diagnostics. 53
Subdivisions and their diagnostics 53
Correlation and examples 53

CHAPTER 5. HALOMORPHIC, GLEISOLIC AND ORGANIC SOILS. 55
Introduction. 55
Solonchacks. 55
General considerations 55
General diagnostics. 55
Subgroups and their diagnostics 55
Gypsisols 55
Correlation and examples 55
Solonetzic–planosolic group (soils with eluvial horizons due to Na) 56
General considerations 56
General diagnostics. 56
Subgroups and their diagnostics 56
Correlation and examples 56
Organic soils 58
General considerations 58
Correlation and examples 58
Gleisolic soils 59
General considerations 59
General diagnostics. 59
Subgroups and their diagnostics 60
Correlation and examples 60

CHAPTER 6. FUNDAMENTAL PATTERNS OF SOIL DISTRIBUTION. BROAD SOIL REGIONS OF THE
WORLD 63
Fundamentals 63
The concept of soil region. 63
Broad soil regions of the world 63
Podsolic regions 63

General considerations 63
Sub-glacial desert regions 66
Tundra regions 66
Coniferous forest regions 66
(Temperate) broad-leaf forest regions 68
Cinnamonic regions 69
Mediterranean cinnamonic regions 69
Mediterranean cinnamonic–ando regions 69
Arid cinnanomic regions 70
Chernozemic regions 70
General considerations 70
(Sensu stricto) chernozemic regions 71
Cinnamonic chernozemic regions 71
Sub-antarctic chernozemic regions 71
Deserts 71
Kaolinitic regions (humid tropics) 72
General considerations. Typical regions 72
Subtropical kaolinitic regions 73
Young kaolinitic regions 73
Kaolinitic–ando regions 73
Kaolinitic–alluvial regions 73
Mountains 73
General considerations 73
High latitude humid mountains 74
Mediterranean mountains 74
Dry mountains 74
Dry mediterranean mountains 75
(Humid) tropical mountains 75
Intermediate regions 75
Grey-wooded regions 75
Brunisolic–cinnamonic regions 75
Brunisolic–arid cinnamonic regions 75
Chernozemic–desert regions 76
Mediterranean mountains–dry mountains 76
Mediterranean mountains–chernozemic regions 76
Transitional kaolinitic regions 76
Brunisolic–kaolinitic regions 76
High latitude humid mountains–kaolinitic 76
Kaolinitic–cinnamonic regions 76
Tropical mountains–kaolinitic 76
Kaolinitic chernozemic regions 76
Palaeo–kaolinitic regions 76
Kaolinitic–podsolic regions 76
Kaolinitic podsolic–cinnamonic regions 76
Mediterranean mountains-kaolinitic 76

CHAPTER 7. SOILS OF THE WORLD (COUNTRY BY COUNTRY) 77
Introduction 77
Northern and Central America 79
Canada 79
United States 79
Mexico 91
Central America 91
West Indies 92
South America 94
Venezuela and Guianas 94
Colombia 94
Ecuador 94

Peru 94
Bolivia 105
Chile 105
Argentina 105
Uruguay 119
Paraguay 119
Brazil 119
Europe 122
European Russia 122
Scandinavian countries 122
Great Britain and Ireland 124
France 125
Western central Europe 126
Eastern central Europe 127
Mediterranean countries 130
Africa 134
North Africa 134
West Africa (from the Atlantic to Togo) 136
Nigeria, Niger, Dahomey, Cameroons, Chad and Central African Republic . 140
Ethiopia, Somalia, Sudan 141
East Africa 146
Central Africa 150
Mozambique and Madagascar 150
Southern Africa 152
Asia 155
Asiatic Russia 155
China 155
Japan 157
Indochina 157
Indonesia 159
India, East Pakistan and Ceylon 166
Southwestern Asia 166
Australia and New Zealand 172
Australia 172
New Zealand 172

CHAPTER 8. AGRICULTURAL POTENTIALITIES 177
Soil relations of plants 177
Soil as a neutralizer of toxins 177
Soil as a reservoir of water 177
Soil as provider of nutients: "potential", "actual" and "mineral" fertility . 178
Agricultural potentialities of the various soil groups: A. Soils with degree of leaching 1 and 2 (dominated by 2:1 clays and not strongly eluviated) . . . 179
Rendzinas 179
Dark clays 179
Ando 180
Raw soils 180
Rankers 181
Brunisolic soils 181
Cinnamonic soils 181
Chernozemic soils 181
Agricultural potentialities of the various soil groups: B. Soils with degree of leaching 2–6 (dominated by 1:1 clays and/or podsolized) 181
Kaolisols 181
Podsolic soils 181
Podsols 182
Agricultural potentialities of the various soil groups: C. Halomorphic, organic and gleisolic soils 182

Saline soils 182
Solonetz 182
Planosols 182
Organic soils 182
Gleisolic soils 183
Agricultural potentialities according to soil region 183
Podsolic regions (*P*) 183
Mediterranean cinnamonic regions (*MC*) 183
Arid cinnamonic regions (*AC*) 183
Chernozemic regions (*Ch*) 184
Desert regions (*D*) 184
Kaolinitic regions (*K*) 184
Mountains (*M*) 185
Palaeo-kaolinitic regions. 185

CHAPTER 9. SOIL SURVEY 187
Soil classification in the lower categories. 187
The concept of land type and its use in soil survey 188
Soil survey method 189

ADDENDUM 191

REFERENCES 193

REFERENCES INDEX 199

GEOGRAPHIC INDEX 203

SUBJECT INDEX 205

Chapter 1 | Soil Formation

Types of weathering

Of the various processes involved in soil formation, clay synthesis is certainly the most important. It may be outlined as follows:

Chemical weathering breaks down the crystalline structure of silicates; silica, alumina, iron and bases are released. When leaching is intense and temperatures are high (humid tropics), the bases and a great part of the silica are leached away. The remaining silica combines with alumina, and forms 1:1 clays poor in silica. The clay fraction consists chiefly of 1:1 clays of low cation exchange capacity. This type of weathering has been called allitic, because the clays formed are rich in alumina, and sometimes gibbsite accumulates.

When leaching is slow and/or temperatures low, the majority of the silica remains in the soil; 2:1 clays, rich in silica, are formed. The clay fraction consists chiefly of 2:1 clays of high cation exchange capacity. This type of weathering may be called siallitic, because it forms clays rich both in silica and alumina. Sometimes weathering takes place at very low pH (< 4). Soil is covered by acid peat or raw humus, which release abundant organic acids. In this case the sesquioxides produced by weathering are complexed and leached; concerning silica it seems that a part is leached but another part may precipitate as opal; the soil formed is poor in clay and in sesquioxides; the clay fraction is very small and rich in silica. We may call this type of weathering podsolic because it forms an ashy (in Russian, pod) horizon, which is the characteristic of podsols.

It seems that synthesis of crystalline clays from silica and alumina is a slow process. In the meantime soil contains a mixture of free silica and alumina (allophane or amorphous clays) of a high cation exchange capacity. That is why in young soils the ratio of cation exchange capacity to clay content is very high; and a part of cation exchange capacity is due to the non-clay fraction. However, since weathering is also slow, clay crystallization usually keeps pace with weathering, and no great quantities of amorphous clays can accumulate. But in the case of volcanic ashes, weathering is rapid, because the material is fine and the surface of contact with water, roots, ect., is very extensive. As a consequence a great amount of amorphous clays (allophane) are formed, and for a long time soil has a very high cation exchange

capacity. This is why young soils formed from volcanic ash differ from the others and form a special group, that of the ando soils.

Podsolic weathering is also rapid; moreover acidity does not favour clay crystallization; that is why the illuvial ("spodic") horizon, in which the products of podsolic weathering accumulate, is rich in amorphous clays.

To sum up, we may distinguish four types of weathering:

(*1*) *Allitic;* this takes place under high temperatures and intense leaching (humid tropics); it produces 1:1 clays of low cation exchange capacity and much of the iron released is free (not bound to the clay).

(*2*) *Siallitic;* this takes place under relatively low temperatures and/or slow leaching (cold and/or dry climates); it produces 2:1 clays of higher cation exchange capacity, and the greater part of the iron released is bound to the clay.

(*3*) *Podsolic;* this takes place under an acid raw humus or peat, which release a great quantity of organic acids, that the products of weathering, etc., cannot neutralize. Alumina and iron are leached away from the weathering layer; the horizon formed is poor in clay and more or less bleached; a great part of the clay fraction is free silica (not bound to the clay). The products of weathering usually accumulate in an illuvial ("spodic") horizon which is rich in amorphous clays.

(*4*) *Allophanic;* this takes place when silicate weathering is very rapid (young soils from easily weatherable minerals, as volcanic ashes, vidrium, etc.); it produces amorphous clays (allophane) of very high cation exchange capacity, higher than in case *2*.

According to many authors in the case of phyllosilicates (mica, etc.) the crystal is not broken. The clay formed is not a product of neo-formation, but mica is altered directly into vermiculite or illite. DUCHAUFOUR (1965) admits this possibility, when climate is temperate.

Many authors admit that leaching can transform one clay into another. It may be that intense leaching eliminates silica from 2:1 clays and transforms them into 1:1 clays. It may be that inundation or sub-irrigation with waters rich in silica and bases re-silicates 1:1 clays and transforms them into 2:1 clays.

Many authors sustain that in a very acid medium (raw humus) the crystal of crystalline clays is broken, and silica and sesquioxides are liberated.

However, crystalline clays seem to be more resistant to weathering than most of the other silicates. And trans-

1

formations of one crystalline clay into another seem to be infrequent.

Maybe these opinions are remainders of the time when allitic and podsolic weathering were unknown. Now it has been proven experimentally, and by direct observation, that kaolinitic soils are not the result of leaching of soils with clays rich in silica; they are formed directly by weathering. Also the very poor in iron, and alumina ashy horizon of podsols may be formed directly by weathering (see Pedro's experiment, p.3). On the other hand, there is abundant evidence that crystalline clays are very stable.

Weathering always requires a certain acidity. In the experiments of Pedro it was twenty times more rapid when leaching was done with acetic acid, than when it was done with water. Leaching seems necessary to eliminate the bases formed. This shows the importance of vegetation (including algae, etc.) in accelerating weathering. They create acidity, although transitory, at certain points. It also explains why weathering is slow in soils rich in powdered carbonates of calcium and magnesium, where acids are easily neutralized; and why weathering is slow in the desert.

Soils formed from unconsolidated rocks inherit clays from the parent material. Naturally, these clays have been formed elsewhere by the fore-mentioned processes, and their composition depends on leaching intensity and temperature at the place of their formation. It may be that inherited clay can be changed by intense leaching or re-silication; but as stated before clays seem resistant to alteration. New clay can be formed, but this is rather difficult if the soil is rich in powdery carbonates of calcium–magnesium. Therefore a distinction should be made between soils formed from consolidated rocks and soils formed from unconsolidated materials more or less rich in clay. A distinction should also be made between weathering of a material rich in easily weatherable silicates (basalt, etc.), limestone, or quartz. In the first case, weathering, as described in the foregoing paragraphs, takes place; in the case of limestone it is a solution, which leaves residues, that may weather to such a degree that the presence of carbonates does not impede it; in the case of quartz it is a comminution, accompanied by some solution of silica.

Many authors have tried to study weathering experimentally. The most comprehensive experiments are perhaps those of PEDRO (1964). This author uses an extractor Soxhlet; the mineral is continuously leached with water and the leachate accumulates and is concentrated in the flask. The experiment can be carried out at different temperatures, with various leaching intensities (always high), with pure water, or water containing different volatile substances (carbonic acid, acetic acid, sulphydric acid, etc.).

Pedro's experiments have shown that the weathering of silicate is relatively rapid. Leaching coarse basalt gravel (average weight of one gravel = 11 g) with pure water, responding to an annual rainfall of 547,500 mm, at

70°C, during 22 months, has ferruginized completely a layer of 2–3 mm on the surface of the fragments; this altered layer represents 5% of the total weight of the fragments; 80% of the silicates of this layer have been broken down.

The alteration has been still deeper in the case of volvic lava; the completely ferruginized layer had a depth of 3–4 mm; 35% of the silicates of this layer have been broken down.

In the case of granite a spongious layer resembling pumice stone has been formed; 16% of the silicates of this layer have been broken down. In 22 months 1.9% of the silica of the granite, 7% of the silica of the basalt, and 7.5% of the silica of the lava have been leached away (into the flask). Taking into account the size of the fragments and the exiguity of the surface of contact, such weathering may be considered as extremely rapid.

Naturally leaching was very intense in the experiment. It corresponds to 547,500 mm of annual leaching rainfall. But what is interesting is the rapidity with which silicates break down in pure water; because if weathering itself was not rapid, an increase of leaching would be useless. The experiment has been repeated at 20°C. The rapidity of weathering (SiO_2 and bases content of the water leaving the soil) has been reduced almost four times (from 1,604 γ l to 414 γ l). But still the process was rapid, since 0.45% of the silica of the basalt had been eliminated in 11 months.

An interesting fact is that leaching with water containing acetic acid (pH 2.5) leached away 2.6% of the basalt in 10 days. It is true that soil pH never reaches such low levels, but in the vicinity of roots or decaying organic matter it can. This shows the importance of vegetation in activating weathering.

Pedro's experiments have confirmed that leaching eliminates silica and leaves behind alumina and iron. In the experiment at 20°C with pure water the leachate, contained (in weight) 23 times more SiO_2 than Al_2O_3; 138 times more SiO_2 than Fe_2O_3. It is only at high temperature and intense leaching that some alumina, but not Fe_2O_3, passed in the leachate, but that is certainly due to the high pH of the water under such conditions (7.3–8.0) and excessive drainage through an extremely coarse material. By reducing drainage from 105 mm (equivalent rainfall)/h to 8 mm the alumina content of the leachate has been reduced from 23.2 to 9.4 γ; but a drainage of 8 mm/h corresponds to 70,000 mm/year, while leaching rainfall seldom exceeds 3,000 mm in the most humid climates of the world. So that it is probable that under natural conditions alumina leaching is practically zero.

By leaching with water containing CO_2 (reducing conditions, outgoing water pH 4.8–5.5) the amount of alumina in the leachate has been reduced, but that of iron increased; in this case leaching eliminated 25% more (by weight) SiO_2 than Al_2O_3; nine times more SiO_2 than Fe_2O_3 + FeO. By leaching with water containing H_2S (reducing conditions, outgoing water pH 5.0–5.2) the amount of alumina has been reduced and that of iron

(21,555 in.)

increased. In this case 24 times more SiO_2 has been leached than Al_2O_3; 10 times more SiO_2 than FeO.

As mentioned before, by leaching with acetic acid, weathering has been accelerated; the SiO_2 content of the leachate leaving the soil increased 21 times; its Al_2O_3 content 250 times; and its Fe_2O_3 282 times; comparison is made with carbonic acid. This shows that when mineral silicates are leached with organic acids, as strong as acetic, weathering is very rapid and practically all the products of weathering except perhaps a small part of the silica are leached away. One of the greater merits of Pedro's work is to show us the possibility of podsolic weathering. Earlier it was admitted that organic acids destroy clays. That may or may not be so. What is certain is that organic acids do not permit the formation of clay; the alumina released by weathering is leached away.

Pedro's experiments have also confirmed that the synthesis of crystalline clays requires much time. No crystalline clay has been encountered either in the residue or in the leachate. In the experiment at 20°C a trioctahedric ferromagnesian smectite, probably stenensite, has been identified; but it did not contain alumina; crystalline clay containing alumina has not been identified. That explains allophanic weathering; when weathering is rapid amorphous clays accumulate.

It would be interesting to repeat Pedro's experiments with clay instead of rock fragments in order to see what the resistance of clay minerals is to weathering. Do they lose silica by leaching? Do they break down by acids? How rapidly and under what conditions? It would also be interesting to repeat Pedro's experiment with the interstices between rock fragments filled with different clays, more or less acid, with or without powdered $CaCO_3$. That would permit the elucidation of many questions. Does acid clay accelerate weathering? Does neutral or alkaline clay, or the presence of $CaCO_3$ interfere with weathering? The first supposition is plausible; and abundant evidence supports the second. The repetition of Pedro's experiments, more especially that of acetic acid, with different minerals and rocks, would permit the classification of them according to their weatherability on an experimental basis. It would also permit the determination of the mineral reserve of the different soils.

HUMIFICATION

Types of humification

Next to clay humus is the most important component in soil. The residues left in the soil by plants and animals are attacked by microorganisms and humus is formed. However, the quantity and composition of the humus formed vary extremely from one soil to another.

Organic matter content is the point of equilibrium between its formation and decay. Formation depends on temperature, humidity conditions and soil fertility. Decay is chiefly affected by temperature and oxygen supply.

Given that temperature accelerates more organic matter decay than plant growth, soils of cold climates are usually richer in organic matter than those of warm climates; and given that humidity is more necessary for plant growth than for organic matter decay, soils of humid climates are richer in organic matter than those of dry climates (JENNY, 1941). The writer (PAPADAKIS, 1952a, 1960a, 1961) has devised a "growth index" by which a comparison of climates in their effect on plant growth can be made; and a "humolytic index" (HI), by which one can compare them in their effect on organic matter decay. By dividing the first index by the second, we obtain a "humogenic index" (Hg), which shows the organic matter content that corresponds to each climate; such indexes are available for many hundreds of stations; a few are given in Table I.

Concerning organic matter composition a distinction should be made between plant or animal residues and the humus synthesized from them by microbial action. Moreover the humus produced is more or less polymerized. The products of only slightly polymerized humus are acid, and are water soluble even in an acid medium. They may be so acid as to produce "podsolic weathering" (see p.1). When less acid they do not produce such weathering, but they complex clay, put it in solution and cause clay illuviation (lessivage). On the other hand, highly polymerized humus substances are flocculated in a neutral or acid medium, and act as clay stabilizers impeding clay dispersion.

Soil chemistry has not yet progressed sufficiently to identify all these substances, and we have not, as yet, analytical methods able to separate them. However, highly polymerized humus (grey humic acid) is dispersible in pyrophosphate; it is not dispersible when NaCl is added. Medium polymerized humus (brown humic acid) is soluble in pyrophosphate to which NaCl has been added; it precipitates when an acid is added. On the contrary,

TABLE I

HUMOGENIC (*Hg*) AND HUMOLYTIC (*Hl*) INDICES

Locations	Hg	Hl
Barrow, Alaska	6.42	3.3
Pangerango (high mountain), Indonesia	5.12	16.8
Rio Grande, Tierra del Fuego, Argentina	5.06	12.2
Moscow, Russia	4.24	20.0
Berlin, Germany	3.69	22.2
Puno (high mountains), Peru	2.98	27.5
Des Moines, Iowa, U.S.A.	2.54	38.3
Bismarck, N. D., U.S.A.	2.36	30.5
Quito (high mountains), Ecuador	2.30	45.5
Atlanta, Ga., U.S.A.	2.16	51.7
Rosario, Argentina	1.95	63.7
North Platte, Neb., U.S.A.	1.50	47.9
Athens, Greece	1.26	59.6
Jakarta, Indonesia	1.38	107.9
Nagpur, India	0.48	207.2
Yuma, Ariz., U.S.A.	0.09	169.9

slightly polymerized humus (fulvic acids) does not precipitate when acid is added. This method, an adaptation of the original one of L. V. Tyurin by DUCHAUFOUR and JAQUIN (1963), gives an idea of the composition of the humus.

Slightly polymerized fulvic acids act as acetic acid; they attack silicates and produce "podsolic" weathering (see p.1). More polymerized fulvic and possibly brown humic acids cannot produce "podsolic" weathering, but they complex sesquioxides and clay, put them in pseudo-solution and cause their eluviation (lessivage). On the other hand, highly polymerized grey humic acids form, with clay, stable flocculates, which cannot be dispersed in an acid or neutral medium. That is why humus composition has a paramount importance from a pedologic point of view; it conditions soil formation.

Highly polymerized humus is called *mull* and slightly polymerized *mor*. But there are various classes of mull; grass mull is more polymerized than forest mull; the type called *moder* is intermediary between mull and mor. And there are also various types of mor. An accumulation of non-humified organic matter is peat.

The nature of humus formed depends on various factors:

(*1*) *Climate*. High temperatures accelerate organic matter decay; peat is seldom formed. However, TYURIN et al. (1962) have encountered the lowest ratio humic/fulvic acids in "lateritic" soils of Vietnam and "krasnozems" of China. The ratio was 0.2 in the first case, 0.4 in the second, against 0.6–0.8 in podsols. This fact shows that production of fulvic acids is abundant in the humid tropics; and explains why these soils are often very poor in bases and reddish although containing more than 1 % carbon. It is also interesting that brown or desert steppe soils are rich in fulvic acids (humic/fulvic acids ratio is 0.5/0.7).

(*2*) *Waterlogging*. By slowing organic matter decay it favours the accumulation of peat or slightly polymerized humus.

(*3*) *C/N ratio of organic residues*. It seems that N is necessary for the synthesis of highly polymerized humus. Moreover a wide ratio slows organic matter decay and favours the production of peat or slightly polymerized humus. A narrow ratio, by accelerating organic matter decay has the opposite effect.

(*4*) *Calcium content of organic residues*. It seems that Ca is necessary for the synthesis of highly polymerized humus. Moreover when calcium content is high, the acids produced by organic matter decay are neutralized, decomposition proceeds rapidly and a highly polymerized humus is produced. When the residues are poor in Ca, the decaying organic matter becomes rapidly acid, which interferes with its further decomposition; peat or slightly polymerized humus is produced.

(*5*) *Rapidity of decomposition of organic residues*. Some substances, sugars, starch, and cellulose, decay more rapidly than others, such as lignin. Some substances, e.g., tannins, act as sterilizers. This affects the rapidity of decay and consequently the humus formed. Some substances contained in plant residues (acids, tannins, etc.) can produce podsolic weathering or complex sesquioxides and clay.

(*6*) *Mixing or not of organic residues with the soil*. When soil residues are mixed with soil, they cannot become acid, because of the buffering action of soil colloids, and the bases which many soils contain. As a consequence decay proceeds rapidly, and a highly polymerized humus is produced. The opposite happens when organic residues accumulate on soil surface. Moreover calcium is more abundant in the first case.

(*7*) *Soil base status*. A soil rich in bases neutralizes the acidity of the organic residues mixed with it, activates decay, and produces a highly polymerized humus; the contrary happens when the soil is poor in bases. Moreover the organic residues produced in an acid soil are usually poor in bases, and their C/N ratio is wide. What is especially important is calcium, because calcium is necessary for the synthesis of highly polymerized humus.

(*8*) *Vegetation*. Grass roots are renewed each year; so that grass vegetation incorporates with soil a great quantity of organic matter in fine roots and rootlets, thoroughly mixed with soil; on the other hand, the organic matter accumulated on soil surface is not great; grass roots are rich in cellulose and other substances of rapid decay; grasses usually grow in soil well provided with bases. For all these reasons organic matter decay is rapid and highly polymerized humus is formed.

The residues of forest vegetation, however, accumulate on the soil surface; they are not well mixed with soil; forest often grows on soil poor in bases. As a consequence organic matter decay is slow and slightly polymerized substances are formed.

The influence of species is also great. Legume residues are usually rich in calcium and have a narrow C/N ratio; on the other hand some plants contain tannins; the residues of coniferous are poor in bases and have wide C/N ratio. However, vegetation is, of course, only one factor of soil formation; it is not the only factor. Under certain conditions grasses can form mor or peat; and coniferous forest can form mull.

(*9*) *Cropping*. By maintaining the soil during certain periods without vegetation, and by increasing its maximum temperature, cropping accelerates organic matter decay. Moreover when a soil is cropped organic residues are well mixed with soil; they do not accumulate on the soil surface. Agricultural plants are usually less lignified and have a narrower C/N ratio than native vegetation; they are, on the whole, grasses or legumes, or cruciferous (rich in N). For all these reasons cropping favours the production of highly polymerized humus. Even the transformation of a forest into grass has this effect.

The distribution of organic matter in the profile varies also from one soil to another. In forest soils the organic matter provided by above-ground organisms is more important than that provided by roots; this matter accumulates on the soil surface, and were it not for the action of certain insects and worms, it would not be mixed with the soil; that is why, cet. par., forest soils are poor in organic matter except for a thin horizon near the surface. On the other hand, the roots of grasses provide much organic matter; cet. par., grassland soils are richer in organic matter and the humus horizon is thicker.

LEACHING: ELUVIATION–ILLUVIATION

General considerations

In well-drained soils water moves in one direction only, it descends, and while descending it is absorbed by plant roots; the surplus, if any, is lost by drainage. As a consequence of this downward movement of water various substances are removed from the higher horizons and transported to the lower ones, or eliminated by drainage.

Soil components vary greatly in their solubility in water charged with CO_2; in this respect they may be classified in the following order (PAPADAKIS, 1961):

(*1*) Salts more soluble than gypsum (NaCl, Na_2SO_4, $MgSO_4$, etc.).

(*2*) $CaSO_4$.

(*3*) $CaCO_3$.

(*4*) Cations (Na, Ca, Mg, K) absorbed by soil colloids and consequently difficult to leach.

(*5*) SiO_2.

(*6*) Fe_2O_3, Al_2O_3, clay and humus.

Naturally the more soluble a substance is, the easier it is leached out of the profile, or deeper is the horizon in which it accumulates. Salts more soluble than gypsum are easily leached away, or they accumulate at great depth; $CaSO_4$ and $CaCO_3$ are less easily leached away or they accumulate at less depth; absorbed cations are still more difficult to leach; silica is even more difficult to leach. It is only in very humid and warm climates (silica solubility increases with temperature) that soils are considerably desilicated; in less humid climates silica accumulates at less depth than carbonates. Sesquioxides (Fe_2O_3 and Al_2O_3) are very difficult to leach except by special processes. The case of clay and humus is analogous.

The reason why substances leached from the upper horizons accumulate in the lower ones is that water is absorbed by roots and the substances contained in it precipitate. If the substance is very soluble and some water is lost by drainage, the substance does not accumulate. In the case of clay, iron and humus, absorption also takes place. When a clay dispersion traverses a soil with very fine pores it is absorbed on the surface of pores (HALLSWORTH, 1963). When dispersed, iron traverses a horizon rather rich in clay, and the iron is absorbed by clay. When dispersed organic matter traverses a horizon rather rich in clay or iron it is absorbed by them. Clay, iron and humus are also precipitated by Ca; and the lower horizons are often rich in this cation.

Leaching rainfall

As has been stated, water moves in the soil in one direction only, it descends; and while descending it is absorbed by plant roots. As a consequence the leaching effect of a humid season cannot be undone by a capillary rise during the non-humid season. That is why annual rainfall, or the difference between annual rainfall and annual potential evapotranspiration, are misleading; what is important is the difference between rainfall and potential evapotranspiration during the humid season, or "leaching rainfall" (PAPADAKIS, 1952a). Normal leaching rainfall (Ln) is computed on the basis of normal (average of many years) monthly rainfall and the corresponding potential evapotranspiration. However, since rainfall varies considerably from year to year, the soil is leached even when normal leaching rainfall is 0; that is why, in the case of dry climates, it is necessary to use another parameter, "maximum leaching rainfall" (Lm); it is the difference between twice the normal rainfall and normal potential evapotranspiration during the non-dry season (PAPADAKIS, 1952a). The non-dry season includes the months during which rainfall is more than half the potential evapotranspiration. The writer has given elsewhere (PAPADAKIS, 1961) leaching rainfall, both normal and maximum, for 2,400 stations scattered all over the world. Table II gives a few examples.

Decalcification and calcification

An important effect of leaching is decalcification, better said decarbonation. A part of the carbonates of calcium and magnesium contained in parent material, and formed by weathering, is removed from the higher horizons. These carbonates are leached out of the profile, or accumulate in the lower horizons. Decalcification progresses downwards and can entirely decalcify soils formed from materials rich in carbonates. When the climate is dry and/or the material rich in powdered carbonates, the carbonates leached from the surface accumulate in the lower layers and form calcic horizons, or pans. As climate becomes drier the accumulation of carbonates takes place at less and less depth, and under very dry conditions it reaches the surface.

Weathering releases calcium and magnesium; when climate is humid they are leached away from the profile. When climate is dry, they accumulate. This process is called calcification. Depending on the balance between the amount of Ca and Mg released on the one hand and that leached on the other, such accumulation may take place in the lower or upper horizons of the soil.

Calcification is sometimes the result of flooding or sub-irrigation with waters containing Ca and Mg. This process will be discussed later.

TABLE II

LEACHING RAINFALL *(Ln)* AND MAXIMUM LEACHING RAINFALL *(Lm)* IN MILLIMETERS (mm) OR INCHES

Location	Ln		Lm	
	mm	inches	mm	inches
New York, N.Y., U.S.A.	470	18.5	—	—
Des Moines, Iowa, U.S.A.	120	4.7	850	28.4
Denver, Colo., U.S.A.	0	0.0	60	2.4
Fresno, Calif., U.S.A.	10	0.4	110	4.3
Swift Current, Sask., Canada	30	1.2	170	6.7
Quebec, Canada	600	23.6	—	—
Habana, Cuba	280	11.0	—	—
Belem, Brasil	1360	53.6	—	—
Rio de Janeiro, Brasil	280	11.0	—	—
Nueve de Julio, Buenos Aires, Argentina	15	0.6	621	24.5
Mendoza, Argentina	0	0.0	0	0.0
Corrientes, Argentina	118	4.6	—	—
London, U.K.	160	6.3	650	25.6
Paris, France	130	5.1	620	24.4
Marseilles, France	60	2.4	520	20.4
Brussels, Belgium	380	15.0	—	—
Berlin, Germany	170	6.7	730	28.8
Moscow, Russia	220	8.7	830	32.6
Stalingrad, Russia	50	2.0	220	8.7
Akmolinsk, Russia	50	2.0	190	7.5
Madrid, Spain	60	2.4	330	13.0
Milan, Italy	240	9.5	810	31.8
Corfu, Greece	730	28.7	—	—
Athens, Greece	80	3.1	350	13.8
Ankara, Turkey	50	2.0	200	7.9
Tel Aviv, Israel	190	7.5	650	25.6
Alexandria, Egypt	0	0.0	80	3.2
Teheran, Iran	20	0.8	150	5.9
Lahore, Pakistan	0	0.0	170	6.7
Calcutta, India	840	33.2	—	—
Nagpur, India	520	20.4	—	—
Colombo, Ceylon	1430	56.4	—	—
Mandalay, Burma	40	1.6	700	27.6
Bangkok, Thailand	610	24.0	—	—
Saigon, Vietnam	1220	48.1	—	—
Jakarta, Indonesia	690	27.2	—	—
Bogor, Indonesia	2820	111.0	—	—
Tientsin, China	90	3.5	440	17.6
Lanchow, China	20	0.8	210	8.3
Pakhoi, China	1330	52.4	—	—
Tokyo, Japan	870	34.2	—	—
Okayama, Japan	340	13.4	—	—
Nagasaki, Japan	1270	50.0	—	—
Sydney, Australia	360	14.2	—	—
Adelaide, Australia	70	2.8	390	15.4
Deniliquin, Australia	10	0.4	140	5.5
Honolulu, Hawaii	80	3.2	510	20.0
Monrovia, Liberia	3860	152.1	—	—
Bouaké, Ivory Coast	305	12.0	—	—
Accra, Ghana	147	5.8	858	33.8
Ibadan, Nigeria	467	18.4	—	—
Kano, Nigeria	362	14.8	932	36.7
Kinsjasa, Congo	620	24.4	—	—
Stanleyville, Congo	610	24.0	—	—
Salisbury, Rhodesia	290	11.4	—	—
Johannesburg, South Africa	40	1.6	600	23.6
Auckland, New Zealand	710	28.0	—	—
Waipita, New Zealand	10	0.4	250	9.8

Acidification

Another effect of leaching is acidification of the higher horizons. Absorbed bases are replaced by H and soil becomes acid. That is why in many soils pH increases with depth. Acidification with pure water or water containing CO_2 is a slow process. It becomes rapid when water contains organic acids (see Pedro's experiment on p.2; in the laboratory we never use pure or carbonated water to determine absorbed cations; but acetic acid can be used).

When a soil is poor in bases, the plants that grow on it are also poor in bases, and their decay produces organic acids, which acidify the soil. That is why some soils owe their acidity to parent material, and were acid since the beginning of their formation. In other cases, soil acidity is due to the type of vegetation (see p.4). Climate acts in two ways: directly, by determining leaching rainfall, and indirectly, by determining vegetation.

Biologic migration of substances from the lower horizons to the surface

Plants absorb bases from the whole profile and deposit them with their residues on soil surface. This process tends to enrich the upper horizons in bases, silica and phosphorous, and counteracts leaching. In soils well provided with bases leaching prevails and pH decreases with depth; but in tropical soils, poor in bases, biological migration prevails, and pH decreases with depth; in many tropical soils the upper 25 cm contain more exchangeable bases than the following 100 cm.

Clay illuviation

As was stated on p.4, highly polymerized humus (grey humic acids) forms with clay flocculates that are very difficult to put in dispersion. To obtain dispersion it is necessary to replace an important part of absorbed calcium by sodium. That is why in soils rich in highly polymerized humus, clay eluviation is due to Na.

However, slightly polymerized humus (fulvic acids and to a lesser degree brown humic acids) does not impede clay dispersion; its acts as dispersing agent; and clay is eluviated. There are, therefore, two different processes of clay eluviation: (*a*) solonization or eluviation by Na; it is common in grassland or dry climate soils; and (*b*) eluviation by slightly polymerized organic substances.

The two processes do not exclude one another. Fulvic acids saturated with Na have perhaps a greater dispersing action than unsaturated fulvic acids; but very seldom is the same soil rich in Na and in fulvic acids; fulvic acids are formed and maintained against decay by acid conditions. It is only under forest vegetation and cold dry climate (western Canada, Siberia) that fulvic acids are produced in the acid litter of organic residues that covers the soil, while the underlain soil may be rich

in Na, and a combination of the two processes is observed.

Eluviation by slightly polymerized humus takes place when mor (raw humus) accumulates on soil surface; but it also takes place, less rapidly, under more or less acid forest mull. That is why this eluviation, very common under coniferous forest or heath is also observed under broad leaf forest, when soil is more or less acid. Certain species, which leave residues poor in bases, with wide C/N ratio, and rich in tannins, favour it, while grasses and legumes have the opposite effect. However, even grasses may produce a humus rich in slightly polymerized substances, if summers are short and cool (polar climates) and/or the soil waterlogged.

This type of clay eluviation is responsible for the formation of the B horizon of podsols, grey–brown podsolic, red–yellow podsolic, etc. It takes place even in the tropics when vegetation is forest and soil is poor in calcium. As was earlier noted Tyurin has encountered the richer in fulvic acid humus in "lateritic" soils of Vietnam and "krasnozems" of China.

Solonization is common when a dry climate and/or impeded drainage impede the elimination of Na from the soil profile. It is responsible for the formation of solonetz, soloths and planosols. Consolidated rocks and volcanic materials are usually more or less rich in Na. The amount of Cl or S is negligible compared to Na, so that weathering produces $NaHCO_3$. When, because of dry climate and/or impeded drainage, this Na is not eliminated with sufficient rapidity it remains in soil solution, and reacts with clay.

$$\text{clay Ca} + 2NaHCO_3 \rightarrow \text{clay Na} + CaH_2(CO_3)_2$$

Since calcium carbonate is various times less soluble than the corresponding salt of Na the reaction follows preponderantly the direction of the arrow, and an appreciable part of Ca is replaced by Na. 15% of Na saturation is amply sufficient to cause clay eluviation; a various times lesser saturation is perhaps sufficient. That is why soils formed from materials rich in easily weatherable Na silicates under dry climate and/or impeded drainage suffer solonization (PAPADAKIS, 1963a). How slow leaching should be to produce the process depends on how rapid the release of Na is by weathering. As time passes, release of Na becomes slower and slower, for lack of easily weatherable minerals; Na leaching exceeds Na release, soil looses Na, and clay becomes saturated with Ca or H. That is why many of the soils in which clay eluviation is due to Na (planosols) are now poor in Na, sometimes acid.

Another case of solonization is when a soil has, at little depth, a water table rich in bicarbonate of sodium. Fluctuation of water table and capillarity raise this substance to the surface, and clay becomes saturated with Na. However, this process seldom produces clay eluviation; eluviation takes place later, when the soil is irrigated. In many cases irrigation produces solonization; it raises the water table; Na bicarbonate rises by capillarity to the surface between two irrigations; Na clay is produced; and this clay is illuviated by the ensuing irrigation. Naturally solonization may also be produced when a soil is flooded or irrigated, with water containing appreciable amounts of sodium bicarbonate.

Leaching with neutral Na salts, chloride for instance, may also replace Ca by Na and produce solonization; but this process is much slower. Bicarbonate is far more effective.

Until recently little attention was paid to the fact that the Na that causes solonization is usually provided by weathering of the parent material of the soil. The high frequency of clay eluviation in Argentinian soils formed from easily weatherable materials (volcanic ashes) under a dry climate and a topography that does not facilitate drainage induced the author (PAPADAKIS, 1963a) to advance this assumption.

Clay illuviation is not confined to humid climates. Solonization is very common in the deserts. The only difference is that when "leaching rainfall" is low, the illuvial horizon is formed at shallow depth. On the contrary, in humid climates clay is illuviated to a great depth, and textural change may be so gradual that no textural horizon can be recognized. Impeded drainage has the same influence as a dry climate; it impedes the downward movement of water and causes clay illuviation to a shallow depth. These facts are well known from soil geography (PAPADAKIS, 1961); but some authors still believe that textural B horizons cannot be formed under very dry climates and attribute them to a change of climate.

As the experiments of HALLSWORTH (1963) have shown, clay eluviation depends greatly on soil structure, being considerably easier when texture is light; in soils of medium and heavy texture eluviation takes place through cracks; and it is favoured by an alteration of wetting and drying. Hallsworth's experiments show that when a clay dispersion traverses a soil layer, a part of the descending clay is absorbed, the amount depending on the diameter of the pores. As a consequence, in clayey soils clay eluviation is negligible, and when an eluvial horizon is formed, it is very thin; on the contrary in permeable soils, clay descends a long way before it is absorbed, and clay eluviation takes place to a considerable depth. In the first case textural change is abrupt; in the second gradual.

We shall note that in experiments on clay eluviation continuous leaching should be replaced by frequent alternation of leaching and drying. Under such conditions loamy and clayey soils acquire a structure, water descends between the peds, and clay eluviation advances deeper. Perhaps under such conditions clay eluviation is deeper in loamy soils than in structureless fine sands. Since clay dispersion descends through cracks, pores, etc., the translocated clay tends to form coatings of oriented clay particles on the channels through which water moves. That is why the peds of illuvial horizons are covered with "clay skins" variously called clay films, clay flows, illuviation cutans and tönhäutchen. Also the presence of clay skins is considered as proof of clay illuviation; but if clay eluviation has taken place some centuries ago, and now

the process has stopped, the surface of the peds may have been mixed with the interior of the peds and no clay skins can be observed. For the same reasons we cannot expect clay skins in horizons that are structureless or with peds that are not stable.

Naturally an increase of texture with depth can be due to other causes (soils formed in anisotropic parent materials, truncated soils, etc.), but in these cases the line that separates the eluvial and illuvial horizons is seldom parallel to the soil surface over great distances (e.g., 1 km). In some points the eluvial horizon is very thick and in others it disappears completely.

There are some difficulties encountered when attempting to establish an exact line of demarcation between clay eluviation produced by Na and that produced by slightly polymerized humus. Soil solution always contains both; and organic substances saturated with Na have a high dispersing power. Presence of raw humus and acidity is a sign of podsolization; but presence of Na in the lower horizons, or in ground water, is a sign of solonization. Another sign is the variation of SiO_2/Fe_2O_3 ratio along the profile. Soluble organic matter (chelates, resines, etc.) eluviate iron. Therefore the clay fraction of the illuvial horizons produced by podsolization tends to have a narrower SiO_2/Fe_2O_3 ratio than that of the eluvial horizon.

Solonization takes place under dry climates, or with impeded drainage; a great part of the silica leached from the eluvial horizon is not leached away from the profile and accumulates in the illuvial horizon. On the contrary iron is not leached by Na. That is why, in the case of solonization, the clay fraction of the illuvial horizon has a tendency to be rich in silica, and have a wider SiO_2/Fe_2O_3 ratio than the eluvial horizon. Another sign is soil structure. Organic matter and ferric iron stabilize clay; they do not cause swelling. Na and silica have the opposite effects. That is why illuvial horizons produced by organic matter are more permeable than those produced by solonization; their structure is weaker; this is the main morphologic difference between textural horizons produced by organic matter and those produced by Na. Since the textural B horizon formed by organic matter is permeable it does not halt clay eluviation; the process continues and the textural horizon is gradually destroyed. In solonetz and planosols Na may be eliminated by time, but the textural difference is much more permanent.

Naturally clay eluviation may be compensated for by the more abundant clay formation in higher horizons; that is why in young soils clay content decreases with depth, while in older soils the contrary is the case. The experiments of Hallsworth show that clay translocation increases with pH; it increases also by replacing distilled water with dilute solutions of Na (N/100). However, in these experiments, the effect of Na has not been so great. As Hallsworth points out, the effect of Na is not in increasing the amount of clay that is eluviated, but in decreasing soil permeability; this causes clay absorption at shallow depth, and an abrupt textural change. It must be added that swelling and retracting results in vertical cracks, into which sand from the eluvial horizon enters and thus columns are formed. These are the essential features of solonetz soils; the amount of eluviated clay is not so great, because the eluvial horizon is thin. However, it is assumed that if the experiments of Hallsworth were done with alternation of wetting and drying in a structured soil, the influence of Na in activating clay eluviation would be greater, and the illuvial horizon would be a little deeper.

In the experiments of Hallsworth magnesium did not increase clay eluviation. It may be that Mg acts by causing swelling and reducing soil permeability; as we have said Na acts also in this way. However, it may also be that Mg-saturated humus protects clay from dispersion less efficiently than Ca-saturated humus. It may also be that if the experiment were done with alternative wetting and drying, Mg would activate clay eluviation.

A question arises: Is it clay that is eluviated or are alumina and silica eluviated separately and combined in the illuvial horizon? The afore-mentioned experiments of Pedro show that free alumina can also be eluviated; it may be that in soils originated from easily weatherable minerals (volcanic ashes, for instance) a part of the clay of the illuvial horizon has been synthesized in this horizon with alumina eluviated from the eluvial horizon; but in the opinion of the author, clay eluviation is more frequent.

Iron eluviation

Iron is also complexed by slightly polymerized organic matter (fulvic acids) and descends along the profile. In this way iron accumulation horizons (iron podsolic B) are formed; but iron is absorbed by clay, so that it cannot traverse a clay horizon. That is the reason why iron eluviation is only conspicuous in sandy soils; a horizon of iron accumulation cannot underlie a clay horizon; but it may overlie it.

Ferrous iron is more mobile, so that in waterlogged soils iron moves from one point to another. However, such a process will be discussed later.

Humus eluviation

As was stated earlier (p.4), slightly polymerized humus (fulvic acids) is dispersible even in an acid medium, so that when soil is rich in this type of humus it is eluviated from the higher horizons by leaching. However, this humus seldom accumulates in the lower horizons; the reason is that it decays rapidly and it is destroyed as fast as it is illuviated. In high latitudes, however, subsoil temperature is near 0°C all the year round; under such conditions the illuviated organic matter does not decay and a horizon of organic matter accumulation (humus podsolic B) is formed. The same happens when the subsoil is constantly waterlogged; waterlogging impedes humus decay and favours its accumulation. Thus humus podsolic B can be encountered even in equatorial climates when soil is waterlogged (ground water podsols).

When permafrost is encountered at certain depth, the organic matter that is illuviated at its surface accumulates and a humus podsolic B is formed at this level. This is very common in tundra soils.

Fulvic acids are fixed by clay or iron, and they cannot traverse a horizon rich in these substances, that is why a humus podsolic B is produced only in sandy soils; it never underlies a horizon of iron or clay illuviation, but it may overlie them.

Humus is soluble in weak Na solution; therefore it is also eluviated in solonetz and planosols. However, in these cases, humus is usually more polymerized (grey humic acids), saturated with calcium, and consequently more stable. Moreover, the subsoil of these soils has, in summer, high temperatures, which favour organic matter decay and impede its accumulation.

Formation of pans

The substances that are illuviated in the lower horizons of a profile are deposited on the surface of soil grains and coat them. In horizons of calcium accumulation, soil grains are usually coated with calcium carbonates; in iron podsolic B, soil grains are usually coated with iron and acquire a more or less bright red colour. In humus podsolic B, soil grains are usually coated with organic matter. Silica also sometimes forms coats on soil grains.

When these coats abound, soil grains are cemented one to another, and aggregates (nodules) that do not break down in water are formed. When coats are still more abundant the horizon is transformed into a continuous pan that is very hard and does not break down in water.

Alternation of wetting and drying favours the process. During wet periods the coats are deposited on soil grains and during the dry periods they adhere firmly to them.

All these substances present the peculiarity that they dry more or less irreversibly, so that when the grain is wetted again the coat does not become unstuck. Clay, however, does not dry irreversibly; moreover it swells, so that it cannot permanently cement soil grains; it cannot form aggregates or pans that do not break down in water. Moreover clay impedes the action of adhesive materials. This is well known in the building industry; to make a good concrete you should eliminate clay from sand and gravel. That is why pans are usually formed from materials poor in clay.

Calcium pans are called petro-calcic, and calcic nodules, calcinodes. Silica pans are called duripan, and silica nodules, durinodes. A pan may contain various cements: iron and silica; iron and organic matter; silica and calcium carbonates; it is named according to the dominating cement.

Iron pans abound under conditions that favour iron eluviation (podsols). The same is true for organic pans. However, organic pans are usually formed under waterlogging conditions that slow the decomposition of organic matter and favour its accumulation. Silica pans (duripans) are common when an easily weatherable material (for instance, volcanic ash), weathers and the silica formed cannot be leached with sufficient rapidity, either because the climate is dry, or because the drainage is not free.

Pans have usually a high bulk density, since the pores are filled with a more or less dehydrated cementing material. In the case of pans formed under waterlogging conditions, waterlogging by destroying soil aggregates produces tamping in the lower horizons. Alternative wetting and drying cannot produce a structure, because soil is poor in clay; moreover coatings impede wetting.

PROCESSES ADDING MINERAL SUBSTANCES TO SOIL PROFILE

Capillarity rise of water from a water table

As was stated at the beginning of the last section (p.4), water moves in the soil in only one direction; it descends; and while descending it is absorbed by plant roots. This is the general rule. However, when there is a water table in the rhizosphere or a little below it, water ascends by capillarity into the soil that overlies the water table, and is absorbed from there by plant roots. The "capillarity fringe" becomes rich in different substances contained in the water table.

Thickness of the capillarity fringe depends on the relation between the rapidity with which water ascends and the rapidity with which it is absorbed by plant roots. A scarce vegetation does not absorb water, and permits it to ascend higher; a dense vegetation absorbs the water so rapidly that the fringe is thin. When vegetation consumes little water and/or the water table is shallow the capillarity fringe may reach the surface.

The water table often contains soluble salts; and a saline horizon is formed in the capillarity fringe. In other cases, the water table contains gypsum or carbonates, and a gypsic or calcareous horizon is formed. In a few cases it contains abundant silica; and a duripan may be formed. In tropical countries the water table often contains iron in solution; iron concretions or laterite is formed (see pp. 11, 12).

The stratification of these substances within the capillarity fringe depends on their solubility (see the section on eluviation–illuviation); silica accumulates near the water table, and very soluble salts nearer to the surface. When an accumulation horizon reaches the surface it is called crust; but a crust may also be formed in a desert soil without a water table, because leaching is very shallow in this case.

Flooding and irrigation

Flooding and irrigation are also important. Flooding increases the amount of water that leaches the soil.

However, flooding or irrigation water usually contains various substances (soluble salts, carbonates, silica, iron, clay, silt, sand). If the soil does not lose water by drainage all these substances accumulate in the soil and may form saline or calcareous horizons, pans, etc. When the soil loses an appreciable amount of water by drainage saline horizons cannot be formed. Formation of other horizons depends on the balance between the amount of the corresponding substance brought by irrigation or flooding and that lost by drainage.

Moisture regime

Since water may leach, or bring, various substances, the moisture regime has a great influence on soil formation. Three moisture regimes can be recognized (RODE, 1955) as the following:

(*1*) *Flushing type:* a part of the rainfall is lost by drainage carrying with it substances removed from the soil. In this case no saline soil can be formed, decalcification and acidification are more or less easy.

(*2*) *Non-flushing type:* no water is lost by drainage; the upper horizons are leached, but the substances removed accumulate in the lower ones.

(*3*) *Exsudational type:* a water table exists at some depth and a part of the evapotranspirated water is provided by it; soil, instead of being leached, is enriched in substances (salts, carbonates, gypsum, silica, iron) and horizons of accumulation of these substances are formed.

Lateral drainage

In slopes the lower soils are flooded with waters that have drained soils situated a little higher in the slope. These waters contain various substances and modify soil formation. A conspicuous case is the abundance of brown podsolic and brun acide in slopes. A bleached horizon cannot be formed because iron is added from higher soils. Not all brun acide and red podsolic soils are formed in this way, but in the case of brown podsolic this process is very common.

Another case is the formation of more clayey soils in the lower parts of the slope; clay coming from higher soils is illuviated in them. Very often lower soils have a better base status for the same reason. However, we should not exaggerate and use this interpretation as panacea. Drainage water, when reaching the surface, rapidly forms rivulets, causes erosion, but it does not add substances; when this water reaches the plain it may flood soils but this is another process that has been discussed on p.10. It is the same when drainage waters feed a water table which affects the formation of other soils (see p.10).

Aeolian additions: dust

In certain cases soil receives appreciable amounts of dust. This dust may be rich in calcium, in salts, in easily weatherable materials (volcanic ash), and can affect soil formation considerably. Dust coming from the desert may impede acidification, podsolization or desilification. Volcanic ashes produce ando soils and duripans. Solonetz or planosols are formed when sea winds or wind from a saline lake bring Na salts to the soil.

OTHER PROCESSES INVOLVING IRON

Formation of iron silicates or free iron: ferrugination, rubification, braunification

At the beginning of this chapter it was said that weathering breaks down the crystalline structure of silicates; a majority of the bases and silica are leached away; alumina combines with the remaining silica to form amorphous clays, which later become crystalline; very little free alumina is encountered in soils. But what happens with iron?

A part of the iron enters the crystal of silicates, or is tightly bound with them. Another part remains free. The proportion of free iron is greater in the case of allitic, podsolic or allophanic weathering, because in this case crystallization of silicates is not favoured. It is less in the case of siallitic weathering for the opposite reason. It depends also on parent material; when parent material is very rich in iron and poor in silica, it is difficult for silicates to bind all this iron; when parent material is rich in silica and poor in iron, free iron is less abundant. Free iron is easily reduced or oxidized and changes colour; however, iron bound with silicates does not suffer such alteration. Free iron is dispersed by slightly polymerized humus (fulvic acids) or silica. Dispersion takes place when soil is wet and flocculates when it dries. Repetition of this process enriches soil in finely flocculated iron. When iron has been dispersed by fulvic acids the flocculate is a more or less brown mixture of finely flocculated iron and organic matter (see KUBIENA, 1953, pl.XIX, fig.1). When iron has been dispersed by silica and later the soil dried thoroughly, iron dehydrates irreversibly, the soil plasma contains very fine red crystallites of dehydrated iron, and the soil is consequently red. However, when the quantity of free iron is great (allitic weathering and/or rich in iron parent materials), iron forms big concretions (see KUBIENA, 1953, pl. XIX, fig.5); it precipitates on the surface of soil grains and coats them with a red coat; the soil is ferruginized. Four cases may, therefore, be distinguished:

(*1*) *Wet conditions.* PEDRO's (1964) experiments — at 20°C — show that when soil is continuously wet, no ferruginization takes place; iron combines with silica and bases to form a beige–yellowish product that has been identified as a trioctahedric ferromagnesian smectite, probably stevensite.

(*2*) *Allitic weathering under aerobic conditions.* Much free ferric iron is released and this iron forms concretions and red coats on soil grains. The soil is *ferruginized*.

(*3*) *Siallitic weathering; soil dries thoroughly from time to time.* Soil plasma is rich in fine red crystallites of dehydrated iron in translucent silica. The soil is *rubified*.

(*4*) *Soil rich in fulvic acids.* Soil contains a brown mixture of finely flocculated iron and organic matter. The soil is braunified.

Case *1* is that of soils under wet conditions but not waterlogged; case *2* that of kaolinitic soils of tropical regions and of soils from ferruginous materials everywhere; case *3* that of cinnamonic soils of climates with a long dry season, and case *4* that of brunisolic soils.

Formation of concretionary horizons

Weathering under intense leaching and aerobic conditions releases much ferric iron. This free iron coats soil particles forming concretions and ferruginized rock brash. As long as these concretions are not dried thoroughly they are soft and permeable to water, but when the soil dries out periodically, the coatings dehydrate irreversibly, protect their core from weathering, and the soil becomes rich in gravel-size concretions and ferruginized rock brash. Under an ever-humid tropical climate soil never dries sufficiently to cause this irreversible dehydration of iron sesquioxides which makes them effective protectors against weathering; moreover, organic acids are produced in relative abundance and corrode them. That is why under such climatic conditions we seldom encounter concretions larger than 2 mm and ferruginized rock brash. They are, however, common in tropical climates with a dry season, being encountered in the lower rhizosphere. The deeper horizons are poor in concretions, because they seldom dry out thoroughly, iron coatings are not irreversibly dehydrated, and as the core of concretions is not protected against weathering, large concretions cannot be formed easily. The surface horizons also are poor in concretions, because coatings are dissolved by plant roots and reduction processes that always take place at one point or another as a consequence of localized accumulation of organic matter. In the opinion of the author, the "stone lines" that are encountered in tropical soils are usually caused by this process; the fact that they are always encountered at some distance from the surface, in the lower rhizosphere, and disappear at greater depth indicates that they are of pedologic origin. Stone lines of geologic origin are seldom parallel to the soil surface, or they directly overly weathered rock.

Parent material is also important. Minerals resistant to weathering are easier to protect by a ferruginous coating. On the contrary, when a gravel is easily weatherable, weathering penetrates before it is effectively coated. Moreover soils formed from materials poor in weatherable minerals yield coarser soils, more exposed to oxidation and periodic droughts.

The great majority of tropical soils are rich in iron concretions, but concretionary horizons with abundant concretions larger than 2 mm and ferruginized rock brash are common in tropical climates with a dry season, and where parent materials are consolidated rocks poor in weatherable silicates. Soils formed from unconsolidated rocks seldom have such horizons.

Gleization

When a horizon is waterlogged the iron is reduced and the soil acquires light grey or greenish colours; such horizons are called gley. When they dry out iron is oxidized and rusty mottles are formed. A characteristic of horizons periodically waterlogged is that ferric iron instead of coating soil grains accumulates in some mottles.

Ferrous iron is much more mobile than ferric, and can migrate from the gley horizon. Iron ascends by capillarity from deep gley horizons to higher ones and contributes to the formation of a concretionary horizon or laterite (see p.12).

Iron and clay eluviation by gleization

When the surface of a soil is periodically waterlogged for long periods, iron and clay may be eluviated from this horizon. The process may be outlined as follows: When soil is waterlogged, ferric iron is reduced to ferrous. Later when the soil is dried, organic acids are formed rapidly by decaying organic matter, soil is transitorily acid (BRAMMER, 1965) and iron is leached from the surface horizon. Acid substances produced by decaying organic matter cause also clay illuviation; and clay is also eluviated. Naturally when conditions favour the accumulation of bases (with irrigation rich in bases water) the process is impeded. It is also impeded when, during the period the surface horizon is not waterlogged, water does not move downwards.

Formation of laterite

In tropical countries, leaching rainfall is usually high and the lower horizons, those overlying bed-rock, are usually saturated with water during the humid season; iron is reduced and ascends into the capillarity fringe, thus enriching it in iron. This is an additional reason why large iron concretions and ferruginized rock brash accumulate in the lower part of the rhizosphere, while they disappear from lower horizons. This ascending iron may cement the soil materials of the capillarity fringe, transforming them into an iron pan (laterite). The pan hardens when the horizon dries out, and that happens during the dry seasons or during the brief dry periods that are experienced even in very humid tropical climates. Laterite formation is naturally favoured by an "exsudational" moisture regime. That is why laterites abound in bottom lands and where an abrupt change of slope brings the water table close to the surface. The depth at which laterite is formed depends on the depth of the water table; sometimes it reaches the soil surface. Laterite usually overlies an horizon impoverished in iron and more or less bleached (pallid zone).

PODSOLIZATION

General considerations

As was stated on pp.7 and 9 slightly polymerized humus (fulvic acids) has a multiple effect on soil formation. It can descend through the profile; it can complex iron and clay, put them in solution and produce iron and clay illuviation. In extreme cases it can produce podsolic weathering: all the products of weathering, except perhaps a part of the silica, are leached and the horizon that suffers from this weathering becomes skeletal. Since all these processes are produced by the same agent, fulvic acids, it has been considered as one, and called podsolization. The term comes from the Russian name pod, which means ashes; it is due to the fact that iron eluviation and/or podsolic weathering produce a bleached horizon resembling ash. For the same reason all soils produced by these processes are called podsols or podsolic; humus podsols, iron podsols, brown podsolic, grey–brown podsolic, red–yellow podsolic, etc. However, later it has been made a practice to reserve the term podsol to soils having a bleached ashy horizon, which is usually produced by iron eluviation or podsolic weathering; and use the term podsolic for soils in which the main effect of fulvic acids has been to produce clay eluviation. Then French pedologists advanced still further, they reserve the term podsolization to the process of iron eluviation, and they have adopted a new term for clay illuviation "lessivage". Podsolic soils are called "lessivés".

Recently (7th Approximation) American pedologists have also begun to make a clear distinction between an illuvial horizon formed by clay and/or organic matter illuviation, which they call "spodic" — from the Greek "spodos" which means ash — and an illuvial horizon formed mainly by clay illuviation, which they called "argillic". Accordingly, they divised the group of soils formed by fulvic acids into "spodosols", which correspond to podsols, and "alfisols" which correspond to podsolic or "lessivé".

We shall note that in the American classification no distinction is made between an "argillic" horizon caused by the action of fulvic acids, and an argillic horizon formed by action of Na, but which has lost the majority of this Na. The argument is that we know little about clay illuviation and we classify soils according to their morphology. This is, in the opinion of the author, a retrogression. Podsolic soils have been traditionally separated from soloths and planosols; the fact that in certain cases it may be difficult to fix the limits by a rapid examination in the field, does not justify the confusion of two so different pedogenic processes.

In this book the author considers it convenient to use the term podsolization for all the processes produced by raw humus (fulvic acids), and to adopt special terms for each particular process: clay eluviation; iron eluviation; humus eluviation; podsolic weathering. Soils in which the only conspicuous effect of fulvic acids (raw humus) is clay illuviation, are classified as podsolic; those in which humus eluviation, iron eluviation or podsolic weathering are conspicuous are classified as podsols.

Conditions favouring the accumulation of raw humus

Podsolization is produced by raw humus (fulvic acids), therefore, conditions that favour the formation of raw humus (mor) also favour podsolization. Such conditions are (see p.4):

(*a*) Short, cool summers, which slow the decomposition of organic matter, and/or waterlogging.

(*b*) Vegetation producing residues poor in Ca, with wide C/N ratio, rich in substances that decay slowly (lignins) or act as sterilizers (tannins, etc.). The most important are coniferous forest, heath, and perhaps Eucalyptus.

Podsolic weathering

Podsolic weathering takes place when a consolidated rock weathers under a layer of very acid raw humus or peat. The acids produced by this raw humus or peat should naturally be sufficient to neutralize the bases liberated by weathering, so that if the consolidated rock consists mainly of quartz podsolic weathering may take place even under a climate that does not favour the accumulation of raw humus. However, if the consolidated rock is an easily soluble limestone podsolic weathering is impossible even under tundra conditions; calcium should be first eliminated.

With podsolic weathering all the substances released — bases, iron, alumina, silica — are leached; it may be that a part of silica precipitates again forming opal, which remains in the profile. The result is a horizon very poor in clay; moreover, the clay fraction is very poor in sesquioxides; the colour is light grey, because of the elimination of iron. This horizon is called A_2, "albic", we shall call it "ashy".

When drainage is very easy (upper slopes) all the products of weathering are leached. No illuvial horizon is formed. The ashy horizon rests directly on bed-rock. The soil is called "podsol ranker" (podsol humo-cendreux in French). When drainage is less free, descending alumina, iron, silica and fulvic acids accumulate at a certain depth and form an illuvial horizon. This horizon is rich in amorphous clay (silica and alumina), iron and humus; its cation exchange capacity is high; it is more or less red, according to the relation between iron and humus, and aeration. This horizon is called iron podsolic B, or humus podsolic B, according to the prevalence of iron or humus. In many cases it may be separated into two horizons, one of humus accumulation and the other of iron accumulation;

the former overlies the latter. The soil is called humus podsol, iron podsol, or humus–iron podsol, according to the nature of the illuvial horizon.

Podsolic eluviation of clay

If the parent material is unconsolidated rock rich in clay, and/or the raw humus is not so acid, podsolic weathering cannot take place; the buffering capacity of the clay impedes very low pH. In this case the main effect of raw humus is clay eluviation. Iron eluviation also takes place, but as iron is absorbed by clay, sufficient iron remains in the eluvial horizon to impede bleaching. The upper horizon is impoverished in clay, but is neither skeletal, nor bleached. An illuvial horizon is formed. The clay of this horizon is the same as that of the eluvial horizon, usually crystalline. As a consequence cation exchange capacity per unit of clay is not high. Organic matter does not accumulate in the illuvial horizon, firstly because it is absorbed by clay; and secondly because summer temperatures are usually higher and the organic matter that descends into B, decays and cannot accumulate. These soils are called podsolic; gray–brown podsolic if the clays are of the 2:1 type; kaolinitic podsolic if they are of the type 1:1 of low cation exchange capacity.

Bleaching (iron eluviation)

When the eluvial horizon has been sufficiently impoverished in clay, iron can be eluviated thoroughly, and the eluvial horizon becomes bleached (light grey). The iron accumulates in the surface of the clay illuvial horizon which becomes red. The podsolic soil is transformed into podsol.

In many cases bleaching follows clay illuviation, but it may take place directly if the parent material is poor in clay (sands). In the first case the B horizon is rich in clay; in the second it is poor, although more or less richer than the upper one.

It is to be noted that when the eluvial horizon is impoverished in clay, clay leaching advances downwards; the illuvial horizon is also impoverished little by little. That is why under conditions favouring the accumulation of raw humus there is a tendency for soils to be sandy.

Humus eluviation

When the eluvial horizon has been sufficiently impoverished in clay and iron, humus eluviation gains momentum and a horizon of humus accumulation is formed in the surface of that of iron accumulation. Soil is transformed from iron podsol to humus podsol. However, this horizon is only formed when, at this depth, summer temperatures are sufficiently low or the horizon is sufficiently waterlogged to check organic matter decay.

Many humus podsols are the result of further podsolization of iron podsols, but when the parent material is poor in iron, a podsolic soil can be transformed directly into humus podsol. Also when parent material is poor in clay and iron, a humus podsol can be formed directly. Naturally low subsoil temperature and/or waterlogging are always required.

Gleization

Ferrous iron is much more mobile than ferric; moreover waterlogging favours the formation of raw humus. As a consequence gleization favours podsolization. Some gleisols (low humic gley) are podsolized; there is an eluvial horizon more or less bleached overlying an illuvial horizon, richer in clay, which is usually gley.

Many podsols have bad drainage (ground-water podsols); in high latitudes where podsols abound, soil is frozen in winter, and waterlogging conditions prevail in early spring; in still higher latitudes there is permafrost at a very shallow depth, and this permafrost impedes drainage. As a consequence the two processes of iron eluviation often act together. However, the eluvial horizons of podsols are very poor in clay; those produced by gleization are not so poor. Podsols often have a humus accumulation horizon.

Formation of pans (by podsolization)

Iron and organic matter act as cements and may cause the formation of nodules or pans. The horizons of iron or humus accumulation of podsols are sometimes rich in nodules, or cemented into pans. Sometimes (duraquods of 7th Approximation) the pan is formed in the eluvial horizon; the horizon is cemented by the soluble humus that accumulates in it, because soil is waterlogged to this level. Since iron and humus eluviation are produced in sandy materials, iron and organic pans are usually more or less sandy.

Depth and rapidity of podsolization

The depth at which the illuvial horizon is formed depends naturally on leaching rainfall and drainage conditions. The higher the leaching rainfall the greater the depth at which the eluvial horizon begins. In the tropics the eluvial horizon is often 1 m (40 inches) or more thick. In cold, dry climates it is much thinner. When the water table is high the illuvial horizon is formed above it and thickness of the eluvial horizon depends on the depth of the water table.

Polar soils have permafrost at a shallow depth. The slightly polymerized humus that is illuviated at the surface of permafrost accumulates, because low temperatures check organic matter decay. The humus–podsolic B is formed at the surface of permafrost. Permafrost also has another effect; because of the shallow depth at which the illuvial horizon is formed and because of capillary ascension of iron, when soil is waterlogged in spring, iron eluviation is counteracted; the eluvial horizon is brown or yellow or glei but seldom bleached.

Podsolization is a rapid process. Planting a coniferous or Eucalyptus soil in a sandy soil may podsolize it in a few decades. Podsolization and accumulation of organic matter are perhaps the most rapid pedogenetic processes.

Modification of podsols and podsolic soils by cropping

When forest is cut or burnt and the land is used to grow crops, or as a pasture, the raw humus that covered the soil before is destroyed; the residues of crops and pasture plants produce a humus which is much richer in highly polymerized substances (see p.4), and less acid. Soil often receives additions of manure, lime and fertilizers. Not only the process of podsolization is stopped, but soil evolves under conditions similar to those of grassland (chernozemic) soils. A horizon rich in organic matter, dark in colour and less acid is produced. Soil is transformed into sod–podsol or sod–podsolic. Naturally such a modification is easier in the case of podsolic soils than in that of podsols.

INFLUENCE OF CLIMATE ON SOIL FORMATION

Climate has a great influence on soil formation; it determines leaching intensity; it determines vegetation; it affects organic matter decay; it conditions weathering. The most striking effect of climate is seen in the difference between temperate and tropical climate soils. In the tropics leaching rainfall is high, even when the climate is semi-arid, temperatures are high, and silica leaching is rapid; the greater part of silica released by weathering is eliminated by leaching, and 1:1 clays are formed. In temperate climates leaching rainfall is low, even when climate is humid, and temperatures are low, only a small part of the silica released by weathering is eliminated by leaching, and 2:1 clays are formed.

The differences between tropical and temperate climate soils are often attributed to age; it is true that tropical soils are usually old. However, we know now that soils rich in 1:1 clay are produced directly by weathering; they are not the products of desilication of other soils. Moreover as LENEUF and AUBERT (1960) and the experiments of PEDRO (1964) have shown, silica leaching is so rapid in the tropics that the time required for desilication is not so long.

Another striking difference of climate are the large areas occupied by sensu stricto podsols in climates with cool, short summers; in warm climates sensu stricto podsols are only encountered when waterlogging and sandy materials make conditions very favourable to podsolization. On the other hand, solonetzic soils abound in dry climates, where Na released by weathering is not eliminated by leaching. Chernozemic soils abound in climates favouring grassland (low leaching rainfall, non-dry spring, see PAPADAKIS, 1961).

Soil depth is also affected by climate. In dry climates soil is continuously dry below a certain depth. In cold climates it is continuously frozen or very cold. That is why under dry or cold climate weathering is confined to the upper horizons, and soils formed from consolidated rocks are shallow. On the contrary, in the humic tropics water penetrates deeply, temperatures are nearly 25°C at great depth, and weathering often penetrates, many meters down.

The influence of climate is often misunderstood. Climate is only one factor in soil formation and it does not act separately; its effect depends on parent material, vegetation, time and drainage conditions. For instance, acid soils are generally associated with high leaching rainfall; but due to particular types of parent material, or vegetation types, they may be encountered also in areas of dry climate. Due to parent material, poor drainage, or time, neutral soils are often encountered in areas of high leaching rainfall.

As was earlier stated what is important is leaching rainfall; total rainfall is misleading. The influence of climate on organic matter formation and decay has been discussed on p.3. We shall add that maximum temperatures are more important than mean temperatures. And maximum temperatures depend on soil cover; they are higher in barren soils; higher under deciduous forest than under evergreen forest. At a depth between 1 and 2 m soil temperature is constant and a little higher than mean annual temperature.

INFLUENCE OF DRAINAGE AND MOISTURE REGIME ON SOIL FORMATION

Poor drainage halts leaching and in certain respects it has the same influence as a dry climate; it causes the accumulation of salts, $CaSO_4$, Ca and Mg carbonates, at a lower depth than usual; clay illuviation also takes place at less depth; Na, K, Mg and SiO_2 released by weathering are not leached, and formation of alkaline, solonetzic or planosolic soils is favoured. Another effect of bad drainage is gleying and laterite formation; organic and iron pans, and fragipans are usually formed at the level of a water table.

The presence of a water table has great influence on soil formation; inundation too. To sum up, moisture regime (see p.10) has a powerful influence on soil formation.

INFLUENCE OF PARENT MATERIAL ON SOIL FORMATION

Soil is parent material more or less modified by pedogenetic processes. Unconsolidated rocks are sufficiently comminuted to permit normal plant growth; soils formed from unconsolidated rocks often differ little from the original parent material. Consolidated rocks undergo profound changes. A distinction should be made between easily weatherable rocks (basalt, etc.) and those poor in weatherable minerals (quartzite, granite, etc.). In the first case a more or less deep soil is formed rapidly; if climate is dry or drainage poor the soil is rich in bases. In the second case the soil is for a long time coarse and shallow; acidification and podzolization may begin immediately; even with a relatively dry climate the soil formed is rather poor in bases.

Parent materials rich in carbonates resist acidification and podsolization for a considerable time. However, a distinction should be made between soft and hard limestones. Calcium solution from hard limestones is slow, and except in dry climates it is eliminated as soon as it is solved; the soil formed is poor in active carbonates and slightly acid; for a long time it is coarse and droughty, and such conditions favour rubification. This is why hard limestones produce slightly acid red soils (terra rossa). On the contrary, soils formed from soft limestones are for a long time rich in active carbonates, even when climate is humid; they are less droughty and less subject to rubification; their colours are more or less grey (rendzinas).

Taking as criterion their influence on soil formation, parent materials may be divided into the following groups:

(*1*) Soft limestone (including marls).

(*2*) Hard limestone.

(*3*) Consolidated rock rich in weatherable minerals (basalts, etc.).

(*4*) Consolidated rocks poor in easily weatherable minerals (quartzite, granite, etc.).

(*5*) Clays (unconsolidated rocks rich in clay and consolidated rocks that contain clay minerals).

(*6*) Sands (unconsolidated rocks rich in unweatherable minerals).

(*7*) Aeolian; recently fixed or alive dunes.

(*8*) Alluvial; recent alluviums.

(*9*) Desilicated materials; products of erosion of kaolisols and laterite.

(*10*) Volcanic ash.

Each of these materials can produce a variety of soils according to climate and other conditions. What is produced can best be stated separately for each soil region; Chapter 6 gives the soil formed from each group of parent material in each soil region.

INFLUENCE OF TIME ON SOIL FORMATION: POLYGENESIS

Accumulation of organic matter is a relatively rapid process; equilibrium is reached in a few decades. As was already stated acidification is rapid in the case of consolidated rocks rich in unweatherable minerals and unconsolidated rocks consisting almost exclusively of such minerals; such soils are acid since the beginning of their formation. On the other hand, acidification is slow in the case of rocks rich in carbonates, or consolidated rocks rich in easily weatherable minerals and of clayey sediments.

Podsolization (formation of genuine podsols with a bleached eluvial horizon) is rapid when the material is coarse. In general podsolization is a relatively rapid process.

Desilication (formation of tropical soils) is direct in the case of consolidated rocks poor in easily weatherable minerals; silica is eliminated during weathering (LENEUF, 1959). In the case of consolidated rocks rich in weatherable minerals, the process requires more time or higher leaching rainfall. According to LENEUF and AUBERT (1960) 22,000 to 77,000 years are sufficient to eliminate all the silica contained in 1 m of granite. In general, the time required for the various pedogenetic processes is shorter than is usually thought.

The materials from which a soil is formed have often undergone one or more periods of weathering in the past; they have lost a more or less greater part of their easily weatherable minerals; they contain clays that have been formed under different conditions of weathering and which are now stable (resistant to alteration). Therefore, when dealing with the age of a soil we shall consider not only the time since the soil lies, more or less undisturbed, in its actual site, but also the weathering that the parent material has undergone before its deposition in the site.

Really young soils are those formed from igneous rocks that have recently come out of the interior of the earth. They are also the more fertile. Mankind always gathers around the volcanoes, in spite of the disasters that are frequent in these areas. On the contrary, soils formed from erosion of deeply weathered surfaces, which have lost all their bases and the greater part of their silica, are old, even if the deposition of the material in the site is recent. An alluvial soil which is formed from the erosion of a young volcano, or a young mountain, is young, but an alluvial soil formed from the erosion of deeply weathered and kaolinized areas, is old. Attention should be paid to the history of the material before its deposition in its actual site.

In order to explain an apparent incompatibility of soil type with actual climate, many authors recur to the easy explanation of a change in climate. Although in some cases such an explanation is plausible, usually it reflects a misunderstanding of the influence of climate on soil formation and of the time required to produce a certain effect. Moreover, changes in drainage conditions, flooding, subirrigation, vegetation, are far more frequent than climatic changes. Thus we should be more cautious when advancing or accepting such hypotheses; they usually serve to avoid problems, instead of solving them.

As KING (1962) says, all classes of topographic and geologic data from every realm of the earth's surface emphasize the ubiquity and persistence in time of radial (vertical) tectonic elevations and depressions of the earth's surface. Such elevations and depressions change drainage conditions, the moisture regime of the soil, and such changes have much more important effects on soil formation than a climatic oscillation.

For all these reasons when studying the soils of an area, the geologic history of an often very extensive region and its geochemistry should be considered.

Chapter 2 | Soil Diagnostics

A great part of the confusion existing in soil nomenclature is due to the lack of precision. Definitions are loose and may be interpreted in different ways. The Seventh American Approximation (1960) marks great progress in this respect; but much remains to be done. Soil diagnostics should be based on fundamental soil properties, for instance, clay composition, clay illuviation, organic matter composition and content, base status, etc. Moreover, the criteria used should be precise. For all these reasons use of laboratory data is preferable to that of certain morphologic features, structure, etc., that may be interpreted differently by different persons.

Use of laboratory data is often avoided on the assumption that soil classification should be done in the field and cannot wait for the results of laboratory analysis. However, the adoption of a classification based on criteria determined in the laboratory does not imply that we cannot classify a soil, unless we have analysed it. There is a correlation between certain morphologic and other features determinable in the field and laboratory data. In the majority of cases we do not need laboratory data to know whether soil is dominated by 2:1 or 1:1 clays, whether it has a horizon distinctly richer in clay, whether a horizon is calcareous, neutral or acid, etc. Although definitions are given in laboratory terms, laboratory data are only necessary in certain cases, when a soil falls within the boundary zone between two groups, or when the pedologist is not yet acquainted with the soils of a region. It is the same as in medicine: a mere examination of the patient by the physician is usually sufficient to diagnose a disease, but in case of doubt the question is settled by laboratory data; and diseases are defined in laboratory terms.

Analytical methods vary from country to country and this creates a difficulty. The author had the difficulty in mind when choosing the diagnostics used in this book, and he has tried to minimize its effects. One of the great services the International Soil Science Association and other international organizations can render to world pedologists is the establishment of a laboratory that can analyze soil samples sent to it from any part of the world. For many countries it would be less costly, and safer, to send their samples to such an international laboratory than to maintain their own laboratories for routine analysis. Besides, an international laboratory would gather much valuable information.

To ignore laboratory diagnostics in soil classification would be the same as botanists ignoring characteristics of flowers in plant classification, because flowers cannot be seen in all seasons. Botanists devise keys that permit classification of plants on the basis of their vegetative characteristics; but fundamental classification is based on characteristics of flowers. All scientists now use laboratory diagnostics in their every day work; why should pedologists be an exception?

DEGREE OF LEACHING

Due to differences in solubility between the different substances that are encountered in soils, leaching follows a certain sequence. Taking as a basis the work of POLYNOV (1937) and of the writer (PAPADAKIS, 1962a) we may distinguish the following phases; for each one we give the substances that are leached and chief characteristics.

(*1*) Calcic: leaching of carbonates of Ca and Mg; they have not yet been eliminated and effervescence with HCl takes place.

(*2*) Acidification: leaching of exchangeable bases; carbonates have been eliminated, soil may be neutral or acid, but it is so little desilificated that clay fraction is dominated by 2:1 clays.

(*3*) Incipient desilification: silica leaching has been so intense that 1:1 clays prevail.

(*4*) Advanced desilification: silica leaching has advanced so far that 2:1 clays are virtually absent; the clay fraction consists of 1:1 clays and sesquioxides.

(*5*) Ferrugination: free iron is so abundant, and 2:1 clays so scarce that pH in water is equal or lower than in normal KCl (both pH below 7).

(*6*) Podsolization: leaching of clay and iron complexed with organic substances has taken place; an eluvial horizon has been formed.

Chlorides and sulfates are naturally the first substances that are leached; but given their mobility, it is not convenient to consider desalinization as phase *1* of the leaching sequence. Horizons differ in their degree of leaching; it decreases with depth, since the lower horizons are less leached. However, due to biological migration (see p.7) the upper horizons are sometimes less acid than the lower ones. Naturally leaching affects all soil components at the same time and the differences between phases are more quantitative than qualitative; each phase marks the peak of leaching of a certain group of substances. Leaching itself is less important than the balance between the release of a substance by weathering and its removal by leaching (CROMPTON, 1960); that is why the duration of a certain phase of leaching depends greatly on the parent material. Many materials do not contain carbonates and leaching begins with phase *2*. As the author has already stated many soils are acid, de-silicated, ferruginized or podsolized since the beginning of their formation; others have been formed from materials already de-silicated or ferruginized. Very often podsolization follows acidification, but it may follow any other phase. Some grey-wooded soils are podsolized, although they are neutral. Others are podsolized after being desilicated.

DEGREE OF WEATHERING: MINERALOGICAL COMPOSITION OF THE NON-CLAY FRACTION

Minerals vary considerably in their weatherability. Taking as a basis the work of JACKSON and SHERMAN (1953) and many other authors, the following sequence can be established: (*1*) olivine–hornblende–pyroxenes; (*2*) biotite; (*3*) albite–microcline–orthoclase; (*4*) quartz; (*5*) anatase (also zircon, rutile, ilmenite, corundum, etc.). Gypsum, dolomite, aragonite, etc., are not included, because the elimination of these minerals from the soil profile is produced by leaching rather than by weathering. Vermiculite, illite, kaolin and other clays, allophane and sesquioxides (gibbsite, boehmite, hematite, gnethite, etc.) are not included because these minerals are products of weathering, and their further weathering, if any, is conditioned by leaching.

Degree of weathering is determined by mineralogical analysis of the non-clay fraction. Another very convenient test is the solubility in acids of the non-clay fraction. Such solubility is relatively high in young soils; it is very low in old tropical soils. Hot 20% HCl or $D = 1.41$ H_2SO_4 may be used. Unfortunately, such determinations have not yet entered routine analysis; moreover, in tropical soils many concretions exist in the non-clay fraction, although they are products of neo-formation. The Belgians use the silt/clay ratio. A better index is mineral reserve: the difference between bases soluble in hot 20% HCl and exchangeability. Solubility of the whole soil in $D = 1.41$ H_2SO_4 is also a good index and is used by the Brazilians. Mineral reserve and the difference between solubility in this acid and exchangeability decrease with weathering.

PEDRO's (1964) experiments (see pp.2,3) provide a convenient method to test the weatherability of minerals and rocks. The mineral is put in a Soxhlet extractor and leached with acetic acid. The amount of alumina leached away per liter of leachate permits the classification of the mineral.

MINERALOGICAL COMPOSITION OF THE CLAY FRACTION: ADJUSTED Tca/CLAY RATIO

Clay is certainly the most important mineral fraction; that is why clay composition is of fundamental importance in soil classification. Young volcanic soils are rich in amorphous clays (allophane, etc.) of very high cation exchange capacity; temperate climate soils are rich in 2:1 clays of rather high cation exchange capacity; tropical soils contain 1:1 clays and sesquioxides of low cation exchange capacity. A part of cation exchange capacity is due to silt; but the silt/clay ratio decreases in the same order. That is why the cation exchange capacity per unit of clay can be taken as a measure of soil age and degree of leaching.

Clay mineralogy has advanced considerably in recent years. However, we cannot rely entirely on mineralogical analysis; in purifying clay for analysis the amorphous clay is lost and it is the most active clay; perhaps other changes also take place. The cation exchange capacity/clay ratio is a more convenient index. Moreover, cation exchange capacity is the most important property of clay; clay is valued for its cation exchange capacity.

Since cation exchange capacity varies according to the method of its determination the method used shall be specified; to be preferred is that of calcium acetate at pH 7; Tca is cation exchange capacity in milliequivalents determined by acetate of calcium at pH 7. Both Tca and the Tca clay ratio are important; clay is clay percentage; that is why, and after many trials, the author uses the adjusted Tca / clay ratio; it is equal to:

$$Tca/clay + Tca/100$$

"Adjusted Tca/clay ratio" is higher than 0.50 in soils rich in 2:1 clays (soils of temperate regions with degree of leaching 1 or 2, e.g., rendzinas, dark clays, chernozems, brunisolic, etc.); it is lower than 0.50 in soils rich in 1:1 clays, but still containing some proportion of 2:1 clays (tropical soils with degree of leaching = 3, e.g., red–yellow podsolic, krasnozems, tropical ferruginous, terre de barre, etc.); it is lower than 0.15 in soils containing almost exclusively 1:1 clays and sesquioxides (tropical soils with degree of leaching = 4, e.g., ferralitic sensu stricto, latosols of Brazil, etc.). Naturally the Tca/clay ratio is significant when the horizon is poor in organic matter (C/clay ratio below 0.1); horizons rich in organic matter may have high Tca/clay ratios, although their clay fraction is dominated by low cation exchange capacity clays.

Horizons rich in organic matter may be treated with H_2O_2, before determining their cation exchange capacity, but we have not, as yet, sufficient experimental data on a possible effect of the treatment of clays cation exchange capacity.

When soil is dominated by amorphous clays (allophane, etc.), apparent density is low; moisture tension at 15 atm. is high in relation to clay content; cation exchange capacity is high. Therefore we have three criteria to distinguish "allophanic" horizons (horizons dominated by amorphous clays):

(*a*) Cation exchange capacity (Tca) is higher than clay content.

(*b*) Moisture tension at 15 atm. is higher than clay content.

(*c*) Apparent density of fine earth (less than 2 mm), air-dry and ready for analysis, is *usually* less than 1.1.

Organic matter also causes a decrease of apparent density, an increase of moisture tension at 15 atm., and an increase of cation exchange capacity. Therefore the horizons tested should not be rich in organic matter (use lower horizons). Iron sesquioxides also cause an increase of moisture tension at 15 atm. but ferruginous soils poor in amorphous clays have low Tca/clay ratios, and their bulk density is not so low. (See p.3 for the definition of "allophanic horizons".)

Ferruginous horizons are usually rich in free iron; Al_2O_3/Fe_2O_3 ratio is low; moisture tension at 15 atm. is rather high; the difference between cation exchange capacity at pH 8.2, determined by the summation method, and Tca is great.

BASE STATUS: ADJUSTED S/CLAY RATIO

Base status is a very important soil characteristic and it is extensively used in soil diagnosis. Unfortunately, pH is not always sufficient; base saturation (S/Tca) also has some shortcomings, more especially in tropical soils. That is why we are using, as a supplementary diagnostic, "adjusted S/clay ratio"; it is equal to S/clay $+ 2S/100$, where S is the sum of absorbed cations in mequiv./100 g of soil, and clay percentage.

"Neutral" and "eutrophic" soils should have S/Tca ratio above 0.50 and adjusted S/clay ratio above 0.25. "Acid" soils have S/Tca ratio below 0.50. "Dystrophic" soils have adjusted S/clay ratio below 0.25. Practically all acid soils are "dystrophic"; but some "dystrophic" soils rich in 1:1 clays and iron oxides may not be "acid". That is why we prefer to use the terms "neutral" or "acid" in the case of "illitic" horizons; and "eutrophic" or "dystrophic" in the case of kaolisols.

In the case of peaty or organic horizons we use S/C ratio; S in mequiv., C is carbon in percent of dry soil. Eutrophic soils have ratios above 1; dystrophic below 1. In this case, base saturation is not taken into consideration.

DIAGNOSTIC HORIZONS

Soil classification and description is greatly facilitated by using diagnostic horizons. For instance, ando soils are characterized by an "allophanic" horizon; chernozem soils by a "dark humic" horizon more than 25 cm (10 inches) deep; kaolisols by a "kaolinitic" or "super-

TABLE III

SYMBOLS OF DIAGNOSTIC HORIZONS

A_2	ashy	*hd*	dark humic
a	albic (bleached)		
ac	acid	*il*	illitic
ae	aeolian		
al	allophanic	*k*	kaolinitic
		kk	super-kaolinitic
Bh	humus podsolic B		
Bhm	organic pan	*l*	slightly lessivé
Bir	iron podsolic B	*la*	laterite
Birm	iron pan		
b	brown	*m*	duripan
br	braunified	*mg*	magnesic
ca	calcareous	*n*	black (niger)
cac	calcareous accumulation	*na*	natric
cam	petro-calcic	*ne*	neutral
ci	cinnamonic		
co	concretionary		
cr	desert crust	*or*	organic
crm	calcareous crust	*pa*	desert pavement
cs	gypsic	*pe*	peaty
		po	positive
dy	dystrophic	*ps*	pseudogley
e	eluvial	*r*	red
eu	eutrophic	*ra*	light red (albic red)
		ro	bed-rock
ff	highly ferruginous	*sa*	salic
fm	fragipan	*sp*	spodic (podsolic B)
G	gley	*t*	textural B
g	semi-gley		
gn	dark gley	*v*	vertisolic
h	humic	*y*	yellow
ha	bleached humic		

kaolonitic" horizon and so on. Diagnostic horizons have been introduced as a basis of soil classification by the Seventh American Approximation. The diagnostic horizons (see Table III) used in this book follow.

Horizons according to clay mineralogy

According to clay mineralogy we may distinguish several diagnostic horizons:

(*1*) *Super-kaolinitic (kk):* adjusted Tca/clay ratio below 0.15; it denotes a clay fraction composed almost exclusively of 1:1 clays and characterizes ferralitic sensu stricto soils; horizons with a clay fraction composed exclusively of 1:1 clays and sesquioxides can have ratios above 0.15 if they are rich in organic matter, but such horizons are not termed "super-kaolinitic", because they cannot be used as diagnostics.

(*2*) *Kaolinitic (k):* adjusted Tca/clay ratio between 0.15 and 0.50; it denotes prevalence of 1:1 clays and characterizes red–yellow podsolic, krasnozems, tropical ferruginous, terres de barre, etc. (see Table IV). Naturally

TABLE IV

<small>CLAY CONTENT THAT IS NECESSARY FOR A HORIZON TO BE DIAG-NOSED AS KAOLINITIC (ADJUSTED Tca/CLAY RATIO BETWEEN 0.15 AND 0.50) ACCORDING TO Tca. IF CLAY CONTENT IS HIGHER, THE HORIZON IS SUPER-KAOLINITIC (ADJUSTED Tca/CLAY RATIO BELOW 0.15)[1]</small>

Tca	Clay %	Tca	Clay %	Tca	Clay %	Tca	Clay %
1.0	2.0– 7.1	2.8	5.9– 23.0	10.0	>25.0	19.0	>61.4
1.2	2.5– 8.7	3.0	6.4– 24.0	11.0	>28.2	20.0	>66.7
1.4	2.9–10.3	4.0	8.7– 36.4	12.0	>31.6	21.0	>72.5
1.6	3.3–11.9	5.0	11.1– 50.0	13.0	>35.1	22.0	>78.5
1.8	3.7–13.6	6.0	13.6– 66.6	14.0	>38.9	23.0	>85.0
2.0	4.2–15.4	7.0	16.2– 87.4	15.0	>42.8	24.0	>92.1
2.2	4.6–17.2	7.5	17.6–100.0	16.0	>47.0	25.0	100.0
2.4	5.0–19.1	8.0	>19.0	17.0	>51.5		
2.6	5.5–21.0	9.0	>22.0	18.0	>56.1		

[1] For instance, when Tca is 3.0 the horizon is diagnosed as kaolinitic if clay percent is included between 6.4 and 24.0; if clay per cent is higher than 24.0 the horizon is diagnozed as super-kaolinitic. Horizons with Tca higher than 7.5 cannot be diagnosed as super-kaolinitic; and horizons with Tca above 25 cannot be diagnosed as kaolinitic. Tca in mequiv./100 g soil.

horizons with a higher ratio can also be rich in kaolinitic clays, if they are rich in organic matter; but such horizons are not termed kaolinitic, because they cannot be used as diagnostics.

(3) Illitic (il): adjusted Tca/clay ratio above 0.50; the horizon is not allophanic; it denotes a clay fraction dominated by 2:1 clays but rather poor in amorphous clays; it characterizes the soils of cold or dry climates and the young soils of humid tropics except ando (rendzina, raw, brunisolic, cinnamonic, chernozemic, etc.).

(4) Vertisolic (v): clay content higher than 35%; S above 30 mequiv.; it is a special case of "illitic" horizon and characterizes vertisols.

(5) Allophanic (al): moisture tension at 15 atm. greater than clay content; cation exchange capacity greater than clay content; apparent density of air-dry laboratory sample (fraction finer than 2 mm ready for analysis) is *usually* below 1.1. It denotes prevalence of amorphous clays and characterizes ando (volcanic) soils. A horizon rich in organic matter can display these diagnostics and not be "allophanic".

(6) Spodic (sp): cation exchange capacity higher than clay content, and C content higher than 1%; dithion-

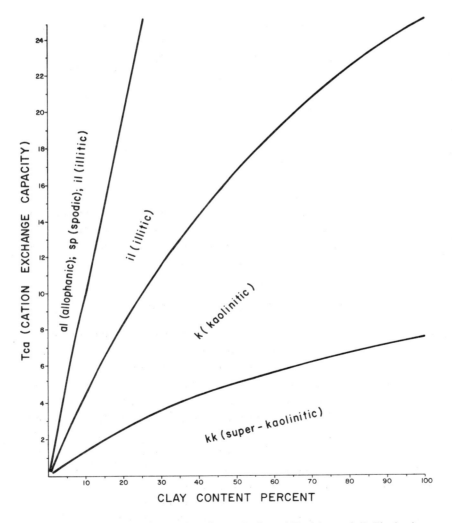

Fig.1. Classification of horizons according to "adjusted Tca/clay ratio". The horizons compared should be poor in carbon (lower horizons). In the case of "spodic" and "allophanic" horizons the ratio permits the supposition that a horizon is not "spodic" or "allophanic", but it is not sufficient to assert that it is.

ate treatment (JACKSON, 1956) eliminates more than half of the alumina of the clay fraction and reduces cation exchange capacity at pH 8.2 by more than 50%; however, dithionate treatment is not necessary, the afore-mentioned diagnostics are sufficient to distinguish a spodic horizon. Since the "spodic" horizon is "humic" and "allophanic" it seems difficult to distinguish it from them; but a "humic" horizon cannot be "illuvial", a "spodic" horizon may be; a "spodic" horizon can be "acid" and "brown" or "red", a "humic" horizon with such characteristics has not such a high cation exchange capacity; in an "ando" soil all horizons are "allophanic" and parent material is "volcanic"; these facts make the distinction easy in the great majority of the cases. The "spodic" horizon characterizes "podsols".

(7) Positive (po): pH in water equal or lower than in normal KCl (pH are below 7); it denotes high content in active sesquioxides and characterizes some terra roxa of Brasil, some hydrol humic latosols, the acrox of the Seventh American Approximation, etc.

(8) Highly ferruginous (ff): free iron content higher than 10% and/or Al_2O_3/Fe_2O_3 ratio, in the fraction soluble in *D* 1.41 H_2SO_4, below 2; it characterizes terra roxa and some cinnamonic and brunisolic soils from ferruginous materials; some "allophanic" and illuvial podsolic horizons are also "highly ferruginous".

Fig.1 gives the classification of horizons according to Tca and clay content; and Fig.2 according to moisture tension at 15 atm. and clay content.

The similarity of highly ferruginous and allophanic horizons in moisture tension at 15 atm./clay ratio and their overlapping show that these soils are related to one another. There is abundant evidence that soils with ferruginous horizons (terra roxa, hydrol humic latosols, acrox) are of volcanic origin and perhaps they were originally ando.

Horizons produced by eluviation–illuviation

Clay eluviation is usually diagnosed on the basis of clay content. When clay eluviation is slight the illuvial horizon is distinctly richer in clay than the eluvial one, but due to higher organic matter content the cation exchange capacity is higher in the eluvial horizon. When clay eluviation is pronounced the illuvial horizon is not only richer in clay, it has also a higher cation exchange capacity. That permits distinction to be made between an "eluvial" horizon which is strongly eluviated, and a slightly lessivé horizon which is little eluviated. Besides clay, iron and organic matter are eluviated; that is why we have "bleached", "humus podsolic B" and iron podsolic B horizons.

The various horizons produced by eluviations may be defined as follows:

(1) Eluvial (e): cation exchange capacity "distinctly" lower than that of some underlying horizon, which is termed *illuvial;* such difference in cation exchange capacity must be accompanied by a "distinctly" lower

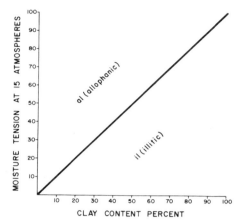

Fig.2. Classification of horizons according to the ratio moisture tension at 15 atm./clay content. The criterion is not sufficient; the horizon should also have other characteristics of "allophanic" or "illitic" horizons ("adjusted S/clay ration", etc). Lower horizons, poor in organic matter, should be compared.

clay and/or C content. A difference is considered "distinct" when the ratio of the figures compared is wider than 8/10. The horizons which are compared should be contiguous; if they are separated by other horizons, total thickness of intercalary horizons must not exceed 25 cm (10 inches); if it is greater, the ratio 8/10 must be wider in the same proportion. For instance if total thickness of intercalary horizons is 50 cm (20 inches) the ratio of cation exchange capacity must be wider than 8/20; the intercalary horizons are considered as forming part of the "illuvial" horizon. The "eluvial" horizons characterize podsolic, solonetzic and planosolic soils. A rapid, but not always decisive diagnostic of eluvial horizons is their sedimentation volume. Twenty-five g of air-dry soil, ready for analysis, are put in a cylinder with approximately 100 ml of 5% Na_2CO_3, well mixed and left overnight; the following day sedimentation volume is determined; that of the eluvial horizon is "distinctly" lower than that of the illuvial. An "ashy" (A_2) horizon is "eluvial" even if the profile does not include an "illuvial" horizon and consequently the fore-mentioned comparisons are impossible.

(2) Slightly lessivé (l): clay content distinctly lower than that of some underlying horizon, but cation exchange capacity is not distinctly lower; a "slightly lessivé" horizon should extend to the soil surface; it cannot be overlain by a horizon richer in clay and which, on the basis of its clay figures, is not "slightly lessivé".

(3) Ashy (A_2): clay content less than 5%; free iron content is so low that it does not turn red on ignition; if the horizon is "neutral" it should be "albic" and underlied directly or indirectly by a "textural" *(t)* or an "iron podsolic B" (*Bir*) horizon; if it is "acid", it should be "albic" or "black" and either rest directly on "bed-rock" or overlie, directly or indirectly, a textural *(t)* or "iron podsolic B" (*Bir*) horizon. It characterizes podsols.

(4) Textural (t): an illuvial horizon with distinctly higher clay content and cation exchange capacity than those of the "eluvial" horizon; it characterizes podsolic,

TABLE V

CLAY CONTENT THAT CORRESPONDS TO "ADJUSTED S/CLAY RATIOS" 0.25 AND 0.50, ACCORDING TO S; IF CLAY CONTENT IS LOWER, THE RATIO IS HIGHER; IF CLAY CONTENT IS HIGHER, THE RATIO IS LOWER[1]

S (mequiv.)	Adj.S/clay ratio 0.25	0.50	S (mequiv.)	Adj.S/clay ratio 0.25	0.50	S (mequiv.)	Adj.S/clay ratio 0.25	0.50	S (mequiv.)	Adj.S/clay ratio 0.25	0.50
	Clay %			Clay %			Clay %			Clay %	
1.0	4.4	2.1	2.4	11.9	5.3	7.0	63.6	19.4	13.0		54.1
1.2	5.3	2.5	2.6	13.1	5.8	8.0	88.9	23.4	14.0		63.7
1.4	6.3	3.0	2.8	14.9	6.3	8.33	100.0	25.0	15.0		75.0
1.6	7.3	3.4	3.0	15.8	6.8	9.0		28.2	16.0		88.9
1.8	8.4	3.9	4.0	23.5	9.5	10.0		33.3	16.67		100.0
2.0	9.5	4.4	5.0	33.3	12.5	11.0		39.4			
2.2	10.7	4.8	6.0	46.1	15.8	12.0		46.2			

[1] Thus, when S (in mequiv.) is 5, if clay percent is higher than 33.3 "adjusted S/clay ratio" is below 0.25; if clay percent is included between 12.5 and 33.3, the ratio is between 0.25 and 0.50; if clay percent is lower than 12.5, the ratio is higher than 0.50. Horizons with S higher than 8.33, have "adjusted S/clay ratio" above 0.25; and horizons with S above 16.67 have ratios above 0.50, whatever clay content may be.

solonetzic, and planosolic soils; it cannot be "spodic" (it cannot have C content above 1% and at the same time cation exchange capacity higher than clay content, except if the soil is ando).

(5) *Humus podsolic B (Bh):* an "illuvial" horizon with "distinctly" higher C content and cation exchange capacity than that of the "eluvial" horizon; in this case not only the ratio of C contents must be sufficiently wide, but the difference must be greater than 0.5%; it characterizes humus-podsols.

(6) *Iron podsolic B (Bir):* a brown or red horizon with two times more free iron than some overlying horizon;

the two horizons should be contiguous or separated by horizons the total thickness of which is less than 25 cm (10 inches).

(7) *"Calcareous accumulation" (cac):* a horizon with 10% more carbonates than the underlying horizon (if the underlying horizon contains 10% carbonates the calcareous accumulation horizon must contain 20% or more).

For the determination of eluvial and illuvial horizons, it is convenient for the horizons to be thinner than 25 cm (10 inches). Many pans are also produced by illuviation but they will be discussed later.

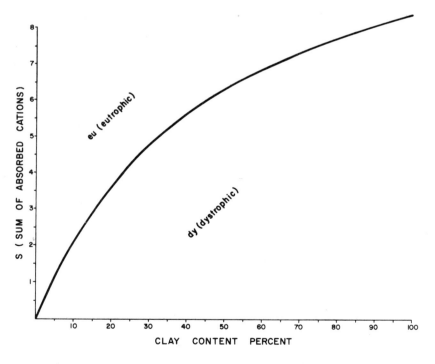

Fig.3. Classification of horizons according to "adjusted S/clay ratio". A "eutrophic" horizon should, in addition, have a base saturation higher than 50%.

Horizons according to base status and salt content

According to base status the following horizons can be recognized (see also Table V and Fig.3):

(*1*) *Acid (ac):* base saturation (S/Tca ratio) lower than 50%.

(*2*) *Neutral (ne):* base saturation above 50%; it cannot be dystrophic.

(*3*) *Dystrophic (dy):* adjusted S/clay ratio below 0.25; in the case of "organic" (including "peaty") soils S/C ratio lower than 1; an "allophanic" or "highly fer-ruginous" horizon cannot be dystrophic; it is "acid" or neutral.

(*4*) *Eutrophic (eu):* adjusted S/clay ratio above 0.25; in the case of "organic" (including "peaty") horizons S/C ratio above 1; an "acid" horizon cannot be "eutrophic"; an "allophanic" or highly ferruginous horizon is "eutro-phic" if it is neutral. The terms "eutrophic" and "dystro-phic" are used in the case of soils dominated by 1:1 clays because in this case the most important criterium is the ra-tio of S to clay content; many soils are dystrophic in spite of having a base saturation higher than 50%. Acid and neutral are used in the case of soils dominated by 2:1 clays, because in this case base saturation is the deciding factor; many soils have adjusted S/clay ratio above 0.25 but they are acid.

(*5*) *Calcareous (ca):* it gives effervescence with HCl; effervescence should be due to active (fine) carbon-ates, not to coarse rock fragments.

(*6*) *Gypsic (cs):* more than 1% of $CaSO_4$.

(*7*) *Salic (sa):* more than 2% salts more soluble than $CaSO_4$.

(*8*) *Natric (na):* more than 15% of the absorbed bases are Na (Na/S ratio higher than 0.15); it characterizes solonetz and alkaline soils.

(*9*) *Magnesic (mg):* absorbed Mg + Na exceed absorbed Ca; more than 7% of absorbed bases is Na; it characterizes magnesium solonetz soils.

Horizons according to colour and iron form (Fig.4)

Soil colour depends chiefly on organic matter and iron. A high content in highly polymerized humus gives soil a black colour. A high content in slightly polymerized humus (fulvic acid) results in high chromas (brown-red); this colour is perhaps due to the iron perplexed by this type of humus. A high content in irreversibly dehydrated iron oxides gives a red colour; on the contrary iron oxides less dehydrated give colours that vary from red to beige according to degree of hydration. Ferrous iron gives light grey–greenish colours. Iron bound with silicates has no great influence on colour; it may cause beige colours. Soils poor both in free iron and humus have the colour of the parent material. The high importance attributed to colour in soil classification is due to its relation with the type of iron and organic matter. However, in many cases a micro-pedologic examination of soil plasma or a chemical test is preferable.

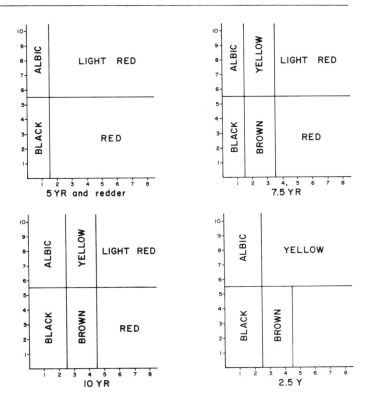

Fig.4. Classification of horizons according to colour.

According to colour and iron form the following horizons can be distinguished:

(*1*) *Black or niger (n):* value 5 or less; chroma 2 or less if hue is 10 *YR* of yellower, 1 or less if 7.5 *YR* or redder; both moist and dry colours must lie within these limits.

(*2*) *Brown (b):* value 5 or less; chroma 2–3, if hue is 7.5 *YR;* 3–4 if 10 *YR* or 2.5 *Y;* both moist and dry colours must lie within these limits.

(*3*) *Red (r):* value 5 or less; chroma 2 or more, if hue is 5 *YR* or redder; 4 or more if 7.5 *YR*, and 5 or more if 10 *YR*; both moist and dry values must lie within these limits; insofar as chroma is concerned it is sufficient that one of them lies within the limits.

(*4*) *Bleached or albic (a):* dry colours with value 6 or more, chroma and hue as black.

(*5*) *Yellow (y):* dry colour with value 6 or more; and chroma 2–3 if hue is 7.5 *YR;* 3–4 if 10 *YR*, and 3 or more if 2.5 *Y*.

(*6*) *Light red (ra):* dry colour with value 6 or more, and chroma 2 or more if hue is 5 *YR* or redder; 4 or more if 7.5 *YR*, and 5 or more if 10 *YR* or yellower.

(*7*) *Gley (G):* rather continuously saturated with water; colour as follows: moist value 4 or more; chroma 1 or less if hue is 2.5 *Y* or redder, 2 or less if hue is 5 *Y* or 7.5 *Y;* hues 10 *Y* and bluer are entirely included in gley; colour should turn red on ignition; it may have rusty mottles, but they should occupy less than 25% of the area.

(*8*) *Semi-gley (g):* saturated with water at some peri-od of the year; distinct rusty mottles; it cannot be gley (*G*).

(*9*) *Pseudogley (ps):* a "gley" or "semi-gley" horizon with vertical bleached stripes, produced by iron leaching.

(*10*) *Black gley (gn):* a "black" and "humic" horizon, permanently or transitorily waterlogged; it may have rusty mottles.

(*11*) *Braunified (br):* a "brown" horizon 10 cm (4 inches) or more thick; organic matter and iron finely flocculated (see KUBIENA, 1953, plate XIX, fig. 1).

(*12*) *Cinnamonic or rubified (ci):* a "red" or "light red" horizon 10 cm (4 inches) or more thick, in which the red colour is due to irreversibly dehydrated iron oxides; micropedologic examination shows minute iron concretions (see KUBIENA, 1953, pl.XIX, fig.5); but if the amount of free iron is high, large concretions are formed and the horizon may be ferruginized; the horizon from time to time becomes thoroughly dry.

The *highly* ferruginous (*ff*) horizon is also due to accumulation of iron oxides; but this has already been discussed.

It should be noted that in the case of soils that have been artificially drained, it is sufficient to ascertain that the gley or semi-gley horizon was in recent times permanently or intermittently waterlogged. The distinction between "braunified" and "rubified" horizons should be done on the basis of micropedologic examination. Colour is usually sufficient, but it may be deceptive in certain cases.

Horizons according to organic matter content and composition

According to organic matter content and composition the following horizons may be distinguished:

(*1*) *Humic (h):* carbon (C) content above 1%; it cannot be either "illuvial" or "bleached".

(*2*) *Bleached humic (ha):* a "bleached" horizon with more than 1% C.

(*3*) *Dark humic (hd):* a humic horizon that is "black" or "brown".

(*4*) *Organic (or):* carbon (C) content above 10%; C/clay ratio above 1/3; but the horizon should not be peaty.

(*5*) *Peaty (pe):* organic residues form a continuous network in which the mineral fraction is imprisoned.

Concretionary horizons and pans

These horizons are formed by cementation of soil grains by different substances produced by pedogenetic processes, chiefly iron, organic matter, or calcium carbonate. These aggregates are sufficiently cemented not to slake in water even if shaken overnight; and they are more or less hard. The following diagnostic horizons can be distinguished:

(*1*) *Concretionary (co):* more than 15% by weight of iron–manganese concretions, iron pan boulders, ferruginized gravel and ferruginized rock brash not belonging in the fine earth fraction, i.e., larger than 2 mm.

(*2*) *Laterite (la):* a continuous pan formed by cementation by iron of a concretionary horizon.

(*3*) *Petrocalcic (cam):* a continuous pan formed by cementation by calcium carbonate; it must have a hardness of 3 or more (Mohs scale).

(*4*) *Duripan (m):* a continuous pan cemented by silica.

(*5*) *Organic pan (Bhm):* a hardpan formed by cementation of a humus podsolic B.

(*6*) *Podsolic iron pan (Birm):* a pan formed by cementation of an iron podsolic B (*Bir*), and containing sufficient free iron to justify its consideration as iron podsolic B.

(*7*) *Fragipan (fm):* a pan that does not give effervescence with HCl, has a base saturation below 100% of cation exchange capacity at pH 7, and underlies one or more horizons with base saturation below 50% of cation exchange capacity at pH 7.

Miscellaneous horizons

(*1*) *Aeolian (ae):* so low in clay content that it is impossible to make a ball to measure "ball resistance" (PAPADAKIS, 1941).

(*2*) *Desert pavement (pa):* a pavement of stones and gravel coarser than 2 mm, that covers the soil surface and impedes further wind erosion; it characterizes some desert soils.

(*3*) *Desert crust (cr):* a surface horizon loosely cemented with products of weathering (soluble salts, gypsum, carbonates, silica); cementation must be sufficient to impede wind erosion; it is encountered in some desert soils.

(*4*) *Calcareous crust (crm):* a "petrocalcic" horizon, that is formed on soil surface; it usually has a hardness less than 3 (Mohs scale).

(*5*) *Bed-rock (ro):* Parent material that has not suffered appreciable weathering. Fine earth is less than half (by weight) of the fraction coarser than 2 mm. Clay content less than 1% of total weight (coarse + fine earth); clay + silt content less than 2% of total weight.

Horizons according to texture

The following classes can be recognized. They are based on the Seventh American Approximation Supplement dated 22 June 1964.

(*1*) *Clay:* more than 35% clay. (*a*) *fine clay:* 35–60% clay; (*b*) *very fine clay:* more than 60% clay.

(*2*) *Sandy:* more than 70% sand; less than 15% clay. (*a*) *sand:* more than 85% sand, less than 10% clay; (*b*) *loamy sand:* 70–85% sand, 10–17% clay.

(*3*) *Loamy:* neither sandy nor clay. (*a*) *coarse loamy:* loamy with less than 18% clay; (*b*) *fine loamy:* loamy with more than 18% clay; (*c*) *silty loamy:* loamy with less than 15% sand.

(*4*) *Skeletal:* more than 60% by weight coarser than 2 mm. All classes may be more or less *skeletal* according to the percentage by weight of particles larger than

2 mm; 10% skeletal means that the fraction above 2 mm is 10% by weight; skeletal without reference to percentage means more than 60% skeletal.

Horizons nomenclature: meaning of some terms

In order to rationalize horizon nomenclature, without departing too much from tradition, the following rules should be used:

(*1*) An *A* horizon must be humic (*h*), slightly lessivé (*l*), or eluvial (*e*); it cannot be kaolinitic (*k* or *kk*).

(*2*) A *B* horizon must be textural (*t*), or humus podsolic (*Bh*), or iron podsolic (*Bir*).

(*3*) A *C* horizon cannot be kaolinitic (*k* or *kk*), humic (*h*), eluvial (*e*), slightly lessivé (*l*); the limit between the *B* and *C* horizons is often arbitrary.

All horizons means to a depth of 50 cm, or to bedrock, or to permafrost whichever is shallower. *Upper limit of effervescence* is the depth at which effervescence with HCl begins. *Lower limit of desalinization* is the depth at which conductivity of the saturation extract at 25°C surpasses 1 mmho. When we say that a soil is 20 cm *eluvial*, we mean that the depth at which the illuvial horizon begins is 20 cm.

OTHER METHODS OF SOIL DIAGNOSIS

The diagnostics proposed in the preceding paragraphs are based on methods that are more or less routine in soil analysis. However, this does not mean that these are the better methods. For instance, the relation between cation absorbing capacity and total alumina of the clay fraction may be a better criterium than T/clay ratio; and S/clay alumina may be better than S/clay ratio. A diagram showing the variation of total alumina of the clay fraction along the profile is very interesting. Also the variation of Al, Fe, and Si soluble in concentrated HCl or H_2SO_4; it is unfortunate that such determinations have been abandoned in many countries.

The effect on cation exchange capacity of treating soil with dithionate-citrate-bicarbonate (JACKSON, 1956), which removes organic matter and iron and alumina bound with it is a promising method. When soil is rich in fulvic acids the reduction of cation exchange capacity at pH 8.2, is enormous. Another interesting index is the effect on cation exchange capacity of heating to 200°C in a nitrogen atmosphere (JACKSON, 1956); it seems that this treatment destroys fulvic acid, but not humic acids. The effect of destroying organic matter on cation exchange capacity is another possibility.

As stated before, apparent density of laboratory sample ready for analysis is a useful index; and sedimentation volume in Na_2CO_3 too; colour of the liquid obtained in this determination is also useful; it shows the amount and composition of organic matter (fulvic acids give brown liquids, humic acids black liquids). Fractionation of humus according to the Tyurin method, modified by DUCHAUFOUR and JAQUIN (1964) is very useful. Micropedologic examinations permit the determination of the form of iron. Moisture tension at 15 and 3 atm., and all the other methods that determine soil absorbing surface permit the detection of amorphous clays.

Much attention is given to soil structure. It is dependent on soil composition. High clay content produces large aggregates, and low clay content granular structures. At equal clay content absorbed Na increases the size and consistency of aggregates. Ceteris paribus 2:1 clays give greater aggregates than 1:1 clays. Organic matter and iron oxides produce small aggregates (crumb structure). Rubified (cinnamonic) soils become hard when dry (in relation to their clay content).

Change in colours after treatment by sodium hydrosulfite ($Na_2S_2O_7$) is also interesting; it denotes free iron content; 1 g $Na_2S_2O_4$ is mixed with 4 g soil; water is added; the container is closed immediately; the mixture is shaken overnight and soil colour is determined the following day.

Variation of cation exchange capacity with pH is also a useful index of the composition of the absorbing complex.

Chapter 3 | Soil Classification: Introduction. Soils with Degree of Leaching 1–2 (dominated by 2:1 clays and not strongly eluviated)

Natural and artificial classifications

As has been stated elsewhere (PAPADAKIS, 1961), classification is very important in every science. The mere classification of a "species" in an adequate scheme shows its characteristic features and the similarities and differences that exist between it and other species classified in the same scheme. Classification is basic to understanding. Long ago man classified everything of interest to him, even abstract qualities, phenomena, etc., and devised a word for each class; the words we are using today are nothing but the names of the units in this classification. It is to be noted that usually for each word in each language there is a corresponding word in all other languages; when a word is lacking in a certain language it is usually because the corresponding thing does not exist or is of little importance; the correspondence is not perfect, but is sufficient to permit the use of dictionaries. This shows that people classified the things of the world in a very similar manner; the units of classification are practically the same; what changes from one people to another are the terms used. As a consequence we may assume that a natural order of things exists in the world, and man-made classifications are nothing but approaches to this natural order. If classifications were entirely arbitrary, every people, every man, would devise his own classification, no correspondence of words would exist, the meaning of the words would change completely from one man to another, and understanding would be impossible. The task of the scientist who devises a classification is to try to discover this natural order of things, and approach it as closely as possible. Also, given the fact that people, who long ago dealt with the matter, already made, consciously or unconsciously, their classification and devised terms (words) to denote its units, the task of the scientist is usually reduced to defining more accurately and arranging more systematically in an adequate scheme the "units" of classification already recognized and named. All natural sciences, zoology, botany, etc., use natural systems of classification. And many of the species, genera, families, orders, etc. that are recognized — the dog, the cat, the oak, the pine, etc. — had been recognized by people in prehistoric times and given special names. The task of the scientists has been to define these units accurately and arrange them in a convenient system.

However, in the newer sciences, e.g., pedology and climatology, many classifications that have been proposed, are, partly or totally, artificial (arbitrary). Many units of classification are mere creations of the classification system and do not exist as generally observed natural units. Another defect of these classifications is that, being arbitrary, they differ considerably one from another and may be multiplied at will; on the contrary, natural classifications, being approaches to the same natural order, tend more and more to converge.

Characteristics of natural classifications

Natural classifications point out the fundamental characteristics of each grouping. As a consequence, classification criteria vary from group to group. For instance, chernozemic soils are characterized by a deep, dark and neutral humic horizon; dark clays by a high clay content and a high sum of absorbed cations; kaolisols by 1:1 clays of low cation exchange capacity, and so on. We always search for the most fundamental characteristic of each group of high or low category, and we use this peculiar criterium in its definition. On the contrary, in artificial classifications soils are divided first according to one characteristic, then according to a second characteristic, and so on. Natural groups are evident to the great majority of those who are acquainted with them. As a consequence a natural classification merely systematizes what has already been accepted; it respects tradition, departing little from it. All plants and animals that long ago had been recognized as separate groups by people have been adopted as such by botanists and zoologists: a classification that ignores groups recognized long ago by pedologists is not likely to be natural.

Soils form a continuum, soil classification cannot be rigid

There is, however, a fundamental difference between soils and plants or animals. Soils form a continuum; usually there are no gaps between one group and related groups. This is why soil systematics cannot be rigid; in the case of soils it is pretentious to talk about orders, suborders, great groups, subgroups, families, series, varieties, genera, species, etc. This is well shown by the history of pedology. Until recently American classification recognized three orders, ten suborders, 40 great groups; in the Seventh Approximation there are ten orders, more than

34 suborders and more than 120 great groups; the number of series recognized in the United States has been increased various times by subdividing series. All this shows that the concept of order, great group, series, etc., is mostly arbitrary. We may say that chernozemic soils form a subdivision of siallitic (with 2:1 clays); that genuine chernozems are a subdivision of chernozemic soils; but it is difficult to say whether genuine chernozems are a group or a subgroup and any discussion on this subject tends to lead to trivialities.

The names and other terms used should be precisely defined

The greater need in present-day pedology is for a precise definition of the names and other terms used. The definitions are so inaccurate that they are meaningless. In a conference given in Rome in October 1965, Dr. G. Smith, of the U.S. Soil Survey, cited the following example. In a recent map, compiled by some of the best pedologists of today, a soil is defined as: "Soil with an *AC* profile. The *A* horizon is thick, dark in colour, but is relatively low in organic matter". Such a definition is completely meaningless. It implies that the soil lacks a *B* horizon; however, there is no more or less precise definition of *B* horizon in world literature, and one does not know what the concept is of the authors of the map of a *B*; one pedologist can identify a *B* horizon where another pedologist denies its existence.

Still more inaccurate is the sentence "the *A* horizon is thick". What is the meaning of "thick"? Dr. Smith, in a pedologic seminary asked the attendants how thick an *A* horizon should be to be considered as "thick". The replies varied from 2 cm to 60 cm. Moreover in an *AC* profile, what is the exact point where the *A* horizon ends and the *C* horizon begins? The reply will certainly vary substantially from one pedologist to another.

The sentence "relatively low in organic matter" is also meaningless. Dr. Smith asked the attendants of the seminary, how much organic matter an *A* horizon should contain to be considered as "relatively low in organic matter". The replies varied from 0.5 to 5%. Definitions like this are the rule in pedologic literature, even nowadays. I remember some years ago I had the opportunity to read the lectures given by one of the best pedologists of our days in a famous university. The definitions of soil groups were so imprecise that it was impossible to understand their meaning; practically the same definition was given for different groups; it was impossible to understand what the difference was between one group and another.

It may be asked, why pedology today is still so backward. To give a precise definition of the concepts of a science is a very difficult task. The definitions should arise from our theories of soil formation, and they should classify together soils that have been formed in the same way. The groups formed on the basis of the definitions should be natural, and when a group is natural it has already been identified by the people who studied it. Thus

the need for precise definitions should not be used as a pretext to reject tradition and formulate an artificial system of soil classification. We should start with the groups already recognized by world pedologists, and try to define them precisely. Naturally some modifications are necessary. When one tries to define a concept, he is obliged to do a critical examination of it, and this provides the opportunity to improve it; pedologic concepts vary from country to country and from author to author, and it is impossible to give a definition that satisfies everyone. But we shall try to depart as little as possible from tradition. That is why it is preferable to maintain the old terms consecrated by time; the establishment of an entirely new nomenclature is dangerous, it may lead us too far from tradition, and lose a great part of the achievements already made in pedology. This is the course followed by all sciences. Until a few centuries ago botany and zoology were in the stage of pre-science. People had long ago recognized a great number of species, genera, classes, etc.— the dog, the cat, the birds, the fishes, the snakes, etc.— but no system of classification existed that embraced all animals, or all plants, and defined each classification unit precisely. Modern zoology and botany have done this task; the first systems and definitions had many defects but they have been gradually improved, and the improvement continues; and they did not reject tradition. The names used, especially at the beginning, were those used by ancient Greeks and Romans 2,000 years ago. They only defined them precisely and made minor changes; for instance, the whale has been separated from the fishes and classified with mammals, and so on. Conserving the old terminology prevented them from departing from tradition and losing what had already been accomplished by mankind in the pre-science stage of zoology and botany. It has been the same in chemistry. Chemistry devised a new terminology based on chemical theory, on the new concept concerning the constitution of chemical substances. But they conserved the old names (fer, sulphur, etc.); they departed from it as little as possible, only when necessary to comply with the new theory. Pedology is now in the pre-science stage. We cannot pretend to have a science, when we have no precise definitions. The great problem is to find adequate definitions, that arise from theory, respect tradition, and are precise and measurable.

The first attempt to solve this essential and difficult problem has been done by American pedologists under the leadership of Guy Smith and Charles Kellogg. They tried to give a soil classification in which all groups are precisely defined, and they introduced the "diagnostic horizons" which greatly facilitate the task.

This book is an effort in the same direction. The author tries to give a system of classification in which all groups are precisely defined, and to make good use of the "diagnostic horizons" introduced by the Americans. The chief differences are that (*a*) the author's system is more closely related to the theory of soil formation; thus this book is begun with a chapter on soil formation; (*b*) the number of diagnostic horizons is considerably greater, although the definitions are simpler, but still precise and easily memorizable for the student; and (*c*) the author sticks more to tradition; the soil groups that are recognized are those recognized long ago by world pedologists, and the definitions that are given are in agreement with the concept of the group; the only departure from tradition is when the theory of soil formation, the need of systematization, or differences in the concept between various countries make such departure unavoidable.

NOMENCLATURE

As it has been pointed out by SIMONSON (1959) the same processes of soil formation (leaching, podsolization, etc.) have contributed to the formation of all soils; the differences are rather quantitative. In one case process A prevailed, while in another case process B has been dominant. This fact implies that many soils belong, at the same time, to two different groups. For instance, a rendzina, or a dark clay, may be a solonchack at the same time; a podsol gleisol; an ando chernozemic; etc. The better way to point out this fact is by using a term that shows that the soil belongs to two groups, for instance, saline rendzina or saline clay or glei podsol, etc. The usefulness of compound terms is shown by the Seventh American Approximation, in which all names derive from a few elements (aqu, argi, ochr, ert, oll, etc.) put together. Moreover compound names are used, for instance, humodic cryorthent, or haplic durargid, etc., but it is not necessary to be so revolutionary as to devise new names; the terms already in use can be compounded.

In many cases two groups of the same broad group differ in the intensity of certain characteristics. For instance, rendzinas vary in the depth of the "humic" horizon from chernozemic rendzinas to serozems; kaolisols vary in base status from "eutrophic" to "dystrophic", and so on. In this case, compound names that immediately show this difference, for instance, eutrophic kaolisol, or dystrophic kaolisol, etc., are used; but in others we prefer to use a name consecrated by time; for instance, serozem; however, a serozem can also be termed non-humic rendzina or desert rendzina.

In pedology the same soil may belong to two different groups, and in some cases may have two different names. For instance, a soil may be rendzina and chernozem at the same time, or ando and chernozem, or gleisol and podsol and so on. A soil may be named grey–brown podsolic and brown lessivé; desert rendzina or non-humic rendzina, and so on. That may surprise many people, but it is a consequence of the nature of the things that are studied in pedology, and it is common in many sciences.

In chemistry, for instance, $FeSO_4$ belongs to the compounds of Fe, and to the sulfates group; and to highly oxidized substances. In organic chemistry there is a scientific name that is seldom used and a common name.

This does not cause confusion, because each name is precisely defined by its formula. The use of a manifold classification and more than one name is confusing when there are no precise definitions, but when the definitions are precise, there is no inconvenience. The trouble in pedology is not caused by the use of various names, or the classification of a soil in different ways, but by the lack of precise definitions and knowledge as to the difference between one term and another, if they are synonymous, if they are distinct, or if they overlap. In zoology and botany there are no manifold classifications, because there a group derives from another group and the classification may be pictured by a genealogic tree. But soils form a continuum, and many soils are situated in a bridge between two or more branches.

When a decimal system of classification is used — it presents great advantages — groups situated between two broader groups may be attached to one or another of them; but that is purely conventional.

Soil names should immediately give an idea of the main features of a soil and its position in the classification system. Given that soils form a continuum, differences within groups are quantitative, and may be better shown by figures. For instance, instead of dividing chernozems into "thick", "thin", etc., we may say "chernozem so many centimeters humic", and so on. In the nomenclature of lower category groups it is convenient to include textural class — it is that of the surface horizon — but in the case of planosols, solonetz and podsolic soils, it is convenient to also include texture of the illuvial horizon, e.g., fine loamy on fine clay. In many cases it is enlightening to include parent material in the name. No general rules can be given. In a natural classification criteria vary from group to group and what is important in one case is useless in another.

SOIL DESCRIPTION

Considerable abbreviation of soil description is possible by using the diagnostic horizons proposed in Chapter 2. For instance, profile 42 of the Seventh American Approximation may be described as follows: *Ahn, ca* 0–12 inches on *Cca,n* 12–45 inches on *Cca,y* 45–65 inches. This description means that the soil has a "humic" (*h*), black (*n*) and calcareous (*ca*) surface (*A*) horizon from 0 to 12 inches; this horizon overlies a calcareous (*ca*) and black (*n*) C from 12 to 45 inches; this horizon overlies a calcareous (*ca*) and yellow (*y*) C.

HIGHEST GROUPS OF SOIL CLASSIFICATION AND THEIR DIAGNOSTICS

One may distinguish the following highest groups:
Rendzina: all horizons are "calcareous".
Ando: all horizons are "allophanic".
Dark clays: all horizons are "vertisolic" (*v*).

TABLE VI

SYMBOLS OF SOILS[1]

a	alluvial		*gp*	low humic gley–lessivé
ag	gleisolic alluvial		*gr*	rendzina–gley
aj	alluvial with petrocalcic horizon		*gs*	saline gley
as	saline alluvial		*gt*	solonetz–gley
aw	wet[2] alluvial		*gu*	planosol–gley
			gz	paddy gleisolic
b	braunerde and recent brown			
b(cac)	braunerde or recent brown with calcareous crust		*h*	organic
b(l)	braunerde or recent brown, slightly lessivé		*hg*	gleisolic organic
ba	brun acide		*hr*	rendzina organic
ba(l)	brun acide slightly lessivé		*hs*	saline organic
bah	organic brun acide		*hw*	wet[2] organic
bao	peaty brun acide			
bat	arctic brun acide		*k*	kaolisol and kaolinitic
bax	ferruginous brun acide		*kc*	kaolisol–cinnamonic intergrades
bc	cinnamonic brown		*(k–c)p*	(kaolinitic–cinnamonic) podsolic
bd	arid brown		*(k–c) (p–u)*	(kaolinitic–cinnamonic) (podsolic–planosolic)
bg	gleisolic brown or braunerde		*kd*	dystrophic kaolisol
bh	organic brown or braunerde		*kdf*	acid ferruginous
bk	jeltozem		*kdl*	krasnozem
bl	lithosolic brown		*kf*	concretionary kaolisol
bm	brown clays		*kff*	ground-water laterite
bo	peaty brown or braunerde		*kg*	gleisolic kaolisol
bp	brown podsolic		*kga*	kaolinitic low humic–gley
bpg	gleisolic podsolized brown		*kgp*	kaolinitic gley podsolic
br	calcareous brown (old rendzina)		*kh*	kaolinitic rubrozem
brd	para-serozem		*kk*	ferralitic
bt	arctic brown		*kkh*	ferralitic rubrozem
by	chernozemic brown (prairie)		*kl*	latosolic kaolisol
byr	degraded chernozem		*kn*	eutrophic kaolisol
bx	ferruginous braunerde		*knf*	tropical ferruginous
			knf(l)	tropical ferruginous lessivé
c	cinnamonic		*kng*	gleisolic eutrophic kaolisol
c(co)	concretionary cinnamonic		*knl*	terre de barre
c(r)	red cinnamonic		*knl(l)*	terre de barre lessivé
c(l)	slightly lessivé or planosolic cinnamonic		*kp*	kaolinitic podsolic
ca	acid cinnamonic		*kpd*	red–yellow podsolic
cb	brown cinnamonic		*kpdg*	gleyed dystrophic kaolinitic podsolic
cd	reddish desert		*kpf*	lateritic podsolic
ced	reddish desert sands		*kpg*	gleyed kaolinitic podsolic
ck	cinnamonic–kaolisol intergrades		*kpgd*	dystrophic gleyed kaolinitic podsolic
cl	lithosolic cinnamonic		*kphg*	kaolinitic humus gley–podsol
cm	cinnamonic clays		*kt*	kaolinitic solonetz
cr	terra rossa and calcareous cinnamonic		*ku*	kaolinitic planosol
cy	reddish brown		*kx*	terra roxa
cx	ferruginous cinnamonic		*ky*	chernozemic kaolisol
d	arid brown		*l*	lithosol
d'	gypsisol		*ld*	desert lithosol (hamada)
dl	desert lithosol		*lh*	lithosolic ranker
dr	serozem		*lha*	acid lithosolic ranker
du	red desert lessivé		*lhh*	organic lithosolic ranker
			lhn	neutral lithosolic ranker
e	aeolian sands (dunes)		*lho*	peaty lithosolic ranker
ecd	red desert sands		*lk*	kaolinitic lithosol
ed	desert sands (ergs)		*ll*	rock outcrops
ew	wet aeolian sands		*lo*	peaty lithosolic (parano)
			lt	tundra rankers
ff	lateritic rock outcrops		*lv*	andic (volcanic) lithosol
			ly	chernozemic ranker
g	gleisolic			
g(na)	alkaline gley		*m*	dark clays (vertisols)
ga	low humic (acid) gley		*ma*	acid sulfate cat clays
gh	organic gleisolic		*m(mg)*	magnesic alkaline clays

[1] The same symbols are used in the maps. They are composed of letters, which always have the same meaning (see footnote 1, p.32). Additional symbols may be composed, following the same rules.

TABLE VI (*continued*)

mn̄	alkaline clays	*rg*	gleisolic rendzina
m (ac)	acid clays	*rw*	meadow² para-rendzina
md	desert² clays	*rh*	organic rendzina
mg	gleyed clays	*rlh*	proto-rendzina
mm	roof clays	*ry*	chernozemic rendzina
ml	clay-rankers	*r'y*	chernozemic para-rendzina
mr	rendzina-clays		
ms	saline clays	*s*	saline, solonchack
mw	wet² clays	*s (na)*	saline–alkaline
my	chernozemic clays	*sal*	salt accumulations
myw	meadow² chernozemic clays	*se*	aeolian saline (lunettes)
		sh	humic saline
n̄	alkaline	*sr*	saline rendzina
		st	saline solonetz
o	peaty	*stt*	takir
oh	peaty-organic (lenist)		
op	peat on podsol	*t*	solonetz
os	saline peaty	*t (mg)*	magnesium solonetz
		tc	cinnamonic solonetz
p	podsol	*tr*	rendzina solonetz
p (Bhm)	humus-pan podsol		
p (Birm)	iron-pan podsol	*u*	planosol
pb	grey–brown podsolic	*uc*	cinnamonic planosol
pb (Bir)	lessivé	*ud*	planosolic red desert
pb (-Bir)	brun lessivé	*ug*	gleisolic planosol
phg	gleyed grey–brown podsolic	*ur*	rendzina-planosol
pbs	sod grey–brown podsolic	*uu*	clay-pan planosol
pc	non-calcic brown	*uug*	gleisolic clay-pan planosol
pcj	duripan non-calcic brown	*uw*	wet (meadow)² planosol
pf	fragipan podsol		
pdk	red–yellow podsolic	*v*	ando
pg	gley-podsol	*v (l)*	ando slightly lessivé or planosolic
ph	humus podsol	*v (po)*	positive ando
phk	kaolinitic humus podsol	*va*	acid ando
pho	peaty humus podsol	*ve*	aeolian ando
pi	iron podsol	*vg*	gleisolic ando
pih	iron-humus podsol	*vh*	organic ando
pk	kaolinitic podsolic	*vj*	ando with pan
pkd	red–yellow podsolic	*vl*	lithosolic ando
pl	ranker-podsol	*vn*	eutrophic ando
po	peaty podsol	*vp*	ando lessivé
ps	sod podsol	*vu*	planosolic ando
psb	sod podsolic	*vw*	hydrol and wet² ando
pt	tundra nano-podsol	*vy*	chernozemic ando
pu	planosolic (solodized) podsol	*vx*	ferruginous ando
pv	allophanic (volcanic) podsol		
pw	grey-wooded	*y*	chernozemic
pwl	lithosolic grey-wooded	*y'*	para-chernozem
pwn	neutral grey-wooded	*y' (l)*	para-chernozem slightly lessivé or slightly planosolic
py	prairie lessivé	*yb*	brunisolic para-chernozem (prairie)
		yc	cinnamonic para-chernozem
q	palaeo-kaolisol	*ye*	aeolian para-chernozem
qf	palaeo-lateritic	*yg*	humic gley
qfr	calcareous palaeo-lateritic	*ygu*	planosolic humic-gley
ql	lithosolic palaeo-kaolisol	*y'j*	para-chernozem with petrocalcic horizon
		yl	grassland ranker
r	rendzina	*yp*	prairie lessivé
r'	para-rendzina	*yr*	calcareous chernozemic
r (cac)	para-rendzina with calcareous accumulation	*yrw*	meadow² calcareous chernozemic
r (cr)	calcareous desert crust	*ys*	saline chernozemic
r (h)	forest (humic) rendzina	*yu*	planosolic para-chernozem
rb	braunified rendzina	*yug*	gleisolic planosolic para-chernozem
rc	cinnamonic rendzina	*yuw*	meadow² planosolic para-chernozem
rd	serozem	*yw*	meadow² para-chernozem
rdc	cinnamonic–serozem	*ywr*	rendzina meadow² para-chernozem
re	aeolian rendzina		

Raw: all their horizons are "illitic"; they do not have "braunified", "cinnamonic", "chernozemic" or "eluvial" horizons; when they have a "humic" (*h*) horizon, this horizon is not directly underlain by "bed-rock" or permafrost.

Rankers: they have a "humic" horizon (*h*), which rests directly on "bed-rock" or permafrost.

Brunisolic: all their horizons are "illitic" (*il*); they do not have "eluvial" (*e*) horizons; they have a "braunified" and/or an "acid" horizon.

Cinnamonic: all their horizons are "illitic"; no horizon is "eluvial" (*el*); they have one or more "cinnamonic" horizons.

Chernozemic: all their horizons are "illitic" and "neutral"; they do not have "eluvial" horizons; they have a "dark humic" horizon 25 cm (10 inches) or more deep.

Kaolisols: one or more horizons are "kaolinitic" (*k*), "super-kaolinitic" (*kk*) or "positive" (*po*); this horizon begins at a depth not lower than 2.50 m (100 inches).

Solonetzic–planosolic: they have an "eluvial" horizon produced by Na or a "natric" horizon.

Solonchaks: soluble salts content is sufficiently large to cause a change in the type of vegetation or cause serious damages to crops.

Podsolic: they have a textural and/or an iron podsolic B horizon; but they do not have any of the following diagnostics: "ashy" horizon, "humus podsolic B", "spodic" horizon, "iron" or "organic pan". Clay illuviation is due to slightly polymerized organic substances. For reasons of convenience soils having "laterite" that intercepts drainage at a depth less than 50 cm (20 inches) are also included in this group.

Podsols: they have one or more of the following diagnostics: (*a*) an "ashy" horizon; (*b*) a "humus podsolic B"; (*c*) an "organic pan"; (*d*) an "iron pan"; (*e*) a "spodic" horizon.

Organic: they have an "organic" or "peaty" horizon.

Gleisolic: they have a "gley" horizon at a depth less than 75 cm (30 inches), or a "semi-gley" horizon at a depth less than 50 cm (20 inches).

---

[1] Letters are used with the following meanings: *a* = alluvial; when following another letter acid; *b* = brunisolic; recent brown is included; *c* = cinnamonic; *d* = desert; in the case of *k* (kaolinitic soils) dystrophic; *e* = aeolian; *f* = ferruginous concretions; in the case of *p* (podsol or podsolic soils) fragipan; *g* = gleisolic; *h* = organic (but not peaty); *i* = iron; *j* = petrocalcic horizon or duripan; *k* = kaolinitic; *l* = lithosol; when following *k* latosolic; *m* = clay; *n* = natric; when following another letter neutral or eutrophic; *o* = peaty; *p* = podsol or podsolic; *q* = palaeo-kaolisol or palaeo-lateritic; *r* = rendzina; *s* = saline; *t* = solonetz; *u* = planosol; *v* = ando; *w* = wet, meadow (environmental term); *x* = highly ferruginous; *y* = chernozemic; *z* = paddy (rice).
[2] Environmental, not systematic.

32

SOIL SYMBOLS

Brevity is necessary and abbreviations are extensively used in our time; moreover they are necessary in mapping. Table VI gives the abbreviations we are using in the maps of this book. They have been coined according to the following rules:

(*a*) As far as possible one letter has always the same meaning[1].

(*b*) When more emphasis should be given to the characteristic shown by a letter it is written before the others.

(*c*) Symbols in parentheses show characteristic horizons, according to Table VI; for instance *b*(*cac*) means a brunisolic soil with a *cac* (calcareous accumulation) horizon.

These symbols present the advantage that, although brief, they show the fundamental characteristics of a soil; and related soils have related symbols; for instance all rendzinas have symbols beginning with *r*; all ando with *v*; all dark clays with *m*; all brunisolic with *b*; all cinnamonic with *c*; all chernozemic with *y*; all kaolisols and kaolinitic with *k*; all podsolic and podsols with *p*; all saline with *s*; all solonetz with *t*; all planosols with *u*; all organic include *h*; all peaty include *o*; all lithosolic include *l*; all gleisolic include *g*.

RENDZINA GROUP (SOILS WITH DEGREE OF LEACHING 1)

General considerations

These soils contain active carbonate and give effervescence with HCl in all their horizons. Since carbonates favour the formation of dark humus, their "humic" horizons are deep and dark, but as climate becomes drier the "humic" horizon becomes shallower and desert rendzinas (serozems) lack a humic horizon. Parent materials are usually powdery or soft limestones (marl or chalk), calcareous loess or sand, etc. When leaching rainfall is low, soils containing carbonates in all their horizons can also be formed from hard limestones, calcareous schists, basalt, etc.; release of Ca and Mg by rock weathering is more rapid than leaching, even in the upper horizons, and they become "calcareous"; naturally in this case carbonate content is lower and red colours are more common, because much of the iron released by weathering is "rubified"; these soils are called "para-rendzinas". Soils containing carbonates in all their horizons can also be formed by subirrigation with water rich in carbonates when the moisture regime is "exsudational"; these rendzinas are called meadow rendzinas; they are often gleyed and may have calcareous accumulations. When the climate is dry, carbonates leached from the upper horizon accumulate in the lower ones, and horizons richer in carbonates are formed. Such horizons are usually soft; the depth at which they are formed depends on the amount of leaching rainfall; it depends also on drainage, which impedes the descent of calcium to deeper levels; in meadow rendzinas and para-rendzinas they may reach soil surface.

General diagnostics

All horizons are "calcareous"; effervescence with HCl is due to fine carbonate coating the soil grains, and to soft concretions, not to coarse fragments of hard limestone.

Subgroups and their diagnostics

Forest rendzinas (r(h)): formed under high leaching rainfall; they have thick "dark humic" horizons; the carbonate leached from the surface is leached away; no horizon of "carbonates accumulation" (*cac*) is formed.

Rendzina with calcareous accumulation (r(cac)): an horizon of "calcareous accumulation" (*cac*) is encountered at some depth.

Serozem (rd): formed under desert conditions they lack entirely a "humic" horizon.

Para-rendzinas (r¹): formed from non-carbonaceous materials, or from hard limestones under dry climate, they have thinner "humic" horizons and redder colours.

Calcareous desert crusts (r(cr)): desert para-rendzinas, the surface of which is loosely cemented with products of weathering.

Meadow para-rendzinas (rw): formed from non-calcareous materials; they owe their carbonates to subirrigation or inundation with calcareous waters, or to bad drainage, which impeded leaching away of the carbonates produced by weathering.

Proto-rendzinas (rlh): the "humic" horizons rest directly on bed rock; usually they have been formed from hard limestone; they are also called "rendzina rankers".

Organic rendzinas (rh), para-rendzinas (r'h) or proto-rendzinas: they have an "organic" horizon.

Chernozemic rendzinas (ry): they have a "dark humic" horizon 25 cm (10 inches) or more deep; they belong both to rendzina and to chernozemic groups.

Cinnamonic para-rendzinas (rc) and serozems (rdc): they have "cinnamonic" horizons.

Aeolian rendzinas (re): all horizons are aeolian.

Rendzinas that have "gley" or "semi-gley" horizons will be discussed later. Some rendzinas are simultaneously dark clays, saline, etc.; they are called rendzina clays, saline rendzinas, etc.

N.B. It is convenient to specify the thickness of the "humic" horizon; for instance, forest rendzina, 20 cm humic; the afore-mentioned terms do not necessarily exclude one another; for instance an organic rendzina can be gleyed, an aeolian rendzina may also be forest rendzina, and so on.

Correlation and examples

The concept of rendzina does not vary considerably from country to country. Some authors extend it to soils the surface of which has been decalcified and does not effervesce with HCl; others exclude soils which owe their high lime content to climate (chernozemic rendzinas, serozems, etc.), and then the problem to fix limits becomes difficult. It is considered more convenient to include in rendzinas all soils, all the horizons of which are "calcareous"; and the problem of limits, which often leads to intense discussions, is solved by admitting that a soil may be simultaneously rendzina and chernozemic, rendzina and dark clay, rendzina and solonchak, etc.; for the reasons already revealed, a soil may belong to various groups at the same time and it is better defined by pointing out this fact. The concept of serozem varies considerably from author to author, and the definitions given are usually meaningless. But all serozems are rich in lime and poor in organic matter; so that we use this term for the rendzinas that do not have a humic horizon. Our rendzinas correspond to "rendolls" of 7th American Approximation, and to "sols calcimorphes" of the French classification (DU-CHAUFOUR, 1965). But the author's concept is wider than that of the 7th American Approximation.

Profile 41 of the 7th American Approximation, classified as *orthic orthustent*, is cinnamonic serozem.

Profile 42 of the 7th American Approximation, classified as *orthic grummaquert*, is chernozemic rendzina clay; it may be gleisolic.

Profile 26 of the 7th American Approximation is saline rendzina clay.

Profile 57 of the 7th American Approximation, classified as *orthic camborthid*, is serozem.

Profile 59 of the 7th American Approximation, classified as *orthic calcorthid*, is serozem with calcareous accumulation at 12 inches.

Profile 60 of the 7th American Approximation, classified as *orthic calcorthid*, is rendzina solonetz natric at 27 inches with gypsic horizon.

Profile 1 of the 7th American Approximation is rendzina 8 inches humic, perhaps gleisolic.

Profile 4 of DUCHAUFOUR (1957), classified as *sol alluvial calcaire*, is rendzina 5 cm humic.

Profile 8 of DUCHAUFOUR (1957), classified as *rendzine forestière*, is rendzina more than 40 cm humic, perhaps rendzina ranker.

Profile 9 of DUCHAUFOUR (1957), classified as *rendzine grise de pelouse*, is rendzina ranker 25 cm deep.

Profile 12 of DUCHAUFOUR (1957), classified as *sol brun calcaire*, is rendzina 8 cm humic.

Profile 79 of São Paulo (ANONYMOUS, 1960a), classified as *lithosol*, is protorendzina 50 cm deep.

The Congolese profile No. 4 of SYS (1960), classified as *rendzina on calcareous crust*, is rendzina clay 30 cm dark humic.

The Australian *skeletal* profile B of STACE (1961) is aeolian rendzina with limestone at 12 inches.

The Australian *aeolian sand* profile A, B and C of STACE (1961) is aeolian rendzina.

The Australian *brown forest*, profile A (STACE, 1961) is chernozemic rendzina 14 inches humic.

The Australian *grey calcareous*, profile A (STACE, 1961) is rendzina 6 inches humic.

The Australian *grey–brown and red calcareous desert soils*, profile A, is rendzina 9 inches humic; profile B cinnamonic serozem with petrocalcic pan at 4 inches; profile C serozem.

The Alaskan *rendzina* profiles (UGOLINI and TEDROW, 1963) are protorendzinas.

The Belgian profile 2 (PAHAULT and SOUGNEZ, 1961) is rendzina 85 cm humic.

The Greek profiles 80, 81, 3, 35, 36, 29, 75, 79, 25, 12 and 16 (ZVORYKIN, 1960) are rendzinas, many of them gleisolic.

The Moroccan *crust tirs* (DEL VILLARS, 1944) are rendzinas with petrocalcic pan, sometimes chernozemic, some of them are rendzina clays.

The Chadian *calcareous slightly natroné* profile Moto (PIAS and GUICHARD, 1959) is rendzina 25 cm humic; the calcareous natroné koona rendzina more than 45 cm humic.

The Turkish *alluvial and youthful* profile (OAKES, 1957) Tarsus silty clay loam is rendzina; Firat river valley, rendzina; and Cumra valley, rendzina.

The Turkish *reddish brown* profile (OAKES, 1957) Konya is rendzina with calcium accumulation at 37 cm.

The Turkish *sierozem* (OAKES, 1957) profile Çakmak is serozem.

The Turkish *rendzina* (OAKES, 1957) profile Yassiviran (2 km from) is rendzina intergrade to rendzina clay.

The Pakistani profile (KARIM and QUASEM, 1961) Rasulpur silty loam is slightly lessivé slightly alcaline rendzina.

The Indian profile (AGARWALL and MUKERJI, 1951) is rendzina.

ANDO GROUP (SOILS RICH IN AMORPHOUS CLAYS)

General considerations

Young soils from volcanic materials are usually rich in amorphous clays (allophane) of high cation exchange capacity. Moreover allophane interferes with organic matter decay and favours organic matter accumulation. This is why these soils have deep "humic" horizons. Their apparent density is usually low, and moisture tension at 15 atm. is higher than their clay content.

Typical ando soils are acid and more or less red, but those formed under drier conditions are neutral and they have low chromas (white–grey–black). The latter differ little from chernozemic or brown soils and are usually classified with them. In Argentina a great part of the materials from which chernozemic soils have been formed is of volcanic origin. This is why these soils have very high cation exchange capacity and are very fertile. Hitherto little attention has been paid to ando soils of dry climate, but they abound in the world and their economic importance is great.

Many ando soils have a "latosolic" profile (texture varies little from horizon to horizon) and have been classified as latosols, although the difference between these two groups is enormous. However, volcanic ashes are rich in wheatherable minerals, weathering releases great amounts of Na, K, Mg and SiO_2, and when a dry climate or poor drainage does not permit their elimination by leaching, "textural" horizons are formed; the high frequency of planosolic and solonetzic soils in Argentina is chiefly due to the volcanic origin of parent materials. For the same reasons the end product of the weathering of volcanic materials is often dark clays (margallitic soils) when the climate is dry or drainage is poor.

Some ando soils of humid climate are very rich in iron, having "highly ferruginous" (*ff*) horizons. It is also true that the origin of many highly ferruginous soils (terra roxa, etc.) is volcanic, sometimes they are ando.

General diagnostics

All horizons are "allophanic" (their moisture tension at 15 atm. is greater than their clay content; Tca is higher than clay content; apparent density of air-dry soil ready for analysis is usually below 1.1).

Subgroups and their diagnostics

Acid ando (va): one or more horizons are "acid".

Chernozemic ando (vy): they have a "dark humic" horizon 25 cm (10 inches) or more deep.

Positive ando (v(po)): one or more horizons are "positive".

Slightly lessivé or planosolic ando (v(l)): A is slightly lessivé (*l*).

Ando lessivé (vp): they have an "eluvial" horizon; they belong both to the ando and to the podsolic groups.

Planosolic ando (vu): they have an eluvial horizon produced by solonization; they belong both to the ando and to the planosolic groups.

Ando with pan (vm): they have a pan cemented with silica.

Organic ando (vh): they have "organic" horizons.

Hydrol-ando (vw): the allophane forms a gel; when dried to air it hardens irreversibly into sand and gravel-sized aggregates; usually they have thick humic horizons; sometimes they are organic.

Ferruginous ando (vx): all their "allophanic" horizons are "highly ferruginous".

Ando with "gley" or "semi-gley" horizons will be discussed later.

Correlation and examples

As has been stated earlier ando soils have been recently identified in humid countries (Japan) and many of them, more especially those of dry and desert climates, are still classified in other groups. Moreover, considerable confusion reigns concerning the concept and consequently the diagnostics of ando soils. The main feature of ando soils is their richness in amorphous clays which results in unusual capacities of absorbing water and ions; and their diagnostics can be fixed in such terms; unfortunately moisture tension at 15 atm. and apparent density of air-dry soil ready for analysis are not always mentioned in the literature, and that makes the separation of ando soils difficult. Ando soils are often called volcanic.

The *andepts* of the 7th American Approximation correspond rather well to the author's ando; their *hydrandepts* to his hydrol-ando; and their *mollandepts* to his chernozemic ando.

In the French classification (DUCHAUFOUR, 1965) ando corresponds to *andosols*.

Profile 5 of the 7th American Approximation, classified as *orthic umbrandept*, is acid ando.

Profile 52 or the 7th American Approximation, classified as *mollic umbrandept*, is chernozemic ando.

Profile 53 of the 7th American Approximation, classified as *hydrandept*, is hydrol positive ferruginous ando.

DARK CLAYS (SOILS RICH IN 2:1 CLAYS, WELL SATURATED WITH BASES)

General considerations

A high content of expanding clays well saturated with bases and rather poor in iron slows leaching and clay eluviation and keeps the soil young for a long time. Moreover, it gives the soil some special characteristics: the soil swells when moistened and shrinks with drying; wide and deep cracks, gilgai and slickensides may be formed; permeability is low. Red clays are not included in this group, because a high iron content impedes clay dispersion and swelling, and soil lacks the characteristics of dark clays.

These soils are so dark that some time ago they were confused with chernozems. However, while some dark clays have deep "dark humic" horizons and may be classified as chernozemic, others are poor in organic matter; their dark colour is due to their poverty in free iron and their richness in Mg, Ca, and Na. Many dark clays have surface horizons that effervesce with HCl and may be classified as rendzinas; in others the upper limit of effervescence is encountered at shallow depth; some dark clays lack entirely a "calcareous" horizon. Related to Ca and humus content is structural stability and self-mulching; according to this characteristic dark clays (vertisols) are divided in the Seventh American Approximation into grumaquerts and grumusterts on the one hand, and mazaquerts and mazusterts on the other.

Dark clays are usually rich in Mg; many of them contain considerable Na; this fact confers to this group some of the features of solonetzic soils. Interior drainage is poor in these soils; exterior drainage is also poor in many cases, because clayey materials are usually deposited in depressions. This is why dark clays are often gleyed, but the horizons formed are usually "dark gley" and it is difficult to decide if a soil is gleisolic or not.

General diagnostics

All horizons are "vertisolic" containing more than 35% clay and have *S* above 30 mequiv.; the upper 5 cm may be less clayey; no horizon is "red".

Subgroups and their diagnostics

Rendzina clays (mr): all horizons are "calcareous"; they belong at the same time to the dark clays and rendzina groups.

Chernozemic clays (my): they have a "dark humic" horizon 25 cm (10 inches) or more thick.

(Sensu stricto) *Dark clays (m):* they are neither rendzinas nor chernozemic.

Cat clays (ma): marine clays contain often sulphides and become acid after drainage; the soil formed is called cat clay; it is a degradation rather than a subdivision of dark clays; cat clays are encountered in polders, mangrove areas, etc.

Roof clays (mm): they are so impermeable that they are used for roofing; in the Mediterranean area roofs are flat and they are made of a mat of reeds covered with these clays.

Ranker clays (ml): the "humic" horizon rests directly on "bed-rock".

Saline, natric or magnesic clays (ms, mn, m(mg)): one or more horizons are "salic", "natric" or "magnesic"; it is convenient to specify at what depth this horizon begins.

Dark clays with "gley" or "semi-gley" horizons will be discussed later.

N.B. These terms do not necessarily exclude one another; e.g., a soil may be a natric rendzina clay or a saline gleyed clay, etc.

Correlation and examples

Dark clays have been called regur, black cotton, margallitic, black clays, black earths, grumusols, vertisols. The concept also varies from author to author. By fixing high limits of S and excluding red colours many soils that are not typical are eliminated from the group. But in each group, except kaolisols and ando, a special subgroup is established for soils with more than 35% clay in all horizons (acid clays, cinnamonic clays, etc.). The problem of limits between dark clays on the one hand, and chernozemic, rendzinas, saline, alkaline on the other, is solved by admitting that a soil may belong at the same time to two groups: chernozemic clay, rendzina clay, and so on; the reasons have been repeatedly stated.

The *vertisols* of the 7th American Approximation correspond rather well to the dark clays of the author as is also the case with the vertisols of French classification (DUCHAUFOUR, 1965).

Profile 42 of the 7th American Approximation, classified as *orthic grummaquert*, is chernozemic rendzina clay.

Profile 43 of the 7th American Approximation, classified as *orthic mazaquert*, is rendzina clay; it may be gleisolic.

Profile 45 of the 7th American Approximation, classified as *orthic grumustert*, is para-rendzina clay.

Profile 46 of the 7th American Approximation, classified as *orthic mazustert*, is dark clay.

Profile 26 of the 7th American Approximation is saline rendzina clay.

Profile 7 of DUCHAUFOUR (1957), classified as *tir à croûte zonaire*, is chernozemic rendzina clay with petro-calcic horizon at 40 cm.

The Australian *rendzinas*, profiles A, B and C (STACE, 1961) are chernozemic rendzina clay.

The Australian *grey of heavy texture*, profiles A and B (STACE, 1961), are rendzina clays; profile C gleyed dark clay.

The Australian *desert loam*, profile B (STACE, 1961), is rendzina clay.

The Australian *black earths*, profiles A, B and C (STACE, 1961), are chernozemic clays.

The American grumusol profil Summit I (JARVIS et al., 1959) is chernozemic clay.

The Santo Tomé *black tropical, with lime concretions* profile 31 (SACADURA and CARVALHO, 1960) is dark clay, perhaps chernozemic.

The South African *subtropical black clay* (VAN DER MERWE and HEYSTEK, 1955) is chernozemic clay; the *intrazonal black clays* Barkly east, Bethal and Sataha chernozemic clays.

The Ghanaian *tropical black earth* profile Akuse (H. Brammer, in: WILLS, 1962) is dark clay.

The Moroccan *gley tirs* (DEL VILLARS, 1944) are rendzina clay-glei; the *broken land tirs* chernozemic rendzina clays sometimes with calcareous accumulation.

The Tchadian *tropical black earth* Tagaga (PIAS and GUICHARD, 1959) is chernozemic clay more than 80 cm humic.

The Ivory Coast *black earth* profile Boudoukou No. 3 (DABIN et al., 1960) is chernozemic clay more than 30 cm humic.

The Dahomeian *tropical black clay* profile (WILLAIME, 1959) H 31, H 15, H 3 and 18 are dark clays.

The Turkish *alluvial and youthful* profile (OAKES, 1957) Carsamba delta is dark clay 20 cm humic.

The Turkish *hydromorphic alluvial* profile (OAKES, 1957) 2 A is rendzina clay gley.

The Turkish *hydromorphic alluvial* profile (OAKES, 1957) Tarsus is rendzina clay gley.

The Turkish *brown* profiles (OAKES, 1957) Kallipinar village, Ankara clay and Yogat are rendzina clay.

The Turkish *reddish brown* profile (OAKES, 1957) Altinova clay is rendzina clay; and Malaya province is dark clay.

The Turkish *grumusol* (OAKES, 1957) profile Çorlu is dark clay; and Havsa chernozemic clay.

The Turkish *rendzina* (OAKES, 1957) profile "7 km from Yassiviran" is rendzina clay; and "2 km from Yassiviran" rendzina intergrade to rendzina clay.

The Birmanese *dark compact* (ROZANOV and ROZANOVA, 1962) profile 520B is dark clay; 12B recent brown intergrade to dark clay; and 111-P alkaline dark clay.

RAW (UNDIFFERENTIATED) SOILS (UNDIFFERENTIATED SOILS WITH DEGREE OF LEACHING 2)

General considerations

Undifferentiated soils consist of parent material, which has been little or not at all modified, except for a certain leaching of bases, which has not yet produced an "acid" horizon; and addition of organic matter, which has not produced a "dark humic" horizon 25 cm or more thick; no horizon is "braunified", "cinnamonic", "gley", "semi-gley", "peaty" or "organic". However, young soils from clayey, carbonaceous, volcanic or kaolinitic materials do not belong to this group, because such materials confer peculiar characteristics. Since undifferentiated soils lack the special characteristics of kaolisols, ando, rendzinas, dark clays, brunisolic, cinnamonic, organic and gleisolic soils, they cannot have "kaolinitic", "allophanic", "vertisolic", "acid", "braunified", "cinnamonic" or "eluvial" horizons, and their upper horizon cannot be "calcareous". The distinction between "raw" soils on one hand, "brunisolic" and "cinnamonic" on the other is often difficult, because micropedologic examinations are not usually included in pedologic routine, and we cannot know if the soil has a "braunified" or "cinnamonic" horizon. A distinction should be made between soils formed from recent alluvium, those formed from dunes and those from consolidated rocks. A classification according to parent material is useful in the case of these soils.

Subgroups and their diagnostics

Lithosols (1): so shallow and/or so skeletal that if all fractions coarser than 2 mm were eliminated the fine earth would not suffice to form a soil 25 cm (10 inches) thick; better said the weight of fine earth per square meter is less than 350 kg. A distinction should be made between desert lithosols and humid climate lithosols; the first are entirely lacking a "humic" horizon.

Desert sands (ed): all horizons are "aeolian"; they

cannot have "humic" horizons; the upper limit of salinity is usually high.

Recent dunes (e): same as desert sands, but they may have "humic" horizon.

Recent alluvial (a); they often have an "exudational" moisture regime.

Recent brown (b): all horizons are "neutral", but the upper limit of effervescence does not reach the surface; they do not have "braunified" or "cinnamonic" horizons. Since such horizons are seldom determined in pedologic examinations, it is often difficult on the basis of published data to distinguish recent brown from braun erde, or from not distinctly cinnamonic soils.

Arid brown (d): recent brown of drier climate; naturally, they cannot have a "humic" horizon more than 25 cm (10 inches) thick.

Brown with calcareous accumulation (b(cac)): same as recent brown, but a "calcareous accumulation" is encountered at some depth.

Brown with petrocalcic horizon (bj): same as recent brown, but a "petrocalcic" horizon is encountered at some depth.

Old rendzinas (br): former rendzinas, the surface of which has been decalcified and does not effervesce with HCl; naturally their "humic" horizon is less than 25 cm thick and they lack the "braunified" horizon of braun-erde.

Para-serozems (brd): same as arid brown, but they lack humic horizons.

Brown clays (bm): recent brown with clay content higher than 35%, but S (sum of absorbed bases) less than 30 mequiv. Naturally they cannot be acid.

Slightly lessivé recent brown (b(l)): recent brown with "slightly lessivé" horizon.

Polygonal: recent brown with the characteristic polygons of arctic or high altitude soils.

Correlation and examples

The concept of raw soils is very old. They were called "azonal" in the older classifications, but in the deserts, horizon differentiation seldom takes place, raw soils abound and they form the "zonal" soil of this climate. Since an azonal soil cannot be zonal, the raw soils of the desert were separated from those of other climates and formed an artificial group; many authors included in it solonchaks, alkaline, solonetz, planosolic soils that have nothing in common with raw soils. It is preferable to follow the example of French authors (DUCHAUFOUR, 1965) and classify desert soils in the group of raw soils (sols minéraux bruts). Desert is not a type of soil, it is a type of environment, a "soil region".

The *entisols* of the 7th American Approximation correspond roughly to our raw soils. The correspondence is better with the "minéraux bruts" of the French classification.

The Argentinian profile classified as *regosolic prairie* by BONFILS et al. (1959) is recent dune. That classified as *prairie intergrade to brown and regosolic* is arid brown; and that classified as *brown*, arid brown.

The Argentinian profile described by GOLLAN et al. (1936) and GOLLAN and LACHAGA (1939) under number 178 is slightly planosolic brown 15 cm humic.

The Antarctic profile described by MOLFINO (1956) under number I is lithosol.

The Brasilian *latosol* profile C-4 (COSTA LIMA, 1953) is recent brown.

The Argentinian profiles 2 and 9 (PIÑEIRO and ZUKARDI, 1959) are arid brown.

The Greek profile 21 (ZVORYKIN, 1960) is recent brown; and profile 24 recent alluvial.

The Santo Tomé *alluvial* profile 112 (SACADURA and CARVALHO, 1960) is recent alluvial; and the *lithosolic from sandstone* profile 45 V.C. recent brown.

The Tanzanian *dark grey* profile 5 (MUIR et al., 1957) is recent brown intergrade to dark clay 10 cm humic.

The Tchadian *steppe brown* profiles 781–782 (PIAS and GUICHARD, 1959) is arid brown.

The Ivory Coast *brown earth* profile 12 (DABIN et al., 1960) is recent brown more than 10 cm humic.

The Ivory Coast *black earth* profile No. Z 2 (DABIN et al., 1960) is recent brown more than 10 cm humic.

The Dahomeian *alluvial* profile (WILLAIME, 1959) H 23 and H 32 are recent alluvial.

The Dahomeian profile (WILLAIME, 1959) JBO 10 is recent brown more than 10 cm humic.

RANKERS (SOILS WITH DEGREE OF LEACHING 2 CONSISTING OF A HUMIC HORIZON RESTING ON BED-ROCK OR PERMAFROST)

General considerations and diagnostics

Rankers are soils in which the "humic" horizon rests directly on "bed-rock" or permafrost. They abound in polar climates and high mountains.

Subgroups and their diagnostics

Lithosolic rankers (lh): the humic horizon rests on "bed-rock". They abound in mountains. Many of them are at the same time lithosols and rankers.

Tundra rankers (lt): the humic horizon rests on permafrost.

Alpine rankers: same as lithosolic, but they often have "organic" or "peaty" horizons above it (*lhh* or *lho*).

Neutral rankers (lhn): all horizons are neutral.

Acid rankers (lha): some horizon is "acid".

"Chernozemic rankers" (ly): they have a neutral "dark humic" horizon 25 cm (10 inches) or more thick.

N.B. These terms do not necessarily exclude one another. It is useful to specify the thickness of the "humic" horizon, e.g., lithosolic ranker 15 cm humic; when a "peaty" or "organic" horizon is present, it is useful to specify its thickness, e.g., tundra ranker 15 cm peaty, 8 cm humic.

Correlation and examples

Soils of cold mountains, shallow but rich in organic matter, have always been considered as a special group and given a special name as "humus alpine", etc. The 7th American Approximation made a special sub-order of them, that of umbrepts; but the suborder includes soils that are much deeper than rankers. In the French classification (DUCHAUFOUR, 1965) rankers correspond to "rankers alpins", and "rankers pseudo-alpins".

Profile 2 of DUCHAUFOUR (1957) classified as *ranker d'érosion à moder* is acid ranker 20 cm deep on granite.

Profile 3 of DUCHAUFOUR (1957) classified as *ranker à mull actif* is ranker 30 cm deep on granitic colluvium.

Profile 11 of DUCHAUFOUR (1957) classified as *rendzine brunifiée* is clay ranker 35 cm deep on limestone.

Profile 81 and 82 of São Paulo (ANONYMOUS, 1960), classified as *lithosol* are rankers 20 cm deep on argilito or fohelho.

The Antarctic profiles described by MOLFINO (1956) under numbers III and V are lithosolic rankers.

The Australian *alpine humus* (profile C, STACE, 1961) is ranker on weathering granite 25 inches deep, 6 inches dystrophic organic.

The American *alpine turf soils* (RETZER, 1956) are alpine (lithosolic) rankers on various substrata; the last one is proto-rendzina.

The Alaskan *regosol* and *lithosol* (TEDROW et al., 1958) are tundra rankers which are more or less deep.

The Colombian *andino humus soil* profile 41 (JENNY, 1948) is alpine ranker 10 in dystrophic organic, 8 in humic; it may be ando; the *parano humus soil* profile Q10 is also alpine ranker 18 in dystrophic organic and 29(?) in humic.

BRUNISOLIC GROUP (SOILS WITH DEGREE OF LEACHING 2 AND "BRAUNIFIED" OR "ACID" HORIZONS)

General considerations

When the climate is cool and the vegetation woody, slightly polymerized humus (rich in fulvic and brown humic acids) is produced. This humus complexes iron, and a finely granulated brown plasma (see KUBIENA, 1953, pl.XIX, fig.1) is formed. Soil has a "braunified" horizon. With a more cold-humid climate, or more time, clay eluviation takes place, an "eluvial" horizon is formed and the soil becomes podsolic. When the soil dries thoroughly from time to time, iron oxides dehydrate irreversible and small concretions are formed (see KUBIENA, 1953, pl.XIX, fig.5); a "cinnamonic" horizon is formed and the soil becomes cinnamonic. Thus brunisolic soils grade to podsolic on the one hand, and to cinnamonic on the other.

Besides climate, parent materials and vegetation are important. Materials rich in iron are difficult to podsolize and produce brunisolic soils under conditions favourable for podsolization. Rendzinas are braunified before their podsolization. Deciduous forest favours braunification, while coniferous forest favours podsolization. Brunisolic soils may be neutral (eutrophic braunerde) or acid (brun acide or oligotrophic braunerde). The first are usually formed from materials rich in ferromagnesian minerals (basalt, etc.). The second form the transition to podsolic soils. Brunisolic soils grade also into chernozemic, in the transition zone between forest and grassland, or on materials rich in carbonates; the soil becomes richer and richer in organic matter and darker in colour; these soils are called chernozemic braunerde; some prairie soils lacking an "eluvial" horizon belong to this subgroup. It is to be noticed that under tropical humic climates humus is rich in fulvic acids; according to TYURIN et al. (1960) the ratio humic fulvic acid is 0.6–0.8 in Russian krasnozems, the same as in podsolic, 0.4 in Chinese krasnozems, and 0.2 in Vietnamese lateritic; it is 1.5–2.5 in sensu stricto chernozem. Therefore the brown soils of warm humid climates with 2:1 clays are probably braunified.

General diagnostics

Brunisolic soils have a "braunified" horizon (iron oxides finely flocculated with organic matter forming a brown plasma) 10 cm (4 inches) or more thick; but soils having an "acid" horizon and lacking "allophanic", "vertisolic", "cinnamonic", "eluvial" or "kaolinitic" horizons belong to the brunisolic group, even if they lack a "braunified" horizon. Since micro-pedologic examinations have not yet entered in routine analysis, it is often difficult to distinguish a neutral brunisolic soil (braunerde) from a "recent brown". For the same reason the distinction between "cinnamonic" and "brunisolic" soils is not easy on the basis of existing soil data.

Subgroups and their diagnostics

Braunerde (b): they have a "braunified" horizon; all horizons are "neutral". As we have said the distinction from recent brown is often difficult.

Brun acide (ba): one or more horizons are "acid", but they do not have an "iron illuvial" (*Bir*) or "spodic" horizon; they usually have braunified horizons.

Calcareous braunerde (br): braunerde with a calcareous horizon. They are usually old rendzinas.

Chernozemic braunerde or prairie (by): their humic horizon is 25 cm (10 inches) deep; they belong to both the brunisolic and the chernozemic groups.

Degraded chernozems (byr): they combine the characteristics of chernozemic and calcareous braunerde ("humic" horizon more than 25 cm thick and "calcareous" horizon at some depth).

Organic braunerde or brun acide (bh or bah): the upper part of the "humic" horizon is "organic"; they are encountered in high mountains and polar climates.

Peaty braunerde or brun acide (bo or bao): the upper part of the humic horizon is "peaty".

Slightly lessivé brown or brun acide (b(l) or ba(l)): the upper horizons are "slightly lessivé".

Arctic brown and arctic brun acide (bt or bat): they are encountered in polar regions; many of them have "red" horizons and might be classified as "cinnamonic" but it may be their red colour is due not to dehydrated iron oxides but to fulvic acids.

Ferruginous braunerde or brun acide (bx or bax): they have a highly ferruginous horizon; their colours are often bright red.

Acid clay (m(ac)): one or more horizons are "acid"; all horizons below 5 cm from the surface contain more than 35% clay.

Brunisolic soils with "gley" or "semi-gley" horizons will be discussed later.

N.B. These terms do not necessarily exclude one another; for instance a braunerde may be simultaneously organic and slightly lessivé. It is useful to specify depth of the "humic", "peaty" or "organic" horizon.

Canadians (STOBBE, 1962) divide brunisols (the term is Canadian) into: forest (mull type A_1) and wooded (mor type A_1); under mor type A_1 podsolic weathering and iron illuviation take place and soil is usually brown podsolic.

Correlation and examples

Brunisols have long ago been identified in Germany by Raman, and European authors have always considered them as a special group. Later the concept traversed the Atlantic and brown forest or brun acide soils have been recognized in the United States. In the 7th American classification brunisolic soils correspond very roughly to ochrepts. In the French classification (DUCHAUFOUR, 1965) they correspond rather well to *"bruns tempérés"*; they also include *"bruns eutrophes tropicaux"*. The *brunisolic* soils of Canada (STOBBE, 1962) correspond rather well to our brunisolic; but they probably include brown podsolic.

The *brown forest soils* of the United States of THORP and SMITH (1949) correspond rather well to our braunerde.

Profile 55 of the 7th American Approximation, classified as *orthic haplumbrept*, is brun acide 32 inches humic.

Profile 14 of DUCHAUFOUR (1957), classified as *sol brun acide*, is brun acide.

Profile 13 of DUCHAUFOUR (1957), classified as *sol brun à mull calcique*, is old rendzina more than 40 cm humic.

Profile 15 of DUCHAUFOUR (1957), classified as *sol brun melanisé (jeune)*, is chernozemic braunerde more than 50 cm humic.

The Australian *skeletal* profile A (STACE, 1961) is

brun acide with weathering micaschist at 11 cm.

The Australian *brown forest*, profile B (STACE, 1961) is old rendzina.

The American profile Holyoke series from Hampten, Mass., Hartford, Conn., and New Haven, Conn. (TAMURA et al., 1959) are brun acide.

The Californian profile Hugo fine sandy loam (ULRICH et al., 1959) is brun acide, but only slightly acide.

The Belgian profile 1 (PAHAULT and SOUGNEZ, 1961) is braunerde 31 cm humic.

The Santo Tomé non-calcareous psammo-regosol profile 113 (SACADURA and CARVALHO, 1960) is brun acide.

The South African *intrazonal black clay* Komgha (VAN DER MERWE and HEYSTEK, 1955) is acid clay.

The Ugandan Buwekula brown profile (RADWANSKI and OLLIER, 1960) is brun acide 3 inches humic.

The *arctic brown* loamy sand (TEDROW and DOUGLAS, 1964) is arctic brown 18 inches humic.

The Alaskan *podsol-like* profiles O-11, A-1 and A-3 (BROWN and TEDROW, 1963) are arctic brun acide.

The English *ferritic brown earth* (STORRIER and MUIR, 1962) is ferruginous braunerde.

CINNAMONIC GROUP (RUBIFIED SOILS WITH DEGREE OF LEACHING 2)

General considerations

When the soil dries thoroughly from time to time, iron oxides are dehydrated irreversibly and form minute concretions that give the soil a reddish (cinnamon) colour. Such soils are called cinnamon (Korichnevie in Russian). Cinnamonic soils abound in climates with a dry season; parent materials rich in ferromagnesian minerals (basalts, etc.) favour their formation. On the contrary soils derived from unconsolidated materials, that are poor in iron in relation to clay content, are seldom red. The thin mantle of soil that is formed on the surface of hard limestone is exposed to drought even when the climate is humid. This is why hard limestones give rise to cinnamonic soils (terra rossa) under an extended gamut of climates. Cinnamonic soils are usually neutral; but some are acid (acid cinnamonic); these seem to be intergrades to kaolisols (krasnozems). In some cinnamonic soils (red cinnamonic) all horizons are "red"; in other (reddish brown) the upper horizons are brown; others (pale cinnamon) have pale colours. Some acid cinnamonic soils (cinnamonic rubrozems) have a "dark humic" horizon 25 cm (10 inches) or more thick. Some cinnamonic soils have a "slightly lessivé" surface, being solonized or podsolized. In cinnamonic soils, iron concretions are usually minute (see KUBIENA, 1952, pl.XIX, fig.5); but a few cinnamonic soils contain iron concretions, ferruginized gravel and ferruginized rock brash larger than 2 mm.

General diagnostics

An horizon 10 cm (4 inches) or more thick is "cinnamonic" ("red" or "light red"). No horizon is "kaolinitic" (*k* or *kk*) or eluvial (*e*). The upper limit of effervescence does not reach the surface.

Subgroups and their diagnostics

Terra rossa (cr): cinnamonic soils formed from limestone.
Red cinnamonic (c): all horizons are "red".
Reddish brown (cy): the upper horizons are "brown"; the "cinnamonic" horizon begins at some depth.
Pale cinnamon (c): neither red, nor reddish brown, but cinnamonic.
Reddish desert sands (ced): all horizons are "aeolian"; no horizon is "humic".
Cinnamonic clays (cm): clay content above 35% throughout the profile.
Cinnamonic rubrozems (ch): they have an "acid", "dark humic" horizon 25 cm or more thick.
Acid cinnamonic (c(ac)): one or more horizons are "acid".
Slightly lessivé or slightly planosolic cinnamonic (c(l)): they have a "slightly lessivé" horizon.
Reddish desert (cd): they lack entirely a "humic" horizon; the "cinnamonic" horizon reaches the surface.
Cinnamonic lithosol (cl): lithosol with cinnamonic horizon.
Ferruginous cinnamonic (cx): "all horizons" are "highly ferruginous".
Concretionary cinnamonic (c(co)): they have a "concretionary" horizon.
N.B. These terms do not necessarily exclude one another; e.g., a reddish brown may be slightly lessivé, and so on.

Correlation and examples

Until recently, neutral, more or less red soils of dry climates were classified with chernozemic as steppe soils. But cinnamonic soils differ from chernozemic not only in the quantity of humus, which is much lower, but also in its quality; their humus is much richer in fulvic acids than that of chernozemic soils. Quite recently Russians formed a special group of these soils and called it cinnamon (korichnevie). Approximately at the same time Kubiena showed that the form in which iron is encountered differs considerably from a brunisolic soil to a cinnamonic one and separated southern braunerde (cinnamonic) from braunerde (brunisolic). Long before, Americans had recognized as a special group the "non-calcic brown" of the west, which are cinnamonic and more or less lessivé; they also had separated southern chernozemic from northern chernozemic and called them "reddish prairie", "reddish chestnut" or "reddish brown"; all these soils are cinnamonic or cinnamonic–chernozemic. In their 7th Approximation, suborder ustalf corresponds to these soils (cinnamonic slightly lessivé or lessivé).

In the French classification (DUCHAUFOUR, 1965) cinnamonic corresponds to "*rouges méditerranéens non lessivés*" and "*châtains rouges et bruns rouges subtropicaux*".

The distinction of cinnamonic soils from chernozemic is easy on the basis of the depth of the dark humic horizon; the distinction from kaolinitic is also easy on the basis of the relation between cation exchange capacity and clay content. The distinction from brunisolic should be based on micro-pedologic criteria.

The *reddish brown* soils of the United States of THORP and SMITH (1949) correspond rather well to our reddish brown; and many *non-calcic brown* are cinnamonic.

Profile 41 of the 7th American Approximation, classified as *orthic orthustent*, is cinnamonic serozem.

Profile 7 of the 7th American Approximation is acid cinnamonic.

Profile 10 of the 7th American Approximation, classified as *rhodustalf*, is terra rossa 22 inches humic.

Profile 95 of the 7th American Approximation, classified as *orthic rhodochrult*, is acid cinnamonic 14 inches humic.

Profile 102 of the 7th American Approximation, classified as *idox*, is ferruginous cinnamonic; perhaps an old ando.

Profile 25 of DUCHAUFOUR (1957), classified as *terra rossa brunifiée*, is cinnamonic clay.

Profile 28 of São Paulo (ANONYMOUS, 1960), classified as *red–yellow mediterranean*, is red cinnamon.

Profile 76 of São Paulo, classified as *lithosol*, is ferruginous red cinnamonic 30 cm humic with basalt at 35 cm.

Profile 80 of São Paulo, classified as *lithosol*, is red cinnamonic 15 cm humic with arenito at 25 cm.

Profiles 20 and 21 of Rio de Janeiro (ANONYMOUS, 1958), classified as *red–yellow mediterranean*, are red cinnamonic.

The Australian *red podsolic*, profile B (STACE, 1961), is slightly lessivé cinnamonic.

The Australian *terra rossa*, profiles B and C (STACE, 1961), are terra rossa.

The Australian *brown earth*, profile B (STACE, 1961), is reddish brown.

The Australian *desert sand plain soils*, profiles A and B (STACE, 1961), are desert sands.

The Australian *stony desert table land soil*, profile A (STACE, 1961), is cinnamonic clay with desert pavement; and profile C, cinnamonic rendzina clay.

The Australian *desert sand hill soil*, profiles A, B and C (STACE, 1961), are reddish desert sands.

The Brazilian profile BA-13 aluviao de gloria (SILVA CARNEIRO, 1951) is cinnamonic; A1-2 (Tana) red cinnamon; and PE-4 (Pajehu) slightly lessivé cinnamonic.

The Argentinian profile 3 (PIÑEIRO and ZUKARDI, 1959) is slightly lessivé reddish brown; profiles 4 and 5 are cinnamonic; profile 6 reddish brown 10 cm humic.

The Greek profile 48 (ZVORYKIN, 1960) is terra rossa 9 cm humic.

The South African *intrazonal black clays of the subtropical region* Sphinx and Sethlers (VAN DER MERWE and HEYSTEK, 1955) are cinnamonic clays.

The South African profile Kenhardt (C. R. van der Merwe and Weber, 1963) is desert cinnamonic lithosol; the *subtropical brown lowveld* profile Blyde river red is cinnamonic on granite, and Shangone red cinnamonic.

The Tanzanian Kando sandy loam profile (MILLER and MEHLICH, 1960) is red cinnamonic slightly lessivé.

The Ivory Coast *tropical ferruginous* profile (DABIN et al., 1960) Bouna No.2 is pale cinnamonic; and Bondoukou (N) No.10 red cinnamonic.

The Turkish *terra rossa* (Oakes, 1957) profile Renyali is terra rossa.

The Turkish *non-calcic brown* (OAKES, 1957, p.108) is reddish brown.

The Turkish *reddish prairie* (OAKES, 1957) is cinnamonic clay intergrade to terra rossa.

The Birmanese tropical cinnamon brown (ROZANOV and ROZANOVA, 1961) profiles 3 F, 103 F, 99 F, 145 F, 179 C, 1-P and 247 W are slightly lessivé reddish brown.

The Birmanese *red–brown savannah* (ROZANOV and ROZANOVA, 1961) profiles 506 and 211 W are red cinnamonic.

CHERNOZEMIC GROUP (SOILS WITH DEGREE OF LEACHING 2 AND DEEP NEUTRAL DARK HUMIC HORIZONS)

General considerations

Grasses enrich the soil with dark humus; and this humus, instead of being accumulated on the soil surface, is well distributed in the upper decimeters. As a consequence, grassland soils have deep "dark humic" horizons and form a natural group recognized long ago by pedologists. Grassland is the climax vegetation where climate is not continuously hot, leaching rainfall is low (less than 20% of annual potential evapotranspiration) and the spring season non-"dry" (PAPADAKIS, 1961). Such conditions are only encountered, over extensive areas, in Russia and the Danubian Basin, Canada, the United States and Argentina. In Russia, where chernozemic soils were first identified, they have been formed from materials rich in carbonates. Moreover leaching rainfall is very low in the Russian steppes. As a consequence Russian chernozemic soils have a "calcareous" horizon, that immediately underlies, or overlaps with the "humic" horizon. Clay illuviation seldom takes place; as the climate becomes drier, the humic horizon becomes less dark and less thick, while the calcareous horizon approaches the surface and gypsum and other salts appear in the lower part of the profile. For all these reasons Russians classified their chernozemic soils on the basis of these criteria into chernozem of greater or lesser thickness, kastanozems and brown soils.

In the United States, Canada and Argentina, many chernozemic soils have been formed from non-calcareous materials; moreover leaching rainfall is often higher. As a consequence the "calcareous" horizon may be lacking entirely, or it is separated from the "humic" horizon by a "non-calcareous" layer. Absence of carbonates and richness in Na, Mg and SiO_2 produced by silicate weathering favour clay illuviation; thus a variety of soils not encountered in Russia have been formed. For all these reasons attempts to apply the original Russian classification to other countries have not been so successful; it is necessary to make a distinction between calcareous chernozemic soils formed from materials rich in carbonates, and para-chernozems formed from silicates, or intensely leached.

General diagnostics

Chernozemic soils have a "dark humic" horizon 25 cm (10 inches) or more thick; no horizon is kaolinitic (*k* or *kk*), or "eluvial" (*e*); all horizons are "neutral".

Subgroups and their diagnostics

Calcareous chernozemic (yr): the "humic" (*h*) horizon rests directly on a "calcareous" (*ca*) horizon, or overlaps with it.

(Sensu stricto) *Chernozems (yr):* calcareous chernozemic with the "humic" horizon entirely "black".

Kastanozems (yr): calcareous chernozemic with the lower part of the "humic" horizon "brown".

Chernozemic "brown" (yr): calcareous chernozemic with the "humic" horizon entirely "brown".

Para-chernozems (y'): between the lower limit of the "humic" horizon and the upper limit of "effervescence" there is a layer which does not effervesce with HCl; many para-chernozems entirely lack "calcareous" horizons.

Reddish para-chernozems (yc): para-chernozems with cinnamonic ("red" or "light red") horizons immediately underlying the "humic" horizon, or overlapping with it.

Prairie: (slightly lessivé para-chernozems) symbol *yb(l):* the upper horizon is "slightly lessivé"; clay eluviation is due to the action of organic substances. Not all soils usually called "prairie" belong to this group; those lacking a textural B are "para-chernozems" (*y'*) "brunisolic para-chernozems" (*yb*) or "chernozemic braunerde" (*by*); those with a strong textural B are "prairie lessivé" (*yp*).

Petrocalcic para-chernozems (y'j): para-chernozems with a "petrocalcic" horizon (calcareous hardpan).

Aeolian chernozems or para-chernozems: all horizons are "aeolian" (very sandy).

Slightly planosolic para-chernozems: the upper horizon is "slightly lessivé" (*l*); clay eluviation is due to Na.

Meadow chernozems or para-chernozems (yw): chernozemic soils with "exsudational" moisture regime; they often contain carbonates or salts due to ground water; they are often gleyed and sometimes alkaline (natric or magnesic); this term is environmental rather than systematic.

Grassland rankers (yl): they have more than 5 tons of non-decayed roots per hectare; *the "dark humic" horizon may be thinner than 25 cm.*

N.B. These terms do not necessarily exclude one another; e.g., a soil may be slightly planosolic petrocalcic para-chernozem. It is useful to specify thickness of the humic horizon.

Correlation and examples

The concept of chernozemic soils is as old as pedology. But, outside Russia, it has been extended to include soils that have humic horizons that are too thin, or colours which are too red; many cinnamonic soils have been classified as chernozemic. That has been unfortunate because the difference between chernozemic and cinnamonic soils is not only a question of humus quantity, but humus quality; the humic/fulvic acids ratio is much lower in cinnamonic soils.

Chernozemic soils correspond to *mollisols* of the 7th American Approximation; but mollisols include some soils that have not sufficiently thick humic horizon to be classified as chernozemic in our scheme.

In the French classification (DUCHAUFOUR, 1965) chernozemic soils correspond to "chernozems", "châtains" and "bruns steppiques".

The *chernozemic* soils of Canada (STOBBE, 1962) correspond rather well to our chernozemic.

The *dark-coloured soils of semi-arid sub-humid and humid grasslands* of THORP and SMITH (1949) correspond rather well to our chernozemic; and the *reddish chestnut* and *reddish prairie* to our cinnamonic chernozemic.

Profile 42 of the 7th American Approximation, classified as *orthic grummaquert*, is chernozemic rendzina clay; it may be gleisolic.

Profile 2 of the 7th American Approximation is para-chernozem 10 inches humic.

Profile 3 of the 7th American Approximation, classified as *orthic argustoll*, is chernozem slightly lessivé 14 inches humic.

Profile 15 of the 7th American Approximation is chernozem 11 inches humic.

Profile 33 of the 7th American Approximation, classified as *orthic hapludoll*, is para-chernozem 15 inches humic.

Profile 68 of the 7th American Approximation, classified as *orthic haplaquoll*, is chernozem 14 inches humic.

Profile 69 of the 7th American Approximation, classified as *orthic argaltoll*, is para-chernozem slightly planosolic 12 inches humic.

Profile 70 of the 7th American Approximation, classified as *orthic calcaltoll*, is rendzina chernozem 10 inches humic.

Profile 71 of the 7th American Approximation, classified as *orthic vermudoll*, is rendzina chernozem 34 inches humic.

Profile 72 of the 7th American Approximation, classified as *orthic argudoll*, is prairie (slightly lessivé para-chernozem) 17 inches humic.

Profile 73 of the 7th American Approximation, classified as *orthic haplustoll*, is para-chernozem 22 inches humic.

Profile 74 of the 7th American Approximation, classified as *orthic argustoll*, is planosolic para-chernozem 11 inches eluvial, 8 inches humic.

Profile 75 of the 7th American Approximation, classified as *orthic calcustoll*, is rendzina chernozem 12 inches humic.

Profile 5 of DUCHAUFOUR (1957), classified as *chernozem*, is chernozem.

Profile 6 of DUCHAUFOUR (1957), classified as *sol châtain de steppe* is kastanozem.

The Argentinian profile classified as *prairie* by BONFILS et al. (1959), are para-chernozems.

The Argentine profiles described by GOLLAN and LACHAGA (1939, 1944) under numbers 185, 208, 120, 158, 150, 101, 133, 191, 204, 212 and 126 are para-chernozems some of which are slightly lessivé.

The Australian *alluvial* profiles A, B and C of STACE (1961) are chernozemic brown.

The Australian *prairie*, profiles A, B and C (STACE, 1961) are chernozemic clays.

The Californian *prairie* soils described by BARSHAD (1964) may be classified in our scheme as follows: Sweeney, para-chernozem 30 inches humic; Sheridan, para-chernozem 18 inches humic; Colma 2, prairie (para-chernozem lessivé) 30 inches humic; Cayoucos, planosolic? para-chernozem 12 inches eluvial, 12 inches humic; Colma I, slightly lessivé para-chernozem 12 inches humic; Gleason, prairie 24 inches humic.

The American *brunizem* profile, Bates loam (JARVIS et al., 1959), is prairie (para-chernozem slightly lessivé) 14 inches humic; and Bateslike loam, prairie intergrade to planosolic para-chernozem 12 inches humic.

The American *red prairie* profile, Newtonia silt clay loam (JARVIS et al., 1959), is reddish prairie 17 inches humic.

The American *chernozem* profiles, Barnes loam S53 ND-11-1 and S54 ND-10-2 (MCCLELLAND et al., 1959), are slightly lessivé chernozems; Aastad loam 554 ND-2-3 is prairie (para-chernozem slightly lessivé) 10 inches humic.

The Californian profiles, Sweney sandy clay loam 555 Cal-41-10 and Denison loam (ULRICH et al., 1959), are para-chernozem.

The American *brunizem* profile, Down silt loam (WHITE and RIECKEN, 1955), is prairie (slightly leached para-chernozem) 12 inches humic.

The Argentinian profile 1 (PIÑEIRO and ZUKARDI, 1959) is chernozemic brown 40 cm humic.

The Asian profiles 548a and 25a (ZONN, 1962) are

chernozemic brown; and profiles 863c and 116c rendzina chernozems.

The Russian profile 1 (AKHTYRSTEV, 1962), classified as *gray* is prairie (slightly lessivé para-chernozem) 30 cm humic; profiles 2, 3, 4 and 5, classified as *dark gray*, are also prairie; profiles 6 and 8, classified as *gray*, are also prairie, the second is perhaps lessivé; profile 10, classified as *podsolized chernozem*, is also prairie 90 cm humic; profile 11, classified as *dark gray*, is also prairie 70 cm humic.

Chapter 4 | Soil Classification: Soils with Degree of Leaching 3–6 (dominated by 1:1 clays and/or podsolized)

KAOLISOLS (TROPICAL GROUP) (SOILS WITH 1:1 CLAYS)

General considerations

Leaching is very intense in the tropics. Normal leaching rainfall (*Ln*) is for example, 3,860 mm (152.1 inches) in Monrovia (Liberia); even in the arid Sudanian savanna at Kano, Nigeria, it is 362 mm (14.8 inches). By comparison it is 160 mm (6.3 inches) in humid London and 220 mm (8.7 inches) in Moscow. Moreover, the solubility of silica increases with temperature. For all these reasons desilication is rapid in the tropics.

As has already been stated, desilication takes place during weathering. The soil enters directly into phases 3–5. Desilication and consequently poverty in 2:1 clays is the principal characteristic of tropical soils. This is why we call them kaolisols, a term introduced in pedology by the Belgians (SYS, 1961). The term latosol has often been used for soils (ando, cinnanomic, etc.) that have high cation exchange capacity and contain amorphous or 2:1 clays. The term lateritic comes from laterite, and implies the presence of laterite, which is only encountered in certain tropical soils. That is why the term kaolisol is preferable because it points out the fundamental characteristic of tropical soils.

Another peculiarity of tropical soils is that weathering penetrates very deeply; soils are usually many meters deep. This is due partly to the high leaching rainfall and partly to the clays formed, which do not swell and do not interfere with water penetration.

Tropical soils are usually poor in bases. But as their colloids have a low cation exchange capacity and are rich in iron sesquioxides, their pH is not so low and does not give a satisfactory idea of their base status. A better criterion is *S*/clay ratio. As we have said, base status and pH decrease with depth in tropical soils. In many cases more than half of the absorbed bases of a profile 1 m deep is encountered in the upper 25 cm.

Another characteristic of kaolisols is their high degree of weathering; they are poor in weatherable minerals; the silt/clay ratio is low. Non-clay fractions consist almost exclusively of unweatherable minerals and iron concretions; the core of concretions is sometimes formed from unweathered minerals and such minerals increase the base reserve, and base reserve/clay ratio. Kaolisols are usually rich in iron concretions. In some cases such concretions are so abundant and large that a "concretionary" horizon is formed. In soils with "exudational" moisture regime the "concretionary" horizon is often cemented with iron and "laterite" is formed.

The profile of tropical soils varies from "concretionary" to "latosolic". A "concretionary" profile consists of an upper part poor in gravel that may have a "humic" or "eluvial" horizon; a part rich in large concretions, ferruginized gravel and rock brash; and a "pallid" zone poor in gravel. A "latosolic" profile is rather uniform in texture and any change is gradual.

In kaolisols cation exchange capacity decreases rapidly with depth. As a consequence the degree of clay eluviation which is necessary to create an "eluvial" horizon, as defined in this book, is very high; many kaolisols that are considered as lessivé are only slightly lessivé according to our definitions.

Kaolisols are very permeable and leaching rainfall is high in the tropics. As the experiments of HALLSWORTH (1963) have shown, under such conditions pure water can produce clay eluviation. This is why in many kaolisols clay content increases with depth. But contrary to what happens in solonetz, planosols and podsols, clay illuviation takes place at great depth and textural change is gradual. This is why only when a "distinct" change in cation exchange capacity takes place within the upper 50 cm (20 inches) is a soil with "kaolinitic" or "superkaolinitic" horizons considered as podsolic, planosolic or solonetzic. Kaolisols that have an "eluvial" horizon belong at the same time to the kaolisolic and podsolic or planosolic–solonetzic groups.

Kaolisols are deep and may serve as parent material for new soils in another erosion cycle. Some soils of dry or desert climate are formed from kaolinized materials. These soils are called palaeo-kaolisols and palaeo-lateritic. Due to actual conditions they may be rich in bases or lime; sometimes they are re-silicated.

Classification

Kaolisols vary in degree of leaching. According to it, they may be divided into two groups (Table VII):

(*1*) *Leaching degree 3:* some horizons are "kaolinitic" (*k*); no horizon is "super-kaolinitic" (*kk*); krasnozems, tropical ferruginous, terres de barre, etc.

(*2*) *Leaching degree 4:* some horizons are "super-kaolinitic" (*kk*); (sensu stricto) ferralitic.

According to base status, kaolisols may be divided into two groups:

TABLE VII

CLASSIFICATION KEY OF TROPICAL SOILS (SOILS WITH ONE OR MORE KAOLINITIC HORIZONS)[1]

(*A*) "Re-basified" and/or "re-calcified"	
(*I*) Without "laterite" or fragments of it	palaeo-kaolisol
(*II*) With "laterite" or fragments of it	palaeo-lateritic
(*B*) Neither "re-basified" nor "re-calcified"	
(*I*) "Laterite" in upper 50 cm (20 inches)	ground-water laterite
(*II*) No "laterite" in upper 50 cm (20 inches)	
(*1*) "Dystrophic" "dark humic" horizon 25 cm (10 inches) or more thick	rubrozem
(*2*) "Humus podsolic B"	humus–kaolinitic podsol
(*3*) One or more horizons are "highly ferruginous"	terra roxa
(*4*) None of these horizons	
(*a*) "Eutrophic"	
(*a1*) "Latosolic" profile	terre de barre
(*a2*) "Concretionary" or "cuirassé" profile	tropical ferruginous
(*b*) "Dystrophic"	
(*b1*) "Distinct" increase with depth of cation exchange capacity in the upper 50 cm (20 inches)	
(*b1.1*) "Latosolic" profile	red–yellow podsolic
(*b1.2*) "Concretionary" or "cuirassé" profile	lateritic podsolic
(*b2*) No "distinct" increase with depth of cation exchange capacity in the upper 50 cm (20 inches)	
(*b2.1*) Latosolic profile	
(*b2.1.1*) One or more horizons are "super-kaolinitic"	ferralitic
(*b2.1.2*) No horizon is "super-kaolinitic"	krasnozem
(*b2.2*) "Concretionary" or "cuirassé" profile	acid ferruginous

[1] Palaeo-lateritic include desert lateritic sands. Ground-water laterites may be "kaolinitic" or "ferralitic" (super-kaolinitic). Terres de barre, tropical ferruginous, ferralitic and krasnozems may be "slightly lessivé" or "lessivé"; many do not show clay eluviation. Tropical ferruginous and acid ferruginous may be "concretionary" or "cuirassé". Ferrisols are krasnozems nearly ferralitic. Soils having "neutral" "dark humic" horizons 25 cm or more thick are chernozemic kaolisols; and those having "eluvial" horizons due to Na, kaolinitic solonetz or planosols.

(*1*) *Eutrophic:* average "adjusted *S*/clay ratio" of the upper 25 cm (10 inches) above 0.25 and base saturation (*S*/Tca ratio) above 50%; tropical ferruginous, terres de barre, etc. However, a terra roxa is "eutrophic" when base saturation is above 50% even if adjusted *S*/clay ratio is not sufficient (in terra roxa much of the clay fraction is not clay but iron oxides).

(*2*) *Dystrophic:* average "adjusted *S*/clay ratio" of the upper 25 cm (10 inches) below 0.25 and/or base saturation (*S*/Tca ratio) below 50%; krasnozems, acid ferruginous, ferralitic, etc.

According to profile kaolisols may be divided into the following groups:

(*1*) *Latosolic profile:* no horizon is "concretionary" or "laterite" or "eluvial"; krasnozems, ferralitic, terra roxa, etc.

(*2*) *Concretionary profile:* one or more horizons are "concretionary" containing more than 15% of large concretions, ferruginized gravel and ferruginized rock brash, in size above the fine earth fraction; tropical ferruginous, acid ferruginous, etc.

(*3*) *Cuirassé profile:* they have a "laterite" horizon (indurated iron pan); tropical ferruginous cuirassé, acid ferruginous cuirassé, etc.

Some kaolisols have an eluvial horizon produced either by podsolization or solonization. They are in the same time kaolisols and podsolic or planosolic–solonetzic. Thus we have:

(*1*) *Kaolinitic podsolic:* they have an eluvial horizon due to podsolization; red–yellow podsolic, lateritic podsolic, etc.

(*2*) *Kaolinitic planosols:* they have an "eluvial" horizon due to solonization; kaolinitic solonetz are seldom encountered; these soils belong at the same time to the planosolic–solonetzic and kaolisolic groups.

Some kaolisols have a "highly ferruginous horizon" (ff); terra roxa, etc.

General diagnostics

Kaolisols have one or more "kaolinitic" (*k*) or "super-kaolinitic" (*kk*) horizon, but usually the upper horizons do not satisfy the definition of "kaolinitic" horizon; due to organic matter, their cation exchange capacity is too high.

Subgroups and their diagnostics

Latosolic (kl): they lack both "concretionary" horizons and "laterite".

Cuirassé: they have "laterite", but it begins at 50 cm (20 inches) or more below the surface, otherwise the soil is ground-water laterite.

Eutrophic (kn): average "adjusted *S*/clay ratio" of the upper 25 cm (10 inches) above 0.25, and base saturation (*S*/Tca ratio) above 50%.

Dystrophic (kd): average "adjusted *S*/clay ratio" of the upper 25 cm (10 inches) below 0.25 and/or base saturation below 50%.

Kaolinitic (k): no horizon is "super-kaolinitic".

(Sensu stricto) *Ferralitic (kk):* they have one or more "super-kaolinitic" horizons.

Palaeo-kaolisols (q): they are encountered under climates with low leaching rainfall and have been re-basified; sometimes they are re-silicated; their pH is usually above 7.

Slightly lessivé(k(l)): the upper horizons are slightly lessivé.

Krasnozems (kdl): dystrophic, latosolic kaolinitic.

Terre de barre (knl): eutrophic latosolic.

Tropical ferruginous (knf): eutrophic concretionary or cuirassé.

Acid ferruginous (kdf): dystrophic, concretionary or cuirassé.

Terra roxa (kx): one or more horizons are "highly ferruginous" (*ff*); the profile is latosolic; sometimes they have "positive" horizons.

Aeolian krasnozems (ke): their horizons are "aeolian" (very sandy).

Kaolinitic or ferralitic rubrozems (kh or kkh): kaolinitic or ferralitic with a "dark humic" horizon 25 cm or more thick; they are dystrophic.

Chernozemic (ky): they have an "eutrophic dark humic" (*hd*) horizon 25 cm (10 inches) or more thick.

Kaolisols with "eluvial" horizon or "laterite", which begins at less than 50 cm from the surface will be discussed later.

Correlation and examples

Pedologists have long ago recognized that tropical soils differ substantially from those of temperate regions. But it has been very difficult to find out what this difference consists of and to propose a satisfactory diagnostic. The ratio of silica to sesquioxides or to alumina gave rather good results. But cinnamonic and ando soils are rich in sesquioxides and are completely different from tropical soils, having a high cation exchange capacity.

Later, mineralogical analysis advanced and richness in 1:1 clays was used as a diagnostic. But in purifying clay for analysis amorphous clay is lost; and many soils (ando, etc.) with very high cation exchange capacity were included in "lateritic" or "latosols" (kaolisols).

The excessive emphasis given to morphology recently has led to the use of depth and uniformity of profile and colour as a criterion; but many cinnamonic and ando soils have a deep, red uniform profile and they are not at all "lateritic" or "latosols", having high cation exchange capacity. It is only recently that due attention is paid to the relation between cation exchange capacity and clay content.

The terminology used has also changed with time. In the beginning tropical soils (kaolisols) were called lateritic, but as the term has been misused and applied to soils that are not tropical, it has been changed to latosol. This term has also been misused (applied to andosol, cinnamonic, etc.). Belgians (SYS, 1960) proposed the term kaolisol, which has the advantage of pointing out the fundamental characteristic of these soils (1:1 clays). The author considers it the best.

In the 7th American Approximation kaolisols correspond to "*ultisols*" and "*oxisols*"; more precisely ultisols correspond to red yellow podsolic in our scheme and oxisols to terra roxa. But both ultisols and oxisols contain soils that have cation exchange capacities too high to be considered as kaolisols; and some kaolisols, that do not have a textural horizon or are poor in free sesquioxides, are classified in other orders. In the French classification (DUCHAUFOUR, 1965) kaolisols correspond to *tropicaux ferrugineux* and *ferrallitiques*.

Profile 40 of the 7th American Approximation, classified as *ultic quatzopsamment*, is aeolian krasnozem.

Profile 34 of the 7th American Approximation, classified as *orthic dystrochrept*, is krasnozem 7 inches humic.

Profile 88 of the 7th American Approximation, classified as *ultustalf*, is krasnozem.

Profile 27 of the 7th American Approximation, classified as *orthic haplacrox*, is terra roxa 28 inches humic.

Profile 28 of the 7th American Approximation, classified as *udox*, is krasnozem 7 inches humic.

Profile 99 of the 7th American Approximation, classified as *udox*, is terra roxa 14 inches humic.

Profile 100 of the 7th American Approximation, classified as *udox*, is terra roxa 14 inches humic.

Profile 26 of DUCHAUFOUR (1957), classified as *rouge ferrallitic lessivé*, is krasnozem 20 cm humic.

Profile 5 of São Paulo (ANONYMOUS, 1960), classified as *red–yellow podsolic*, is rubrozem 30 cm dark humic.

Profile 7 of São Paulo (ANONYMOUS, 1960), classified as *red–yellow podsolic*, is slightly lessivé krasnozem.

Profile 8 of São Paulo (ANONYMOUS, 1960), classified as *red–yellow podsolic*, is red–yellow podsolic intergrade to terre de barre lessivé 46 cm eluvial.

Profile 10 of São Paulo (ANONYMOUS, 1960), classified as *red–yellow podsolic*, is slightly lessivé krasnozem 23 cm humic.

Profiles 11 and 12 of São Paulo (ANONYMOUS, 1960), classified as *red–yellow podsolic intergrade to yellow latosol*, are ferrallitic 20 cm humic the former, 18 cm the latter.

Profile 15 of São Paulo (ANONYMOUS, 1960), classified as *podsolized with gravel*, is terre de barre 20 cm humic.

Profiles 16 and 17 of São Paulo (ANONYMOUS, 1960), classified as *podsolized with gravel*, are krasnozems.

Profiles 18, 20, 22, 23, 24 and 25 of São Paulo (ANONYMOUS, 1960), classified as *podsolized Lins and Marila*, are terre de barre lessivé.

Profiles 26 and 27 of São Paulo (ANONYMOUS, 1960), classified as *red–yellow mediterranean*, are terre de barre 35 cm humic the former, 33 cm humic the latter.

Profiles 29, 30, 31 and 32 of São Paulo (ANONYMOUS, 1960), classified as *terra roxa estructurada*, are eutrophic terra roxa.

Profiles 33, 35, 36 and 38 of São Paulo (ANONYMOUS, 1960), classified as *terra roxa legitima*, are dystrophic terra roxa.

Profiles 34 and 37 of São Paulo (ANONYMOUS, 1960), classified as *terra roxa legitima*, are eutrophic terra roxa.

Profiles 39, 40, 41, 43 and 46 of São Paulo (ANONYMOUS, 1960), classified as *dark red latosol*, are ferralitic.

Profile 44 of São Paulo (ANONYMOUS, 1960), classified as *dark red latosol*, is kransnozem 15 cm humic.

Profile 45 of São Paulo (ANONYMOUS, 1960), classified as *dark red latosol*, is dystrophic terra roxa.

Profiles 47–63 of São Paulo (ANONYMOUS, 1960), classified as *red–yellow latosols*, are ferralitic.

Profiles 64–66 of São Paulo (ANONYMOUS, 1960), classified as *red–yellow latosols intergrades to red–yellow podsolic*, are ferrallitic.

Profile 67 of São Paulo (ANONYMOUS, 1960), classified as *humic red–yellow latosol*, is ferralitic rubrozem 50 cm dark humic.

Profile 68 of São Paulo (ANONYMOUS, 1960), classified as *humic red-yellow latosol*, is feralitic 240 cm humic.

Profile 69 of São Paulo (ANONYMOUS, 1960), classified as *Campos de Jordao*, is kaolinitic rubrozem 80 cm dark humic.

Profile 70 of São Paulo (ANONYMOUS, 1960), classified as *Campos de Jordao*, is ferralitic rubrozem 85 cm dark humic.

Profile 71 of São Paulo (ANONYMOUS, 1960), classified as *Campos de Jordao*, is ferralitic 10 cm humic.

Profiles 83 and 84 of São Paulo (ANONYMOUS, 1960), classified as regosols, are sandy krasnozems.

Profile 86 of São Paulo (ANONYMOUS, 1960), classified as *regosol intergrade to red–yellow podsolic*, is slightly lessivé sandy krasnozem.

Profile 77 of São Paulo (ANONYMOUS, 1960), classified as *lithosol*, is rubrozem more than 22 cm dark humic with granite at 50 cm.

Profile 78 of São Paulo (ANONYMOUS, 1960), classified as *lithosol*, is krasnozem 19 cm dark humic on phyllite at 23 cm.

Profiles 1–3 of Rio de Janeiro (ANONYMOUS, 1958), classified as *yellow latosols*, are ferralitic.

Profiles 4–6 of Rio de Janeiro (ANONYMOUS, 1958), classified as *red latosols*, are ferralitic.

Profiles 8 and 9 of Rio de Janeiro (ANONYMOUS, 1958), classified as *orange-coloured latosols* are ferralitic.

Profiles 10 and 11 of Rio de Janeiro (ANONYMOUS, 1958), classified as *dark yellow latosols*, are rubrozems.

Profile 14 of Rio de Janeiro (ANONYMOUS, 1958), classified as *red–yellow podsolic*, is slightly lessivé terre de barre 25 cm humic.

Profile 22 of Rio de Janeiro (ANONYMOUS, 1958), classified as *mediterranean red–yellow podsolic*, is terre de barre 35 cm humic.

Profile 23 of Rio de Janeiro (ANONYMOUS, 1958), classified as *red podsolic latosol*, is slightly lessivé ferralitic 25 cm humic, perhaps intergrade to terra roxa.

Profile 24 of Rio de Janeiro (ANONYMOUS, 1958), classified as *red latosolic podsolic*, is slightly lessivé terre de barre 15 cm humic.

Profile 25 of Rio de Janeiro (ANONYMOUS, 1958), classified as podsolic *orange-coloured latosol*, is slightly lessivé ferrallitic.

Profile 27 of Rio de Janeiro (ANONYMOUS, 1958), classified as *latosolic yellow podsolic*, is slightly lessivé krasnozem 15 cm dark humic.

Profiles 16 and 18 of Rio de Janeiro (ANONYMOUS, 1958), classified as *yellow rego-latosol*, are krasnozems.

Profile 17 of Rio de Janeiro (ANONYMOUS, 1958), classified as *yellow rego-latosol*, is terre de barre.

Profile 19 of Rio de Janeiro (ANONYMOUS, 1958), classified as *yellow rego–latosol*, is slightly lessivé ferrallitic, perhaps acid ferruginous.

Profile 28 of Rio de Janeiro (ANONYMOUS, 1958), classified as *alluvial*, is gleyed terre de barre 15 cm humic.

Profile 29 of Rio de Janeiro (ANONYMOUS, 1958), classified as *alluvial*, is terre de barre 30 cm humic.

The Congolese profile No.2 of SYS (1960), classified as *melanic regosol*, is kaolinitic rubrozem 30 cm humic.

The Congolese profile No.5 of SYS (1960), classified as *tropical black clay*, is clayey chernozemic kaolisol 40 cm humic.

The Congolese profile No.6 of SYS (1960), classified as *allophanic brown on volcanic ashes*, is kaolinitic rubrozem 105 cm humic.

The Congolese profile No.8 and 9 of SYS (1960), classified as *ferrisol*, are krasnozem with the first intergrading to terre de barre.

The Congolese profile No.10 of SYS (1961), classified as *humic ferrisol*, is terra roxa rubrozem 28 cm dark humic.

The Congolese profile No.11 (SYS, 1960), classified as *humic ferrisol*, is ferralitic rubrozem 24 cm dark humic.

The Congolese profiles No.12, 13 and 15 of SYS (1960), classified as *ferralsols*, are ferralitic.

The Congolese profile No.14 of SYS (1960), classified as *ferralsol*, is krasnozem.

The Congolese profile No.16 of SYS (1960), classified as *humic ferralsol*, is krasnozem 36 cm humic.

The Congolese profiles No.17, 18 and 20 of SYS (1960), classified as *arenoferrals*, are sandy krasnozems.

The Congolese profile No.22 of SYS (1960), classified as *kaolisol with dark horizon* is ferralitic 67 cm humic.

The Congolese profile No.23 of SYS (1960), classified as *xeroferrisol*, is kaolinitic rubrozem 35 cm humic.

The Australian *lateritic podsolic*, profile B (STACE, 1961), is tropical ferruginous lessivé 14 inches eluvial.

The Australian *lateritic podsolic*, profile C (STACE, 1961), is tropical ferruginous.

The Australian *brown podsolic*, profile B (STACE, 1961), is terre de barre lessivé.

The Australian *red podsolic*, profiles B and C (STACE, 1961), are terre de barre lessivé.

The Australian *krasnozem*, profiles A, B and C (STACE, 1961), are tropical ferruginous.

The Australian *lateritic red earths*, profile A (STACE, 1961), are terre de barre.

The Australian *lateritic red earth*, profile B (STACE, 1961), is palaeo-kaolinitic.

The Australian *lateritic red earth*, profile C (STACE, 1961), is acid ferruginous cuirassé at 70 inches.

The Australian *terra rossa*, profile A (STACE, 1961) is terre de barre.

The Australian *brown soils of light texture*, profiles A, B and C (STACE, 1961), seem to be palaeo-kaolisols.

The Australian *calcareous red earths*, profiles A, B and C (STACE, 1961), are palaeo-kaolisols.

The Australian *desert sand plain* soil, profile C (STACE, 1961), is palaeo-lateritic.

The Australian *calcareous lateritic*, profiles A, B and C (STACE, 1961), are palaeo-lateritic, although the third is dubious.

The American *reddish brown lateritic* soils described by ENGLAND and PERKINS (1959) may be classified in our scheme as follows: Davidson clay loam, Jasper and Oylethorpe Co., krasnozems; Greenville sandy loam, krasnozems; Greenville sandy clay loam, krasnozem; Decatur clay, clay krasnozem; Decatur clay, krasnozem.

The American *red–yellow podsolic*, profile Hagerstown III (HUTCHESON, 1963), is krasnozem intergraded to acid cinnamon.

The American *lateritic* profiles 1, 2, 3 and 4 (NYUN and MCCALEB, 1955) are slightly lessivé krasnozems.

The Brazilian *latosol* profile C-1 (COSTA LIMA, 1953) is krasnozem.

The Brazilian *lateritic clay* profile MG-79 grupo A (FAGUNDES et al., 1951) is krasnozem; and profile MG-88 grupo C terre de barre.

The Brazilian profile AP-1 Macapa (SILVA CARNEIRO, 1953) is ferralitic; AP-3 (Matapi–Colonia) acid ferruginous 35 cm humic; AP-5 (Vicente Monteiro) ferralitic; AP-6 (Campo de Prata) acid ferruginous; AP-7 (Ferreira Gomes) acid ferruginous 30 cm humic; AP-10 (Managano Colonia) acid ferruginous 45 cm humic; AP-11 (Mazacao Taboal) krasnozem 25 cm humic; AP-13 (Amapa) krasnozem; AP-14 (Olapoque) acid ferruginous 90 cm humic; AP-15 (Olapoque) ferralitic 40 cm humic; AP-16 (Olapoque) ferralitic 55 cm humic.

The Congolese *lithosol* profile 1 (HUBERT, 1961) is terre de barre 14 cm humic, granite at 20 cm.

The Angolan *reddish brown semi-arid* profile 1286 (BOTELHO DA COSTA et al., 1959) is terre de barre; the *non-calcic brownish* profile 1931 and 325/55 are terre de barre lessivé; the *non-calcic reddish brown* profile 1964 slightly lessivé terre de barre; the *reddish brown semi-arid* profile 111/54 krasnozem intergrade to cinnamon; and the *non-calcic reddish brown* profile 362/55 krasnozem.

The Portuguese Guinea profile P-30 (DA SILVA TEIXEIRA, 1959) is slightly lessivé ferralitic; and P-27 slightly lessivé krasnozem.

The Santo Tomé *dark brown tropical* profile 9 (SACADURA and CARVALHO, 1960) is terre de barre; *tropical yellow ferralitic* profile 106 is also terre de barre; *yellow ferralitic* profile 157 krasnozem; and *hydromorphic* profile 58 gleyed slightly lessivé terre de barre.

The South African *grey ferruginous lateritic* profile Curlews (C. R. van der Merwe and Weber, 1963) is tropical ferruginous; the *lateritic red earth* Acornhoek terre de barre lessivé; the Piet Retief ferralitic; the Laatsgevorden krasnozem; the Merensky ferralitic; the *lateritic yellow earths* Jessievale ferralitic; the Nelshoogte krasnozem; and the Neshoogte II ferralitic; profiles Pearl and Llandulo of the winter rainfall region are krasnozem.

The Tanzanian *red loam* profiles 47–49 (MUIR et al., 1957) are terre de barre slightly lessivé; *pallid* profile 3 is krasnozem slightly lessivé.

The Ugandan Buwekula red and Buwekula brown profiles (RADWANSKI and OLLIER, 1959) are krasnozems.

The Ghanian *forest ochrosol* profiles Kumasi, Bekwai and Boi (H. Brammer in: WILLS, 1962) are tropical ferruginous the first, the others acid ferruginous; the forest *ochrosol–oxysol* Nzima-Boi is acid ferruginous; the *savanna ochrosol* Mimi krasnozem; the *savanna ochrosol* Toje terre de barre; the *regosolic ground-water laterite?* tropical ferruginous lessivé 20 inches eluvial.

The Ghanian *ground-water laterite* profile B 1334 (OBENG et al., 1963) is tropical ferruginous 3 inches humic concretionary to the surface; the *forest ochrosol* profile B 1424 and B 1171 are tropical ferruginous; the *savanna ochrosol* profile B 382 is acid ferruginous lessivé 11 inches eluvial; the *forest oxysol* profile 387 is acid ferruginous 3 inches humic.

The Ghanian *ochrosol* Nzima 707 (AHN, 1961) is acid ferruginous 25 inches humic; it seems that the other profiles of the series, for which no clay data are given are eutrophic and consequently tropical ferruginous; the *oxysol* profile B 887 is acid ferruginous 2 inches humic.

The Ghanian *ochrosol–latisol* profile Amantin SOB 3 (SMITH, 1962) is tropical ferruginous cuirassé 80 inches deep; the *latisol–oxysol* Bediesi SOB 17 is slightly lessivé terre de barre 9 inches humic; the *latosol–oxisol* Damongo SOB 28 terre de barre; the *ochrosol–latisol* Ejura SOB 8 tropical ferruginous cuirassé 45 inches deep, 19 inches eluvial; EES 2 tropical ferruginous; and SOB 45 terre de barre; the *ground-water laterite* SOB 9 tropical ferruginous cuirassé 61 inches deep; Kpelesawgu PYS 27 slightly lessivé tropical ferruginous intergrade to ground-water laterite 4 inches humic, and Kpelesawgu PYS 28 slightly lessivé tropical ferruginous intergrade to ground-water laterite 3 inches humic; and the *latisol* Santabona SOB 50 tropical ferruginous lessivé 21 in eluvial 3 inches humic.

The Cameroonian profiles described by MARTIN (1959) are slightly lessivé krasnozems and slightly lessivé acid ferruginous.

The Ivory Coast *strongly leached ferralitic* profile

(DABIN et al., 1960) W 8 (Tabou), Aya 8 (Ayamé) and W.C. (Pata) are ferralitic.

The Ivory Coast *ferrisols* profile (DABIN et al., 1960) Bamoro, Hire, Kouassikpo, Tortya and Divo No.11 are tropical ferruginous.

The Ivory Coast *tropical ferruginous* profiles (DABIN et al., 1960) Dabakala No.40, Bouma No.3, Varale No.6, Bouma No.20, and Bouma No.13 are tropical ferruginous.

The Ivory Coast *mineral hydromorphic* profiles (DABIN et al., 1960) BA 30 (Bandena) and W 7 (Cavally) are dystrophic kaolisols more or less gleisolic.

The Dahomeian *typical ferralitic* profile (FAUCK, 1962) JBE 1 is tropical ferruginous; JAS 1 is acid ferruginous cuirassé (laterite at 100 cm); and JBN eutrophic ground-water laterite 40 cm deep.

The Dahomeian *tropical ferruginous lessivé* profile (FAUCK, 1962) JJ 4 is tropical ferruginous lessivé; JAN 2 is acid ferruginous, perhaps cuirassé at 90 cm; TDS 6 is acid ferruginous possibly cuirassé; JAN 9 is tropical ferruginous; JDS 1 is tropical ferruginous; JBE 21 is probably ground-water laterite.

The Dahomeian *slightly ferralitic* profiles MH 13, MH 14, MH 15 and MH 17, are slightly lessivé terre de barre; MH 25, MH 7 and MH 19 are terre de barre lessivé; MH 21 is terre de barre lessivé; MH 2 krasnozem lessivé.

The Dahomeian *tropical ferruginous* profile MH 3 is terre de barre lessivé.

The Dahomeian hydromorphic profile MH 20 is terre de barre lessivé and gleisolic.

The Dahomeian *red slightly feralitic* profiles (WILLAIME, 1959) H 21, H 25 and H 16 are terre de barre.

The Dahomeian *tropical ferruginous* profiles (WILLAIME, 1959) H 28, H 30, H 22 and 2 are terre de barre, the last dubious.

The Dahomeian profile (WILLAIME, 1962) Kounakankouo 1 is acid ferruginous lessivé; Kounakankouo 13 terre de barre lessivé; Kounakankouo 16 tropical ferruginous lessivé; Koumagou 33 acid ferruginous lessivé; JBO 18 terre de barre lessivé; JBO 11 tropical ferruginous lessivé; and JBO 12 tropical ferruginous.

The Pakistani profile (KARIM and QUASEM, 1961) Chorobari silt loam is palaeo-kaolisol; Panch Gancchia fine sandy loam palaeo-kaolisol.

The Celanese profiles A–C (PANABOKKE, 1959) are terre de barre.

The Vietnamese profiles 442 and 17 (FRIDLAND, 1961) are terres de barre.

The Burmese profiles 62-P, 125-K and 5-A (ROZANOV and ROZANOVA, 1961) are terres de barre; 63-A krasnozem; and 4-B cinnamon intergrade to terre de barre.

The Sarawak profile 8 (BECKETT, 1961) is krasnozem intergrade to ferralitic 3 inches humic.

The Japanese red and yellow soil (KANO et al., 1963) is slightly lessivé krasnozem intergrade to cinnamon.

PODSOLIC (LESSIVÉ) SOILS (SOILS WITH MODERATE PODSOLIZATION)

General considerations

As was stated earlier leaching with slightly polymerized humus may have various effects. When leaching is moderate the only effect is illuviation of clay and to a certain extent iron; but neither "ashy" nor spodic horizons are formed. This is the main characteristic of podsolic soils. Usually podsolic soils are formed under broad-leaf forest, because the humus formed under this vegetation is not so rich in fulvic acids as that formed under coniferous forest; it is also less acid. But not all podsolic soils are formed under broad-leaf forest; and broad-leaf forest does not always produce podsolic soils. Climate and parent material can modify the result. For reasons of convenience soils having "laterite" that intercepts drainage, at a depth less than 50 cm (20 inches), are included in the podsolic soils.

General diagnostics

Podsolic soils have one or both of the following characteristics:

(*a*) A textural horizon produced by slightly polymerized humus; in the case of kaolinitic soils (soils with one or more kaolinitic horizons) the textural horizon should begin at a depth less than 50 cm (20 inches).

(*b*) Laterite at a depth less than 50 cm (20 inches).

They cannot have any of the following horizons: "ashy", "humus podsolic B", "organic pan", "iron pan", "spodic". But soils that combine a "textural" and an "ashy" horizon (grey wooded) may be classified either as podsolic or podsols.

Subgroups and their diagnostics

Podsolic soils can be divided into two broad groups:

(*1*) *Illitic podsolic:* all their horizons are illitic, that is, dominated by 2–1 clays of medium to high cation exchange capacity.

(*2*) *Kaolinitic podsolic:* they have one or more kaolinitic or super-kaolinitic horizon, and/or laterite.

Illitic podsolic

Grey–brown podsolic (pb): they have a "textural" horizon; this horizon or part of it may be also iron podsolic B (*Bir*); no horizon is "cinnamonic"; the eluvial (*e*) horizon is not totally "bleached" (*a*) or "black", a part is "brown" or "yellow".

Sod or turf podsolic (ps): grey–brown podsolic with a "dark humic" horizon 10 cm (4 inches) or more thick.

Non calcic brown (pc): they have a "textural" (*t*) and a cinnamonic horizon, which may overlap.

Grey-wooded (pw): they have a "textural" horizon; the eluvial horizon is "bleached" or "black"; usually all their horizons are "neutral", and they often have "calcic" horizons.

Fragipan podsolic (pf): they have a "fragipan" below the textural horizon.

Allophanic podsolic (pv): "all horizons" are allophanic; allophanic podsolic may have a "spodic" horizon because the distinction between "allophanic" and "spodic" horizons is difficult; naturally an allophanic podsolic soil should have a distinct increase of cation exchange capacity and clay content with depth.

Brun lessivé (pb(-Bir)): they have a "textural" horizon; but they have not an "iron podsolic B" (their "indice d'entraînement du fer" is less than 2); this is a special case of grey–brown podsolic.

Lessivé (pb(Bir)): they have an "iron podsolic B", this is a special case of grey–brown podsolic.

Prairie lessivé: they have a "dark humic" horizon "neutral" and 25 cm (10 inches) or more thick, and a "textural" horizon.

Podsolic soils that have "gley" or "pseudo-gley" horizons are treated on pp.59–61.

N.B. These terms do not necessarily exclude one another; a "brun lessivé" may be "sod-podsolic"; an "allophanic podsolic" may be "sod-podsolic" or "grey-wooded", and so on.

Kaolinitic podsolic

Red–yellow podsolic (kpd): the soil is dystrophic; no horizon is "concretionary" or "laterite".

Lateritic podsolic (kpf): same as red–yellow podsolic; but they have "concretionary" horizons; no laterite is present in the upper 50 cm.

Ground-water laterite (kff): "laterite" in the upper 50 cm (20 inches); it may be "eutrophic" or "dystrophic".

Tropical ferruginous or terre de barre lessivé (knf(e)) or *(knl(e)):* tropical ferruginous or terre de barre with "distinct" increase with depth of cation exchange capacity in the upper 50 cm (20 inches). Both are "eutrophic".

Palaeo-lateritic (qf): they contain abundant (more than 50%) laterite fragments, but are encountered under a dry climate; they have been re-basified or re-silicated and their pH is usually above 7.

Correlation and examples

Soils of humid climate with textural horizon have always been called podsolic; the term has even been sometimes extended to soils, in which clay eluviation has been produced by natrium. The distinction between podsolic and podsols is also relatively old. In the United States, since MARBUT (1936), grey–brown podsolic has been separated from podsols. And in the 7th Approximation (1960) the difference has been defined better by establishing the difference between "textural" and "spodic" horizons. In the same time in France a distinction has been established between "lessivé" soils, which have a textural horizon but no high iron eluviation (their "indice d'entraînement du fer" is not so high), and podsols, which show iron eluviation.

In the 7th American Approximation, illitic podsolic corresponds to *alfisols*; and red–yellow podsolic to *ultisols*.

In the French classification (DUCHAUFOUR, 1965) illitic podsolic corresponds to *bruns lessivés* and *lessivés*.

The *non-calcic brown* of the United States of THORP and SMITH (1949) corresponds rather well to our non-calcic brown; and their *grey–brown podsolic* to our grey–brown podsolic.

The *eutroboralfs* of the 7th American Approximation correspond rather well to our grey-wooded; their *normudalfs* to our grey–brown podsolic; and their *normustalf* to our non-calcic brown.

Profile 9 of the 7th American Approximation is grey–brown podsolic 15 inches eluvial.

Profile 11 of the 7th American Approximation, classified as *orthic typudalf*, is grey–brown podsolic 9 inches eluvial.

Profile 22 of the 7th American Approximation, classified as *orthic typaltalf*, is grey-wooded 5 inches eluvial.

Profile 31 of the 7th American Approximation is non-calcic brown duripan at 22 inches.

Profile 32 of the 7th American Approximation is grey–brown podsolic 25 inches eluvial, 8 inches dark humic, gravelly.

Profile 84 of the 7th American Approximation, classified as *orthic fragudalf*, is fragipan grey–brown podsolic 44 inches deep.

Profile 84a of the 7th American Approximation, classified as *glossic typudalf*, is grey–brown podsolic 18 inches eluvial 4 inches dark humic.

Profile 85 of the 7th American Approximation, classified as *orthic fraglossudalf*, is fragipan grey-brown podsolic 33 inches deep, 7 inches dark humic.

Profile 86 of the 7th American Approximation, classified as *durustalf*, is non-calcic brown with duripan at 22 inches.

Profile 12 of the 7th American Approximation is grey–brown podsolic 7 inches eluvial, 5 inches humic, perhaps gleisolic.

Profile 98 of the 7th American Approximation, classified as *fragochrult*, is fragipan grey–brown podsolic 27 inches deep, 10 inches eluvial, 2 inches humic.

The solonetzic-like profile of Canada, described by JANZEN and MOSS (1956) is probably grey-wooded.

Profile 16 of DUCHAUFOUR (1957), classified as *sol brun lessivé*, is grey–brown podsolic (brun lessivé) 6 cm dark humic, 35 cm eluvial.

Profile 18 of DUCHAUFOUR (1957), classified as *lessivé à moder*, is grey–brown podsolic (lessivé) 12 cm dark humic, 44 cm eluvial.

Profile 20 of DUCHAUFOUR (1957), classified as *podsolique* (with pseudogley), is pseudogleyed podsolic 45 cm eluvial, 4 cm dark humic.

The Australian *grey–brown podsolic*, profile A (STACE, 1961), is grey-wooded 7 inches eluvial, 2 inches humic.

The American *grey-wooded*, profiles I, III and IV of HAPSTEAD and RUST (1964) are grey-wooded.

The American *grey–brown podsolic*, profiles Memphis I, Memphis II and Grenada I (HUTCHESON et al., 1959) are grey–brown podsolic.

The American grey-wooded profiles Cen 1, Int 1 and M2 (RADEKE and WESTIN, 1963) are grey-wooded.

The American profile Holyoke series from Essex, N.J. (TAMURA et al., 1959) is grey–brown podsolic.

The American *brunizem* profile Gundy silt loam (WHITE and RIECKEN, 1955) is prairie lessivé.

The American transitional and *grey–brown podsolic* profiles Pershing silt loam P 429 and Weller silt loam (WHITE and RIECKEN, 1955) are grey–brown podsolic.

The Belgian *brown soil* profile 2 (AMERYCK, 1960) is grey–brown podsolic 25 cm humic; the *prepodsol* profiles 5–7 are also grey–brown podsolic.

The Belgian profile 3 (PAHAULT and SOUGNEZ, 1961) is grey–brown podsolic 18 cm eluvial, 8 cm humic.

The British profiles Batcomb flinty silt loam, Winchester flinty loam and Charity flintly silt loam (AVERY et al., 1959) are grey–brown podsolic.

The Angolan reddish brown semi-arid profile 60/54 is non-calcic brown.

The *red–yellow podsolic* of the United States of THORP and SMITH (1949) corresponds rather well to our red–yellow podsolic.

The *normudults* of the 7th American Approximation correspond more or less to our red–yellow podsolic; but clay composition is not taken as a diagnostic in the 7th Approximation.

Profile 96 of the 7th American Approximation, classified as *orthic typochrult*, is red–yellow podsolic.

Profile 97 of the 7th American Approximation, classified as *fragochrult*, is fragipan red–yellow podsolic 32 inches deep, 13 inches eluvial, 31 inches humic; the fragipan horizon may belong to a buried profile.

Profiles 1, 2, 3, 4, 6 and 9 of São Paulo (ANONYMOUS, 1960), classified as *red–yellow podsolic*, are red–yellow podsolic.

Profiles 13 and 14 of São Paulo (ANONYMOUS, 1960), classified as *podsolized with gravel*, are red–yellow podsolic, the former is 40 cm eluvial, 20 cm humic; the latter 30 cm eluvial, and 15 cm humic.

Profile 19 of São Paulo (ANONYMOUS, 1960), classified as *podsolized Lim and Marila*, is red–yellow podsolic 35 cm eluvial.

Profiles 12 and 13 of Rio de Janeiro (ANONYMOUS, 1958), classified as *red–yellow podsolic*, are red–yellow podsolic.

Profile 15 of Rio de Janeiro (ANONYMOUS, 1958), classified as *red–brown podsolic*, is red–yellow podsolic.

Profile 26 of Rio de Janeiro (ANONYMOUS, 1958), classified as *podsolic orange-coloured latosol*, is red–yellow podsolic 40 cm eluvial.

The Australian *lateritic podsolic*, profile A (STACE, 1961), is lateritic podsolic 14 inches eluvial, 3 inches humic.

The Australian *yellow podsolic*, profiles A–C (STACE, 1961) are red–yellow podsolic; C is gleyed.

The American *red–yellow podsolic* profiles Hagerstown I, Maury I, II and III (HUTCHESON, 1963) are grey-brown podsolic intergrading to red–yellow podsolic.

The American *lateritic* profile 5 (NYUN and MC-CALEB, 1955) is red–yellow podsolic.

The Portuguese Guinea profiles P 196, P 392 and P 72 (DA SILVA TEIXEIRA, 1959) are red–yellow podsolic.

The South African profiles *grey ferruginous lateritic* Klipkok and Letoba (C.R. Van der Merwe and Weber, 1963) are ground-water laterites; and the *lateritic red earth* Johannesburg is red–yellow podsolic.

The Ugandan Buwekula shallow profile (RADWANSKI and OLLIER, 1959) is red–yellow podsolic 10 inches humic.

The Ghanian *ground-water laterite* profile B 1532 is eutrophic ground-water laterite 10 inches deep.

The Ghanian *oxysol* profile Tikobo B 863 is red–yellow podsolic 17 inches eluvial, 2 inches humic.

The Ghanian *ground-water laterite* Changnalili SOB 47/A (SMITH, 1962) is eutrophic ground-water laterite 16 inches deep.

The Dahomeian profile (WILLAIME, 1962) Atacora 27 is lateritic podsolic; Koumagou 5 red–yellow podsolic; and JBO 25 lateritic podsolic.

PODSOLS (SOILS WITH ADVANCED PODSOLIZATION)

General considerations

When the humus formed is very rich in slightly polymerized substances and very acid (fulvic and other acids) weathering follows the podsolic type; all the substances released by breakage of the crystal of silicates are leached away; and an "ashy" horizon very poor in clay, and free iron is formed. The soil is podsol. When drainage is very easy (steep slopes) the alumina, iron, etc., are leached from the profile. The soil consists of an "ashy" horizon, resting on "bed-rock". Such soils are called podsol rankers (podsols humo-cendreux of French authors). When drainage is less free, the alumina, silica and iron leached from the "ashy" horizon accumulate at lower depth and form a "spodic" horizon, which is rich in amorphous clays, iron, and organic matter, and consequently has a high cation exchange capacity.

For the reasons explained earlier the accumulation of iron (if any) usually takes place nearer to the surface; and that of humus (if any) still nearer; so that three horizons may sometimes be distinguished: (*1*) a humus podsolic B; (*2*) an iron podsolic B; and (*3*) a textural" horizon; they usually overlap. Naturally podsolic weathering is only possible when the humus formed is sufficiently acid to neutralize the bases formed by weathering, and maintain a low pH. When parent material is soft limestone, this is impossible. It is very easy when the parent material is rich in slowly weathering acid silicates. Podsolic weathering is also difficult in the case of unconsolidated rocks rich in clay, because clay may buffer the acidity of humus. However, not all podsols are produced by podsolic weathering. In many cases the process begins with clay eluviation. When the eluvial horizon has been sufficiently impoverished in clay, iron is also eluviated, and an "ashy" horizon is formed. In other cases the parent material is so poor in clay, that iron eluviation can begin at once.

Under a very cold climate and coniferous forest, heath or tundra vegetation, podsolic weathering prevails. Podsols are usually formed directly. If an illuvial horizon is formed it is rich in amorphous clays. In a less cold climate and broad leaf forest, podsols are often the result of further podsolization of a podsolic soil; and the illuvial horizon formed is usually rich in crystalline clays. A humus podsolic B (*Bh*) or organic pan (*Bhm*) is formed when the decomposition of organic matter is very slow, because of waterlogging or temperatures approaching 0°C, even in summer, at the depth at which these horizons are formed. That is why such soils abound in tundra (permafrost near to the surface), and under waterlogging conditions (ground-water podsols). The formation of fragipan, with high bulk density, also requires waterlogging conditions that favour the formation of iron silicates. Low temperatures and waterlogging also favour the formation of "peaty", "semi-gley" and "gley horizons". That is why many podsols have such horizons, and some belong at the same time to the podsolic and organic or gleisolic groups.

General diagnostics

Podsols have one of the following diagnostics: (*a*) an "ashy" horizon; (*b*) a "spodic" horizon; (*c*) a "humus podsolic B"; (*d*) an "organic pan"; (*e*) an "iron pan"; (*f*) a "fragipan", which is not overlied by a "textural" horizon.

Subdivisions and their diagnostics

Podsol ranker (pl): they have an "ashy" horizon resting on "rock bed"; naturally they do not have any of the following horizons: "spodic", "humus podsolic B", "organic pan", "iron pan" or "fragipan".

Humus podsols (ph): they have a "humus podsolic B".

Iron podsols (pi): they have an "ashy" horizon and an "iron podsolic B".

Humus-iron or iron-humus podsols (pih): they are at the same time humus podsols and iron podsols.

Organic pan podsols (p(Bhm)): they have an organic pan.

Iron pan podsols (p(Birm)): they have an iron pan.

Fragipan podsols (pf): they have a "fragipan" which is not overlied by a "textural" horizon.

Tundra nano-podsols (pt): humus podsols, in which the illuviation of humus has taken place in the surface of permafrost; their depth is usually small and that is why they are called "nano-podsols"; due to that and to churn-

ing by freezing and thawing they often do not have an "ashy" horizon; the eluvial horizon is "brown" or "gley" or "semi-gley".

Giant podsols: podsols in which the illuvial horizons (*t*, *Bh*, *Bir*, etc.) begin at a depth greater than 50 cm (20 inches); they are usually encountered in warm climates.

Brown podsolic (p(−A₂)): they have an "iron podsolic B" which is at the same time "spodic" (C content above 1% and cation exchange capacity higher than clay content), but they do not have any of the following horizons: ashy, humus podsolic B, organic pan. Although called "podsolic", brown podsolic is probably the product of podsolic weathering and should be classified as podsol (see the 7th American Approximation). The simpler symbol *bp* is also used for these soils.

Kaolinitic podsols (phk): one or more horizons are kaolinitic.

Yellow podsols: the "eluvial" (*e*) horizon is "yellow" (*y*).

Sod podsols (ps): they have a dark humic (*hd*) horizon 10 cm (4 inches) or more thick.

Podsols with "gley" or "semi-gley" horizons will be discussed later.

Correlation and examples

The concept of podsol is as old as pedology. But there are differences of opinion concerning the inclusion of certain soils in it. For instance podsol rankers are frequently not considered as podsols, because they have not an illuvial horizon; the eluviated sesquioxides and organic matter have been leached away. On the other hand brown podsolic is frequently not classified as podsol because it has no "ashy" horizon. In tundra nano-podsols illuviation takes place in the upper part of permafrost and this humus podsolic B is often considered as not forming part of the profile; the same happens in kaolinitic podsols, where illuviation is profound; these soils are often called kaolisols with dark horizon.

The *podsolic* soils of Canada (STOBBE, 1962) correspond rather well to our podsols and podsolic.

The *podsols* of the United States of THORP and SMITH (1949) correspond rather well to our podsols; and their *brown podsolic* to our brown podsolic.

The *spodosols* of the 7th American Approximation correspond rather well to our podsols.

In the French classification (DUCHAUFOUR, 1965) our ranker-podsols correspond to "humo-cendreux", our brown podsolic to "ocre podsolique", our iron podsols to "podsols ferrugineux", our humus-iron podsols to "podsols humo-ferrugineux"; our humus podsols are usually hydromorphic and they are included in "*podsols hydromorphiques*", which also include our glei podsols.

Profile 39 of the 7th American Approximation, classified as *spodic quartzopsamment*, is aeolian humus podsol, 12 inches eluvial.

Profile 24 of the 7th American Approximation, classified as *orthic fragochrept*, is fragipan podsol 16 inches deep, 2 inches dark humic.

Profile 29 of the 7th American Approximation, classified as *orthic fragochrept*, is fragipan podsol 16 inches deep, 9 inches dark humic.

Profile 50 of the 7th American Approximation, classified as *cryandept*, is allophanic humus podsol 5 inches eluvial.

Profile 54 of the 7th American Approximation, classified as *entic cryumbrept*, is tundra humus nano-podsol 11 inches eluvial.

Profile 20 of the 7th American Approximation, classified as *orthic typorthod*, is iron podsol $1\frac{1}{2}$ inches eluvial.

Profile 56 of the 7th American Approximation, classified as *andic fragochrept*, is fragipan podsol 20 inches deep.

Profile 19 of DUCHAUFOUR (1957), classified as *brun podsolique*, is brown podsolic 13 cm dark humic.

Profile 21 of DUCHAUFOUR (1957), classified as *podsol ferrugineux subalpin*, is iron podsol 15 cm eluvial, 5 cm dark humic.

Profile 22 of DUCHAUFOUR (1957), classified as *podsol humo-ferrugineux à alios*, is iron pan humus podsol 52 cm deep, 47 cm eluvial, 4 cm dark humic.

Profile 23 of DUCHAUFOUR (1957), classified as podsol *humo-ferrugineux à pseudogley*, is pseudoglei humus iron podsol 35 cm eluvial, 10 cm dark humic.

Profiles 73 and 74 of São Paulo (ANONYMOUS, 1960), classified as *podsols*, are humus-iron podsols.

The Antarctic profile, described by MOLFINO (1956) under number II, is humus nano-podsol 25 cm eluvial, 5 cm humic.

The Australian *alpine humus*, profile A of STACE (1961), is iron pan podsol 18 inches deep.

The Australian *moor podsol peat*, profiles A and B (STACE, 1961), are humus-pan podsols.

The Australian *moor podsol peat*, profile C (STACE, 1961), is glei podsol 26 inches eluvial, 10 inches peaty, 16 inches humic.

The Australian *podsol*, profile A (STACE, 1961), is humus podsol 114 inches eluvial.

The Australian *podsol*, profiles B and C (STACE, 1961), are humus pan podsols 56 and 36 inches deep, respectively.

The Australian *ground-water podsol*, profiles A and C (STACE, 1961), are humus pan podsols 25 and 24 inches deep, respectively.

The Australian *ground-water podsol*, profile B (STACE, 1961), is humus podsol 35 inches eluvial.

The Alaskan meadow soils described by DOUGLAS and TEDROW (1960) may be classified in our scheme as follows: Barrow 1, tundra nano-podsol 23 inches eluvial; Umiai 3 tundra nano-podsol 12 inches eluvial.

The Alaskan *upland* soil, profile Franklin Bluff 2 of DOUGLAS and TEDROW (1960), is tundra humus nano-podsol 19 inches eluvial.

The Colombian *andino podsol*, profile 37 (JENNY, 1948), is podsol ranker 6 inches organic, 3 inches humic; and the *giant podsol* humus-iron podsol 18 inches humic, 56 inches eluvial.

The Colombian *giant podsol*, profile Sabaneta, series 94 (BARSHAD and ROJAS-CRUZ, 1950), is humus-iron podsol 56 inches eluvial, 32 inches humic.

The Belgian *podsol*, profile 8 (AMERYCK, 1960), is humus-iron podsol 38 cm eluvial; the *humus-iron post-podsol*, profile 10, is humus-iron podsol; and the *humus post-podsol*, profile 11, is humus-iron podsol 40–50 cm eluvial.

The Ghanaian ground-water podsol Atuabo (H. Brammer, in: WILLS, 1962) is humus-podsol 23 inches eluvial, 2 inches humic.

The Ghanaian *oxysol*, profile Friedricksburg B 990 (AHN, 1961), is humus podsol 27 inches eluvial, 1 inch humic, but the other members of the series are not podsols; the *ground-water podsol*, profile Atuabo B 999, is humus podsol 23 inches eluvial, 2 inches humic.

The Ivory Coast *mineral hydromorphic*, profile Port Bouet (DABIN et al., 1960), is humus podsol 15 cm eluvial, 5 cm dystrophic organic, 5 cm humic.

The Sarawak profile 3 (BECKETT, 1961) is humus podsol 108 inches eluvial.

Profile 101 of the 7th American Approximation, classified as *ustox*, is kaolinitic humus podsol 40 inches eluvial, 22 inches humic.

Chapter 5 | Halomorphic, Gleisolic and Organic Soils

INTRODUCTION

These soils have nothing in common, except perhaps their high occurrence in badly drained areas. Their degree of leaching varies from 1 to 6. But for reasons of convenience they will be treated in the same chapter.

SOLONCHACKS

General considerations

Solonchacks are soils rich in salts more soluble than $CaSO_4$. Many are rich in carbonate from the surface down and may be considered as saline rendzinas; others have a "natric" or "magnesic" horizon and may be considered as saline alkaline or saline solonetz soils; others have all their horizons "vertisolic" and may be considered as saline clays; and so on. Many solonchacks have a "calcareous accumulation" horizon at some depth; sometimes a petrocalcic horizon.

According to the kind of salts KOVDA (1961) divides solonchacks into: (*a*) soda solonchacks (Na_2CO_3, $NaHCO_3$); (*b*) sulphate solonchacks (Na_2SO_4, $MgSO_4$); and (*c*) chloride solonchacks (NaCl, $MgCl_2$, $CaCl_2$): this distinction is very important, because salts vary in their toxicity, etc.; soda solonchacks correspond to our saline alkaline.

Another interesting distinction is between *active* solonchacks, in which an "exsudational" moisture regime continuously brings salts from a saline water table, and *residual* solonchacks, in which the saline water table lies so deep that there is no communication by capillarity.

Naturally salt concentration is important, but due to the mobility of salts it is more convenient to specify the total amount (of all horizons) in tons per unit area.

General diagnostics

A soil is considered solonchak, when the salt conten in the soil is sufficiently large to cause a change in the type of vegetation, or when crops suffer seriously from the presence of salts. However, it is better to specify the amount of salts per unit surface, which permits a more liberal use of the term saline.

Subgroups and their diagnostics

Saline alkaline (s(na)): they have a "natric" horizon, but lack an "eluvial" horizon.
Saline solonetz (st): they have a "natric" or "magnesic" "textural" horizon.
Sulphate saline (s(SO_4)): salts are chiefly sulfates.
Chloride saline (s(Cl)): salts are chiefly chlorides.
Saline rendzinas (sr): all horizons are "calcareous".
Saline clays (ms): all horizons are "vertisolic".
Gypsum saline (s(SO_4Ca)): they contain two and a half times as much gypsum as other more soluble salts.
Saline chernozemic (ys): they have a "dark humic" horizon 25 cm (10 inches) or more thick.
Peaty saline (os): they have a "peaty" horizon.
Organic saline (hs): they have an "organic" (*or*) horizon.

Solonchacks with "gley" or "semi-gley" horizons will be discussed later.

These terms do not necessarily exclude one another; e.g., a soil may be saline alkaline rendzina or saline rendzina clay, and so on.

Gypsisols

Gypsisols (*d'*) are soils very rich in gypsum; since gypsum is not toxic these soils have not the general diagnostics of solonchacks, except if they are also rich in other more soluble salts. They are gypsisols, but in the same time they belong to other groups: gleisolic, raw, etc.

Correlation and examples

In the 7th American Approximation solonchacks correspond to the great group *salorthids*.

In the French classification (DUCHAUFOUR, 1965) solonchacks correspond to "salins".

The *solonchacks* of the United States of THORP and SMITH (1949) correspond rather well to the author's concept of solonchacks.

Profile 26 of the 7th American Approximation is saline rendzina clay.

The Argentine profile No.186 described by GOLLAN et al. (1936) and GOLLAN and LACHAGA (1939) is saline 8 cm humic.

The Australian *solonchack*, profile A (STACE, 1961), is chlorine saline rendzina; profile B chlorine saline dark clay; and profile "C saline" chlorine saline terra rossa.

The American *calcium carbonate solonchack*, profile Hamerly clay loam (McCLELLAND et al., 1959) is saline rendzina 7 inches humic.

The Turkish solonchack (OAKES, 1957), profile Malya State Farm, is saline rendzina 12 cm humic.

SOLONETZIC–PLANOSOLIC GROUP (SOILS WITH ELUVIAL HORIZONS DUE TO Na)

General considerations

Clay eluviation, under conditions not favourable to podzolization, is due to Na. However, two cases should be distinguished:

(*a*) Those in which the "textural" horizon is "natric" or "magnesic" (over 15% of absorbed bases not including H, is Na; or Na is between 7 and 15%, but Na + Mg exceed Ca); these soils are called solonetz or magnesium solonetz.

(*b*) Those in which the "textural" horizon is neither "natric" nor "magnesic"; they are called planosols.

Some soils of this group do not yet have an "eluvial" horizon, but they have "natric" or "magnesic" horizons; they are called alkaline or magnesium alkaline. Planosols are often considered as former solonetz that have lost a great part of their Na by leaching (solodized solonetz or solods). As was pointed out elsewhere (PAPADAKIS, 1963b), it is very probable that the majority of planosolic soils have never been solonetz; silicate weathering releases much Na, K, Mg and SiO_2 which produce clay illuviation when for climatic or drainage reasons leaching is slow.

General diagnostics

Solonetzic–planosolic soils have an "eluvial" horizon due to Na; the group includes also soils that lack this horizon, but have a "natric" or "magnesic" horizon at a certain depth, that should be specified.

Subgroups and their diagnostics

Solonetz (t): the "textural" horizon is "natric".

Magnesium solonetz (t(mg)): the "textural" horizon is "magnesic".

Planosol (u): the "illuvial" horizon is "textural"; the "dark humic" horizon, if any, is less than 25 cm (10 inches) deep.

Clay pan planosol (uu): a planosol with the "textural" horizon "vertisolic".

Planosolic para-chernozem (yu): they have a "dark humic" horizon 25 cm (10 inches) or more thick; the "illuvial" horizon is "textural".

Alkaline (ñ): they lack an "illuvial" horizon, but have a "natric" horizon.

Magnesium alkaline (ñ(Mg)): same as alkaline, but instead of a "natric", they have a "magnesic" horizon.

Cinnamonic planosol (uc): a planosol with "cinnamonic" horizon.

Planosolic red desert (ud): a planosol lacking humic horizon and in which the "textural B" begins at 15 cm (6 inches) or less from the surface; it is usually covered by a "desert pavement"; colours are usually, but not necessarily, reddish.

Rendzina solonetz or planosol (rt or ru): calcareous to the surface.

Kaolinitic solonetz, planosol or alkaline (kt, ku or kñ): they have a "kaolinitic" or "super-kaolinitic" horizon.

Acid solonetz (ta): solonetz with low pH (lower than 7).

These terms do not necessarily exclude one another; e.g., an alkaline soil may be rendzina. It is useful to specify thickness of the "dark humic" and "eluvial" horizons; and in the case of solonetz and alkaline soils the depth at which the "natric" or "magnesic" horizon begins: e.g., planosol 20 cm eluvial, 10 cm dark humic; or solonetz 12 cm dark humic, natric at 25 cm.

Correlation and examples

There are not many controversies concerning the concept of solonetz. But planosolic soils are considered by Russian authors as leached (solodized) solonetz whereas in the United States they prefer to call them planosols without implications concerning their formation. As has been stated before planosols owe their formation to Na, but in the majority of cases they have been formed directly, they have never been solonetz (PAPADAKIS, 1963b). Under dry conditions and/or impeded drainage Na released by weathering is leached slowly; it accumulates and produces clay eluviation; that is why clay eluviation is so common under dry climates.

In the French classification (DUCHAUFOUR, 1965) the author's solonetz corresponds to "sols à alcalis lessivés"; his alkaline to "sols à alcalis non lessivés"; and the planosols to soloths, but the concept of planosol is ampler.

The solonetzic soils of Canada (STOBBE, 1962) correspond rather well to our solonetzic.

The *solonetz* of the United States of THORP and SMITH (1949) corresponds rather well to the author's concept of solonetz.

The *natrargids, natralbolls, natraquols, natriborolls, natrustolls, natraqualfs, natriboralfs, natrudalfs,* and *natrustralfs* of the 7th American Approximation are solonetz.

The *solodized solonetz* profile of Canada, described by JANZEN and MOSS (1956), is magnesium solonetz.

The *solod* profile of Canada, described by JANZEN

(1962), is magnesium solonetz 11 inches humic, 17 inches eluvial, magnesic at 17 inches.

Profile 25 of the 7th American Approximation is solonetz 27 inches eluvial, natric at 27 inches.

Profile 58 of the 7th American Approximation, classified as *orthic durorthid*, is solonetz 14 inches eluvial with duripan at 21 inches.

Profile 63 of the 7th American Approximation, classified as *natrargid*, is solonetz 5 inches eluvial, natric to the surface.

Profile 64 of the 7th American Approximation, classified as *nadurargid*, is solonetz 6 inches eluvial, natric to the surface with petro-calcic horizon at 21 inches.

Profile 4 of the 7th American Approximation, classified as *natralboll*, is solonetz 6 inches humic, 7 inches eluvial, natric at 9 inches.

Profile 18 of the 7th American Approximation, classified as *natraquoll*, is chernozemic solonetz 17 inches eluvial, 10 inches humic.

Profile 76 of the 7th American Approximation, classified as *natrustoll*, is chernozemic magnesium solonetz 13 inches eluvial, 10 inches humic, magnesic at 13 inches.

Profile 17 of the 7th American Approximation is solonetz natric at 6 inches and 3 inches humic.

Profile 87 of the 7th American Approximation, classified as *natrustalf*, is solonetz 36 inches eluvial, 6 inches humic, natric at 36 inches.

Profile 89 of the 7th American Approximation, classified as *orthic typustalf*, is magnesium solonetz 15 inches eluvial, magnesic at 26 inches.

Profile 30 of Rio de Janeiro (ANONYMOUS, 1958), classified as *hydromorphic*, is kaolinitic solonetz 28 cm eluvial, natric at 28 cm.

The Argentine soil (Rufino), classified as *planosolic prairie* by BONFILS et al. (1959), is solonetz 28 cm eluvial, 24 cm humic, natric at 28 cm.

The Argentine profile described by GOLLAN and LACHAGA (1939, 1944) under No. 176, is slightly lessivé alkaline; No.209 recent brown alkaline at 35 cm; No.216, 181, 187, 192, 196, 140 and 188 are solonetz.

The Australian *brown podsolic*, profile C (STACE, 1961), is magnesium solonetz 12 inches eluvial.

The Australian *solonetz*, profiles A, B and C (STACE, 1961), are solonetz.

The Australian *solodized solonetz*, profiles A, B and C (STACE, 1961), are solonetz; the first magnesium solonetz.

The Australian *soloths*, profiles A, B and C (STACE, 1961), are solonetz; the first is magnesium solonetz and the second cinnamonic solonetz.

The Australian *solonized brown*, profile Aa (STACE, 1961), seems to be solonetz 12 inches eluvial.

The Australian *solonized brown*, profiles Ba and Bb (STACE, 1961), are alkaline serozems.

The Australian *desert loams*, profiles A and C (STACE, 1961), are solonetz, the first cinnamonic, the second gravelly.

The Australian *stony desert tableland* soil, profile B (STACE, 1961), is magnesium solonetz 2½ inches eluvial.

The Californian profile *tierra fine sandy loam* (ULRICH et al., 1959) is chernozemic magnesium solonetz 17 inches humic, 30 inches eluvial, magnesic at 30 inches.

The Brazilian profile AL-1 Inhapi (SILVA CARNEIRO, 1951) is solonetz 90 cm eluvial, natric at 90 cm; PE-3 (serrinha) is also solonetz 38 cm eluvial, natric at 38 cm; and PE-5 (Taboado) is cinnamonic magnesium solonetz 36 cm eluvial, magnesic at 36 cm.

The Russian *meadow solonchack solonetz*, profile 59 (PANIN and ARISTARKHOV, 1962), is slightly saline chernozemic rendzina alkaline to the surface; and profile 60, classified as *solonchack meadow*, is slightly saline rendzina alkaline to the surface, 23 cm humic.

The Greek *solonetz-like* profile 17 (ZVORYKIN, 1960) is alkaline rendzina-glei 20 cm humic.

The Ghanaian *tropical grey earth* (H. Brammer, in: WILLS, 1962) is solonetz 12 inches eluvial, natric at 60 inches.

The Ghanaian *alluvial* Sene SOB36 (SMITH, 1962) is solonetz 10 inches eluvial, natric at 10 inches.

The Pakistani profile (KARIM and KHAN, 1955) bGFSL is acid solonetz 3 inches eluvial, natric at 3 inches; DGFSL acid solonetz 4 inches eluvial, natric to the surface; GFSL/1 acid solonetz 2 inches eluvial, natric to the surface; GFSL/2 solonetz natric to the surface; and gYFSL acid solonetz 3–4 inches eluvial, natric to the surface.

The Burmese *dark compact* (ROZANOV and ROZANOVA, 1962) profiles 415-W and 153-P are solonetz.

The *haplargids*, *normargids*, *argialbolls*, *argiaquolls*, *argiborols*, *argiudolls*, *argiustolls*, and *argixerolls* of the 7th American Approximation are usually planosols; some are slightly planosolic.

The *podsolic solonetzic* profile of Canada, described by JANZEN and MOSS (1956), is planosol.

The *solodic* profile of Canada, described by JANZEN (1962), is planosolic para-chernozem 12 inches humic.

Profile 62 of the 7th American Approximation, classified as *haplargid*, is planosolic red desert 4 inches eluvial.

Profile 13 of the 7th American Approximation is planosolic para-chernozem 10 inches eluvial.

Profile 8 of the 7th American Approximation is a clay-pan planosol 5 inches eluvial, 2 inches humic.

Profile 16 of the 7th American Approximation is clay-pan planosol 7 inches eluvial, 3 inches humic.

Profile 79 of the 7th American Approximation, classified as *orthic albaqualf*, is planosol 17 inches eluvial, 4 inches humic.

Profile 80 of the 7th American Approximation, classified as *orthic glossaqualf*, is planosol 19 inches eluvial, 7 inches humic.

The Argentine soil (Carlos Tejedor), classified as *planosolic prairie* by BONFILS et al. (1949), is planosol intergrade to solonetz 40 cm eluvial, 23 cm humic.

The Argentinian profiles No.139, 195 and 163

(GOLLAN et al., 1936) are planosols; No.167, 150, 154, 148, 156 and 116 are planosolic para-chernozem.

The Australian *grey–brown podsolic*, profile B (STACE, 1961), is planosolic para-chernozem 12 inches eluvial.

The Australian *grey–brown podsolic*, profile C (STACE, 1961), seems to be planosol 16 inches eluvial.

The Australian *brown podsolic*, profile A (STACE, 1961), is magnesic planosol, 9 inches eluvial.

The Australian *meadow podsolic*, profiles A and B (STACE, 1961), are planosols; the latter magnesium planosol.

The Australian *solodized brown*, profile Ab (STACE, 1961), is planosol 38 inches eluvial.

The Australian *red–brown earths*, profiles A, B and C (STACE, 1961), are planosols.

The Australian *brown earth*, profile C (STACE, 1961), is planosol 11 inches eluvial.

The Australian *brown earth*, profile a (STACE, 1961), is planosolic para-chernozem 8 inches eluvial.

The Australian *red and brown hardpan* soils, profile B (STACE, 1961), is planosol 12 inches eluvial.

The American *red prairie* profile Newtonia silt loam (JARVIS et al., 1959) is reddish planosolic para-chernozem 11 inches humic.

The American *reddish prairie* profile Zaneisloan 1, 2, 3, 4 and Kingfisher 1 (MOLTHAM and GRAY, 1963) are cinnamonic planosolic para-chernozems; Kingfisher 2 is cinnamonic planosol 7 inches eluvial.

The Californian profile Elkorn sandy loam (ULRICH et al., 1959) is planosol.

The American *planosol* profiles 1 and 2 (WHITE, 1961) are clay-pan planosols; the first is intergrade to solonetz.

The Brazilian *latosol* profile C-5 (COSTA LIMA, 1953) is planosol 25 cm eluvial.

The Argentinian profiles 7 and 8 (PIÑEIRO and ZUKARDI, 1959) are planosols.

The Greek profile 38 (ZVORYKIN, 1960) is planosol 15 cm humic; and profile 22 planosol 19 cm eluvial.

The Angolan *grey–brown semi-arid* profiles 548/555 and 2055/53 (BOTELHO DA COSTA et al., 1959) are planosols 8 cm eluvial and 26 cm eluvial, respectively; the *reddish brown semi-arid* profile 1198 is cinnamonic planosol 20 cm eluvial.

The Mozambiquan profiles Malan (Comba) and Maquizemane (Cuxane) (GOUVEIA and GOUVEIA, 1959) are both planosols 25 cm eluvial.

The Pakistani profile (KARIM and QUASEM, 1961) Simla fine sandy loam is planosol 6 inches eluvial, 13 inches humic; and Deonara fine sandy loam planosol 16 inches eluvial.

The Indian profiles 2 and 3 (AGARWALL and MUKERJI, 1951) are planosols.

The Burmese *red–brown savannah* (ROZANOV and ROZANOVA, 1962) profiles 412W, 403W and 97P are planosols.

The Burmese *dark compact* (ROZANOV and ROZA-NOVA, 1962) profile 620B is clay-pan planosol intergrade to solonetz.

ORGANIC SOILS

General considerations

Acid organic matter produces podsolization; the mineral particles mixed with it, or the mineral horizon underlying it suffer podsolic weathering or other forms of podsolization (clay eluviation, iron eluviation, etc.). An "ashy" or "eluvial" horizon is formed. Therefore many organic soils are podsols or podsolic; and in this case the peat may be considered as forming part of the O horizon and the organic horizon as part of A_1. The soil may be classified on the basis of its mineral horizons with specification of the thickness and nature of the organic horizons that overlie them.

In some cases the organic horizons lie directly on "bed-rock" or permafrost. In this case it is essential to specify if the organic horizon is "eutrophic" or "dystrophic", its thickness and the nature of the underlying rock: for instance, 50 cm of peat on granite; or 40 cm of organic on limestone, and so on. Sometimes it is necessary to mention two organic horizons, because one part is "peaty" and the other "organic"; or a part is "dystrophic" and the other "eutrophic".

If the organic horizon overlies an unconsolidated material, which did not suffer any modification, an analogous terminology may be used; for instance, 50 cm of organic on clay. But such cases are rare. When podsolization has not taken place, a humic horizon is formed on the top of the parent material, and such a horizon should be mentioned; for instance, eutrophic organic 50 cm and humic 15 cm on clay. Very often below the humic horizon, or overlapping with it, a gley horizon is formed and that horizon should be mentioned; for instance, eutrophic 40 cm of organic on gley (sand).

It may be concluded that organic soils do not need a special nomenclature; it is sufficient to mention the depth and base status ("eutrophic", "dystrophic") of the "peaty" and/or organic horizons and the nature of the underlying soil or substratum. But when the term used should denote a broad group of soils, then one may consider a soil organic when total thickness of the "peaty" and "organic" horizon is more than 25 cm (10 inches), or they rest directly on "bed-rock" or permafrost. Organic soils may also be divided into "eutrophic" or "dystrophic", "peaty" or "organic" according to the prevalent base status and nature of this horizon.

Correlation and examples

In the French classification (DUCHAUFOUR, 1965) our organic soils correspond to "*hydromorphes orga-*

niques"; they are subdivided into "*tourbe eutrophe et mesotrophe*" and "*tourbe oligotrophe*".

The *organic* soils of Canada (STOBBE, 1962) correspond rather well to the author's organic (sensu lato).

The *histosols* of the 7th American Approximation correspond rather well to his organic (sensu lato); their *fibrists* to the author's peaty; and their *saprist* to his organic (sensu stricto); they recognize an intermediary sub-order of lenists.

Profile 1 of DUCHAUFOUR (1957), classified as ranker *pseudo-alpin (sol humique silicaté)*, is dystrophic organic 25 cm organic, more than 15 cm humic on sand.

Profile 10 of DUCHAUFOUR (1957), classified as *sol humique carbonaté à humus brut*, is eutrophic organic 40 cm deep on limestone.

Profile 31 of DUCHAUFOUR (1957), classified as *tourbe oligotrophique*, is dystrophic peat more than 200 cm thick.

Profile 75 of São Paulo (ANONYMOUS, 1960), classified as *organic*, is dystrophic organic 35 cm on peat.

The Australian *moor peat* profiles A, B and C are dystrophic peat.

The Australian *alpine humus* profile B is dystrophic peat 24 inches thick on weathering gneiss.

The Australian *acid swamp* profile A (STACE, 1961) is eutrophic peat 13 inches thick on gravelly sand.

The Australian *acid swamp* profile B (STACE, 1961) is dystrophic 30 inches thick on acid clay.

The Australian *acid swamp* profile C (STACE, 1961) is eutrophic organic with water table at 29 cm.

The Australian *fen* profiles A and B (STACE, 1961) are peat on limestone or calcareous sand.

The Alaskan *bog* profile (TEDROW et al., 1958) is peat on permafrost.

GLEISOLIC SOILS

General considerations

The classification of gleisolic soils is controversial. And this is because many of these soils belong at the same time to other groups. Some gleisolic soils receive, by flooding or subirrigation, waters rich in salts more soluble than gypsum; if the soil does not lose water by drainage these salts accumulate, and a saline gleisolic soil is formed. When the soil receives waters rich in gypsum, and drainage is not sufficient to eliminate this gypsum, a gypsic gleisolic soil is formed. In an analogous way a gleisolic soil with a calcium accumulation horizon or a petro-calcic horizon (*cam*) or a duripan (*m*) can be formed. Some gleisolic soils are alkaline or solonetz because they received water rich in natrium carbonate and drainage was not sufficient to leach away Na. All these soils are well provided by bases. They are encountered under more or less dry climates in which potential evapotranspiration is less than rainfall. Under these climates waterlogging is usually due to flooding or subirrigation.

However, when rainfall exceeds potential evapotranspiration, rainfall alone can produce waterlogging if drainage is not sufficiently rapid to eliminate the water received. In this case the soil, instead of being enriched in bases, is rapidly acidified. It may become podsolic or podsol.

Clay eluviation is very common in gleisolic soils. Waterlogging favours the formation of slightly polymerized humus that produces clay eluviation. In many gleisolic soils Na also accumulates. Consequently, both podsolization and solonization are frequent.

Gleisolic soils have a water table at a shallow depth. This water table fluctuates, rising during the humid season and descending during the dry one. It may be permanent or transient. A distinction should be made between soils in which waterlogging reaches frequently the surface, so that all horizons are "gley" (*G*), "semi-gley" (*g*) or "dark gley" (*gn*); and those in which only the lower horizons are affected by gleization; the first may be called "gley"; the second "gleyed".

Since ferrous iron is relatively mobile, gleization produces transfer of iron. In semi-gley horizons iron concentrates at some points as rusty mottles or concretions; iron usually concentrates near the roots because roots absorb water, dry the soil, and water moves from the surrounding particles to the roots. By an analogous process the water table is impoverished in iron and this iron concentrates in the capillary fringe that overlies it, where it forms rusty mottles or concretions.

When rice is grown the soil is flooded for long periods and little by little it becomes gleisolic. According to the balance between the substances brought by the water and those eliminated by subterranean drainage the soil may be enriched or impoverished in salts, gypsum, calcium–magnesium, carbonates, silica. A frequent result is clay illuviation. Paddy soils differ from the other soils of the region, but paddy soils do not form a natural group; they vary considerably one from another, according to the nature of the original soil and the balance between the substances brought by the water and those leached out by drainage. Very often rice is grown in soils that were waterlogged, and consequently gleisolic, before starting rice growing.

In many soils the formation of a transient watertable is due to internal causes (impermeability). Because of their texture, dark clays are easily waterlogged and "black gley" horizons are very common in them. Many "textural B horizons" cause bad drainage; the horizon becomes "gley" (*G*) or "semi-gley" and the "eluvial" horizon may also become "semi-gley", and bleached vertical stripes may be formed in the textural B horizon (pseudogley horizons).

General diagnostics

A horizon that begins at a depth of less than 75 cm (30 inches) is "gley", and/or a horizon that begins at a depth of less than 50 cm (20 inches) is "semi-gley"; soils that have a dark-gley horizon underlain directly by bed-rock or permafrost are also gleisolic.

Subgroups and their diagnostics

Gley: the higher "gley" or "semi-gley" horizon reaches the surface or is directly overlain by an "ashy" and/or a "dark gley" horizon that reaches the surface; when an "acid" "dark gley" horizon overlies directly "bed-rock" or permafrost the soil is also gley.

Humic gley (yg): gley soils with a "neutral" "dark gley" horizon 25 cm (10 inches) or more thick; they are at the same time chernozemic and gley; they cannot have an "eluvial horizon".

Planosolic humic gley (ygu): same as humic gley but they have an "eluvial" horizon; they are at the same time gley, chernozemic and planosolic.

Low humic gley (ga): gley soils with one or more "acid" or "dystrophic" horizons; they cannot have an "eluvial", "kaolinitic" or "super-kaolinitic" horizon.

Low humic gley lessivé (gp): same as "low humic gley" but they have an "eluvial" horizon; they cannot have the diagnostic horizons of podsols ("ashy", "humus podsolic B", "spodic" horizon, "organic pan", "iron pan", "fragipan" not overlied by "textural" horizon).

Gley podsol (pg): it has one or more of the following horizons: "humus podsolic B", "organic pan", "iron pan", "spodic", "fragipan" not overlain by a "textural" horizon; some horizons are "gley" or "semi-gley", and all the others that overlie it, including the surface horizon, are "gley", "semi-gley", "dark gley" or "ashy".

Gley-planosol (gu): gley soils with an "eluvial horizon due to solonization".

Gley-solonetz (gt): gley soils with a "natric textural" horizon.

Kaolinitic low humic gley (kga): same as low humic gley, but one or more horizons are "kaolinitic" or "super-kaolinitic"; no horizon is "concretionary" or "laterite".

Lateritic low humic gley (kgf): same as kaolinitic low humic gley, but they have a "concretionary" horizon; they cannot have "laterite".

Kaolinitic gley podsolic (kgp): same as low humic gley but they have a kaolinitic or superkaolinitic horizon; they cannot have a "concretionary" horizon or "laterite".

Lateritic gley podsolic (kgpf): same as kaolinitic gley podsolic but they have a "concretionary" horizon; they cannot have "laterite".

Kaolinitic gley-podsols (kpg): same as gley-podsol, but they have a "kaolinitic" or "superkaolinitic" horizon.

Ground-water laterites (kff): all horizons overlying laterite till to the surface are "gley", "semi-gley" or "dark gley".

Rendzina-gley (gr): all horizons are calcareous (ca).

Saline gley (gs): they have one or more saline horizons.

Alkaline gley (gna): they have a "natric" horizon; but they do not have an "eluvial" horizon.

Tundra gley: the "gley" or "semi-gley" horizon is underlied by permafrost.

Pseudogley (gp(ps)): low humic gley with pseudogley horizon.

Marmorized grey–brown podsolic (pb(tg)): grey-brown podsolic with the "textural" horizon "semi-gley".

Stagnogley (pg(eg,BG)) or (gp(eg,BG)): gley podsol or humic gley with the "eluvial" horizon semi-gley and the "illuvial" horizon "gley".

Gleyed soils: they are gleisolic but not gley; the higher "gley" or "semi-gley" horizon is overlain by an horizon that is neither dark-gley nor ashy. They may be rendzina dark clays, ando, planosols, etc.; their symbol is formed by adding *g* to that of the soil.

Pseudogleyed soils: gleyed soils with a pseudogley horizon.

These terms do not necessarily exclude one another; for instance, a soil may be saline rendzina gley or saline tundra gley, and so on.

Correlation and examples

Since waterlogging favours both gleization and the accumulation of organic matter, in many classifications gleisolic soils form only one group with organic soils. In the American classification of THORP and SMITH (1948) gleisolic and organic soils formed the suborder of hydromorphic soils; within this suborder humic gley, low humic gley, and ground-water laterites were recognized as special great groups.

In the French classification (DUCHAUFOUR, 1965) gley and organic soils form the suborder of "hydromorphiques"; within this class "gley humiques", "gley peu humiques", "pseudogley de surface" and "pseudogley de profondeur" form special great groups; hydromorphic podsols belong to the class of podsols ("sols à humus brut") and laterites to the class of "sols à sesquioxides". In this classification a "sol brun lessivé" (grey–brown podsolic) with "semi-gley" "textural" horizon is called "marnorisé"; when the "eluvial" horizon is "semi-gley" and the textural horizon "pseudo-gley" it is called "pseudogley"; and when an "ashy" horizon has been formed "stagnogley".

Since gleisolic soils belong also to other groups, in the 7th American Approximation they have been classified in different orders, forming the following suborders: *aquerts* (gleisolic clays); *aquepts* (gleisolic soils without vertisolic, neutral dark humic or "textural B" horizon); *aquolls* (humic gley and other gleisolic chernozemic); *aquods* (gleisolic podsols); *aqualfs* (gleisolic podsolic planosolic and solonetz); *aquults* (gleisolic kaolinitic podsolic).

The author considers it convenient to discuss together all gleisolic soils; but that does not mean that he considers them as forming a natural unity; moreover, in his conception a soil may belong to more than one classification units at the same time. Humic gley belongs to chernozemic; humic gley lessivé to planosols (and chernozemic); low humic gley to brunisolic; low humic gley lessivé to podsolic; gley podsols to podsols; gley-planosols to planosols; gley-solonetz to solonetz; kaolinitic and lateritic low humic gley to kaolisols; kaolinitic and lateritic gley-podsolic to kaolinitic podsolic; kaolinitic gley podsols to kaolinitic podsols; ground-water laterites to kaolinitic podsolic; rendzina humic gley to rendzina (and chernozemic); saline gley to solonchack; tundra gley to tundra rankers; alkaline gley to alkaline; marmorized grey–brown podsolic to podsolic; stagnogley podsol or podsolic to podsol or podsolic; gleyed and pseudogleyed soils to the corresponding groups.

Profile 35 of the 7th American Approximation, classified as *orthic cryaquent*, is humic gley 3 inches organic, 8 inches dark humic.

Profile 6 of the 7th American Approximation, classified as *umbraquult*, is gleyed, slightly lessivé brun acide, $12\frac{1}{2}$ inches dark humic.

Profile 36 of the 7th American Approximation, classified as *cryaquent–haplaquent intergrade*, is low humic gley 6 inches dark humic.

Profile 37 of the 7th American Approximation, classified as *hydraquent*, is humic gley more than 34 inches humic and gley until the surface.

Profile 38 of the 7th American Approximation, classified as *haplaquent*, is low humic gley.

Profile 44 of the 7th American Approximation, classified as *mazaquert–aquult intergrade*, is low humic gley clay 6 inches dark humic.

Profile 47 of the 7th American Approximation, classified as *fragaquept*, is fragipan gley podsol 22 inches deep, 8 inches dark humic.

Profile 48 of the 7th American Approximation, classified as *fragaquept–ochrept intergrade*, is fragipan gley podsol 16 inches deep, 9 inches dark humic.

Profile 49 of the 7th American Approximation, classified as *ochraquept*, is low humic gley 8 inches organic.

Profile 61 of the 7th American Approximation, classified as *salorthid*, is saline rendzina gley.

Profile 65 of the 7th American Approximation, classified as *argalboll*, is humic gley lessivé 19 inches deep, 13 inches humic.

Profile 66 of the 7th American Approximation, classified as *argalboll–aquoll intergrade*, is humic gley lessivé 13 inches deep, 13 inches humic.

Profile 67 of the 7th American Approximation,

classified as *haplaquoll*, is humic gley 17 inches humic.

Profile 21 of the 7th American Approximation, classified as *humaquod*, is humus pan gley-podsol 17 inches deep.

Profile 79 of the 7th American Approximation, classified as *albaqualf*, is planosol-gley 17 inches deep, 1 inch dark humic.

Profile 80 of the 7th American Approximation, classified as *glossaqualf*, is planosol gley 19 inches deep, 7 inches dark humic.

Profile 81 of the 7th American Approximation, classified as *ochraqualf*, is low humic gley lessivé 15 inches deep.

Profile 82 of the 7th American Approximation, classified as *fragaqualf*, is fragipan low humic gley lessivé 14 inches deep.

Profile 83 of the 7th American Approximation, classified as *natraqualf*, is solonetz-gley 12 inches deep, 8 inches dark humic.

Profile 91 of the 7th American Approximation, classified as *plinthaquult*, is lateritic low humic gley slightly lessivé 13 inches humic.

Profile 92 of the 7th American Approximation, classified as *ochraquult*, is kaolinitic low humic gley lessivé 14 inches deep, 7 inches humic.

Profile 14 of the 7th American Approximation, classified as *orthic argaquoll*, is slightly lessivé humic gley 16 inches humic.

Profile 94 of the 7th American Approximation, classified as *fragaquult*, is fragipan low humic gley lessivé 21 inches deep, 12 inches illuvial, 1 inch dark humic.

Profile 17 of DUCHAUFOUR (1957), classified as *brun lessivé marmorisé*, is gleyed grey–brown podsolic 40 cm eluvial, 3 cm dark humic.

Profile 28 of DUCHAUFOUR (1957), classified as *pseudogley*, is pseudogley.

Profile 29 of DUCHAUFOUR (1957), classified as *alluvial à gley*, is rendzina gley 10 cm dark humic.

Profile 30 of DUCHAUFOUR (1957), classified as *gley à anmoor* (sol humique à gley), is rendzina humic gley 45 cm dark humic.

The American *light coloured planosol* profiles Toloca silt loam (JARVIS et al., 1959) are planosolic humic gley 19 inches and 11 inches humic, respectively.

The Alaskan *tundra* profile (TEDROW et al., 1958) is humus nano–podsol gley.

The Tanzanian profile 137 (MUIR et al., 1957) is planosolic humic gley 30 cm eluvial, 30 cm humic.

The Tchadian *humic lacustrine recent alluvium* Kouloudia of PIAS and GUICHARD (1959) is humic gley 50 cm humic, probably slightly planosolic; and Madirom, humic gley 40 cm humic, probably latosolic.

Chapter 6 | Fundamental Patterns of Soil Distribution. Broad Soil Regions of the World

The concept of soil region

Soils vary considerably within small distances; in the Pampean region of Argentina, one encounters para-chernozems, planosolic para-chernozems, planosols and solonetz very close to one another, sometimes in the same enclosure. In Ghana, western Africa, one encounters tropical ferruginous, tropical ferruginous lessivé, tropical ferruginous cuirassé, and ground-water laterites within short distances (see maps of AHN, 1961). Moreover, within each one of these fundamental types there is a great variation in thickness of the humic and eluvial horizons, depth to laterite, etc. For all these reasons the geographic soil unit is a complex of various types that may differ considerably from one another. This fact has been long ago recognized and soil catenas or soil associations have long been used.

The catena and association concepts have evolved toward a more flexible and geographic concept, that of soil region. A soil region is an area in which soil distribution follows a definite pattern. A soil region is usually characterized by a certain factor or combination of factors of soil formation. For instance, in Argentina (PAPADAKIS, 1963a) broad regions are usually characterized by vegetation (grassland, temperate woodland, subtropical woodland, subtropical forest, desert, etc.); within each region soil distribution is usually governed by parent material and drainage. In western Africa (PAPADAKIS, 1965) some regions are characterized by a combination of factors that result in dystrophic soils, others by parent material, others by drainage, etc. No definite rules can be given.

Soil regions are natural units; their classification should be natural, and artificial schemes should be avoided. Criteria cannot be established a priori; each region imposes its own criteria.

The difference between soil region and soil association is that in the establishment of soil regions emphasis is given to the pattern of soil formation and distribution; on the other hand, soil associations are mapping units, caused by the impossibility to map soil groupings individually. But in fact soil associations are soil regions empirically established.

The distribution of soils follows some patterns that are repeated in many parts of the earth. When one is acquainted with these fundamental patterns it is easier to understand the distribution of soils in each country. That is why these fundamental patterns or broad soil regions of the world shall be dealt with in this section (see Table VIII and IX).

We may distinguish the following groups or broad regions.

(*1*) *Podsolic regions* where temperatures are usually low and weathering is seldom allitic. Vegetation is usually forest, heath, or tundra. Humus rich in slightly polymerized substances and more or less acid (mor, moder, forest mull) is formed. Podsolization takes place; according to its intensity soils tend to be podsol, podsolic, or merely brunisolic.

(*2*) *Cinnamonic regions* where soil is thoroughly dried from time to time; rubification takes place; soils tend to be cinnamonic.

(*3*) *Chernozemic regions* where vegetation is grassland. A deep, neutral, dark humic horizon is formed; soils tend to be chernozemic.

(*4*) *Kaolinitic* leaching rainfall and temperature are high. Weathering tends to be allitic, and kaolinitic soils abound.

(*5*) *Deserts* where pedogenesis is slow and usually raw soils prevail; non-elimination of salts and sodium carbonates produced by weathering results in an abundance of saline and solonetz soils.

(*6*) *Mountains*; since climate differs considerably from one mountain to another, mountainous regions may be considered as mere modifications of other regions. But as various climates are often encountered near one another in the same mountain, it is more convenient to consider them as separate from the others. A common characteristic is the abundance of rankers, lithosols and other raw soils.

PODSOLIC REGIONS

General considerations

Podsolization prevails when humus is rich in acid, slightly polymerized substances. Such humus is produced

TABLE VIII

CLASSIFICATION OF BROAD SOIL REGIONS

(I) Great groups

Prevalent process	Region
Podsolization	Podsolic
Rubification	Cinnamonic
Humification	Chernozemic
Erosion (transport of materials from one point to another)	Deserts
De-silication (allitic weathering)	Kaolinitic
Erosion, lateral drainage, plus some of the other processes	Mountains

(II) Podsolic regions

Prevalent vegetation and other peculiarities	Region
Subglacial desert; erosion	Subglacial desert
Tundra; permafrost at shallow depth	Tundra
Coniferous forest; high leaching rainfall	Atlantic podsol
Coniferous forest; high leaching rainfall. Volcanic materials abound	Atlantic podsol-ando
Coniferous forest; moderate leaching rainfall	Continental podsol
Coniferous forest; same, humification by cropping	Sod podsol
Coniferous forest; gleization	Gley podsol
Coniferous forest; gleization, permafrost	Permafrost gley podsol
Coniferous forest; moderate leaching rainfall, permafrost	Permafrost podsol
Volcanic materials abound	Podsol ando
Broadleaf forest; clay eluviation; neither podsolic weathering nor iron eluviation	Grey–brown podsolic
No clay eluviation; iron mobilization	Brunisolic

(III) Cinnamonic regions

Climate and other peculiarities	Region
Mediterranean with moderate to high leaching rainfall	Mediterranean cinnamonic
Same; volcanic materials	Mediterranean cinnamonic-ando
Arid with low leaching rainfall	Arid cinnamonic
Moderate to high leaching rainfall during a warm season	Cinnamonic

(IV) Chernozemic regions

Prevalent process	Region
Humification	Chernozemic
Humification plus rubification	Cinnamonic chernozemic

(V) Kaolinitic regions

Prevalent process; other peculiarities	Region
De-silication (allitic weathering)	Kaolinitic
Same; volcanic materials abound	Kaolinitic–ando
Same; no de-silicated alluvial materials abound	Kaolinitic–alluvial
De-silication plus podsolization	Subtropical kaolinitic
Young soils abound	Young kaolinitic

TABLE VIII *(continued)*

(VI) Mountains

Climate and other peculiarities	Region
High latitude humid	High latitude humid mountains
Same; volcanic materials abound	High latitude humid mountains
Mediterranean	Mediterranean mountains
Same; volcanic materials abound	Mediterranean mountains
Dry	Dry mountains
Dry Mediterranean	Dry mediterranean mountains
Tropical highland	Tropical mountains
Same; volcanic materials abound	Tropical mountains–ando

(VII) Transition regions

Transition between	Region
Podsolic–chernozemic (forest steppe)	Gray-wooded
Brunisolic–cinnamonic	Brunisolic–cinnamonic
Brunisolic–arid cinnamonic	Brunisolic–acid cinnamonic
Chernozemic–desert	Chernozemic–desert
Mediterranean mountains–dry mountains	Mediterranean mountains–dry mountains
Mediterranean mountains–chernozemic	Mediterranean mountains–chernozemic
Brunisolic–kaolinitic	Mediterranean brunisolic–cinnamonic
High latitude mountains–subtropical kaolinitic	High latitide humid mountains–kaolinitic
Tropical mountains–kaolinitic	Tropical mountains–kaolinitic
Kaolinitic–chernozemic	Kaolinitic–chernozemic

(VIII) Paleo-kaolinitic regions

Prevalent processes	Region
Desilication (illitic weathering)–podsolization	Kaolinitic–podsolic
Desilication (illitic weathering)–podsolization–rubification	Kaolinitic–podsolic–cinnamonic
Desilication (illitic weathering)–rubification, erosion	Mediterranean mountains–kaolinitic

by sub-glacial desert, tundra, heath, coniferous forest and to a lesser extent broad-leaf forest; under very cold climate even grassland can produce it. On the other hand soil should not dry thoroughly from time to time, otherwise rubification prevails. And temperatures should not be too high, otherwise allitic weathering takes place.

Vegetation is to a great extent conditioned by climate. The climatic requirements of these various vegetation types are, according to PAPADAKIS (1961), as follows.

(*a*) *Sub-glacial desert.* Average daily maximum of the warmer month above 0°C; summer too cool for tundra.

(*b*) *Tundra.* Average daily maximum of the two warmer months above 6°C (42.8°F); summer too cool for coniferous forest.

(*c*) *Coniferous forest.* Average daily maximum of the two warmer months above 10°C (50°F); average daily minimum of the same months above 6°C (42.8°F); frost-free period too short for broad-leaf forest.

(*d*) *Broad-leaf forest.* "Available frost-free period" longer than 4½ months.

Naturally vegetation does not depend only on temperature; it depends on humidity, and on soil. Forest is replaced by grassland when conditions are too dry for it and by a mixed vegetation of woody plants and grasses, when climate is too dry for grassland. Moreover soil conditions have a great influence. An acid soil favours coniferous forest or heath, and a neutral one broad-leaf forest or grassland. Moreover soil depends not only on vegetation but on a great number of other factors, more especially leaching rainfall, drainage, and parent material. That is why the limits of soil regions are determined taking account of all factors, and may differ considerably from the climatic limits of the corresponding vegetation types.

Therefore, on the basis of vegetation one may distinguish four podsolic regions: (*1*) *sub-glacial desert*, where lithosolic rankers, often ranker-podsols abound; (*2*) *tundra*, where humus nano-podsols, arctic brown and gleisols prevail; (*3*) *coniferous forest*, where podsols prevail; and (*4*) *broad-leaf forest*, where according to leaching rainfall and parent material podsolic or brunisolic soils are common.

TABLE IX

SYMBOLS OF SOIL REGIONS

A–K	ando–kaolinitic		*K–Ch*	kaolinitic–chernozemic
A–B–M	ando–brunisolic–mountainous		*K–P*	kaolinitic–podsolic
AC	arid cinnamonic		*K–P–C*	kaolinitic–podsolic–cinnamonic
AC–B	arid cinnamonic–brunisolic			
AP	atlantic podsol		*M*	high latitude humid mountains
			M–A	high latitude humid mountains–ando
B	brunisolic		*M–K*	high latitude humid mountains–kaolinitic
B–A	brunisolic–ando		*M'*	mediterranean mountains
B–Al	brunisolic–alluvial		*M'–A*	mediterranean mountains–ando
B–C	brunisolic–cinnamonic		*M'–Ch*	mediterranean mountains–chernozemic
B–K	brunisolic–kaolinitic		*M'–DM*	mediterranean mountains–dry mountains
			M'–K	mediterranean mountains–kaolinitic
CCh	cinnamonic chernozemic		*MC*	mediterranean cinnamonic
Ch	chernozemic		*MC–A*	mediterranean cinnamonic–ando
Ch'	antarctic chernozemic			
Ch–D	chernozemic–desert		*P–A*	podsol–ando
CP	continental podsol		*P–K*	podsolic–kaolinitic
			PGP	permafrost gley–podsol
D	desert		*PP*	permafrost podsol
DM	dry mountains			
DM'	dry mediterranean mountains		*SD*	sub-glacial desert
			SK	subtropical kaolinitic
GB	grey–brown podsolic		*SP*	sod podsol
Gl	glaciers			
GP	gley–podsol		*T*	tundra
GW	grey–wooded		*TM*	tropical mountains
			TM–A	tropical mountains–ando
K	kaolinitic		*TM–K*	tropical mountains–kaolinitic
K–A	kaolinitic–ando			
K–Al	kaolinitic–alluvial		*YK*	young kaolinitic
K–C	kaolinitic–cinnamonic			

Sub-glacial desert regions

In sub-glacial desert plant growth is so little, that lithosols prevail. As a consequence of alternate freezing and thawing many soils of the sub-glacial desert are polygonal. Sub-glacial desert is shown in the maps by the symbol *SD*.

Tundra regions

Leaching rainfall is usually very low in tundra: 30 mm in Barrow (Alaska), 40 in Anadyr (Siberia) and so on. Under such conditions slightly polymerized humus (fulvic acids) may be formed but complexed iron does not descend the profile; soil is brunisolic (arctic brown), sometimes neutral. However, some areas receive higher rainfall. Even in dry areas, lowlands receive runoff from other soils. In these lowlands waterlogging is common, because permafrost is encountered at a shallow depth. Due to low temperatures and waterlogging slightly polymerized humus is formed. This humus is only slightly eluviated because the amount of water that is lost by drainage is very little. But the organic substances that are illu-

viated at the surface of permafrost are conserved, incorporated into its surface by freezing and thawing and a humus podsolic B is formed. On the other hand, iron elimination from the eluvial horizon is seldom total, because the slow downward movement of iron complexed by organic matter is counteracted by the capillary rise of ferrous iron from the lower waterlogged horizons. Therefore, soils are humus nano-podsol with the eluvial horizon varying in colour from brown to gley (gley-podsols); peat is very common. As a consequence of alternate freezing and thawing many soils of the tundra are polygonal. Where parent material is rich in carbonates organic or highly humic proto-rendzinas are formed. Tundra regions are shown in the maps by the symbol *T*.

Coniferous forest regions

In coniferous forest the following broad regions may be distinguished:

(*a*) *Atlantic podsol:* leaching rainfall is high; waterlogging frequent; soil is acid; peaty soils, humus-podsols, iron podsols and gleisolic soils abound.

Many atlantic podsol regions extend into warm climates on acid sandy soils. In this case podsols are often produced by degradation of grey–brown podsolic soils which also abound. It may be these regions should be separated from the colder ones.

(*b*) *Continental podsol:* leaching rainfall is moderate; waterlogging less frequent; peaty soils are less common; iron podsols abound.

(*c*) *Sod podsol:* leaching rainfall is moderate; temperatures higher than the average for podsol regions; cropping has transformed a great part of the soils into sod-podsols or sod-podsolic.

(*d*) *Gley-podsol:* although leaching rainfall is not so high topographic conditions make waterlogging common; waterlogging combined with low temperatures produced many peaty soils and gley podsols.

(*e*) *Permafrost gley-podsol:* although leaching rainfall is low the presence of permafrost at a shallow depth favours waterlogging; peaty and gley podsols abound.

(*f*) *Permafrost podsol:* leaching rainfall is moderate. Because of this and the presence of permafrost at a shallow depth iron leaching is counteracted; bleaching is not complete and many soils are yellow podsols; for the same reasons solodized podsols are encountered.

Some of these regions have an ando variance.

Atlantic podsol regions. As stated before these regions are characterized by high leaching rainfall. Coniferous forest is caused by climatic causes in the colder parts; but in the warmer ones it is due to soil acidity produced by a combination of high leaching rainfall and poor in bases and clay materials.

In the warmer parts soil evolution often begins with a brunisolic soil, which gradually loses its clay and becomes podsol; DUCHAUFOUR (1959) pointed out this fact; he says that in this case podsols are not the climax soil but the product of degradation. The term "atlantic podsol" has been borrowed from this author and it has been extended by the present author (PAPADAKIS, 1964) to all coniferous forest regions with high leaching rainfall, warm or cold.

Soils vary from iron podsol when drainage is deep and climate rather warm, to humus podsols under opposite conditions; organic pans, orstein and fragipans are very common. Atlantic podsols are very acid and "humic" horizons are very thin; we pass almost directly from A₀, the organic residues accumulated on soil surface, or from a peaty horizon, to the "ashy albic" horizon.

Soil distribution in the atlantic podsol regions may be outlined as follows:

(*a*) *Good drainage, warmer climate:* iron podsol.

(*b*) *Poor drainage, colder climate:* humus podsol, organic pan podsol, fragipan podsol; peaty soils.

(*c*) *Conditions intermediate between a and b:* iron-humus podsol.

(*d*) *Transition between these regions and those of brunisolic–grey–brown podsolic:* grey–brown podsolic.

(*e*) *Combination of d and b:* fragipan podsolic.

Atlantic podsol regions are shown on the maps by the symbol *AP*. They are encountered in Alaska, Canada, Russia, Finland, Norway, Denmark, Sweden, Germany, Poland, Czechoslovakia, Belgium, The Netherlands, France, England, Ireland, Spain, Portugal, New Zealand and Australia.

Atlantic podsol–ando regions. In northwestern United States and Kamchatka (Siberia) volcanic materials are frequent; ando soils are common; organic soils too. These regions are shown on the maps by the symbol *AP-A*.

Continental podsol regions. The main characteristics of these regions, from a pedogenetic point of view, are coniferous forest and moderate leaching rainfall; they correspond to *P* (PAPADAKIS, 1960a). Coniferous forest is not due to soil poverty, but to shortness of the "available" frost-free season, which is less than $4\frac{1}{2}$ months (PAPADAKIS, 1961).

Since vegetation was coniferous forest or heath since the beginning of formation, weathering took place in an acid medium, and an "ashy albic" horizon, very poor in clay and in iron, has been formed. But since leaching rainfall is low, these substances accumulated at a shallow depth and a "spodic" horizon rich in amorphous clays has been formed. This horizon is not so poor in bases; it may contain considerable reserves of unweatherable minerals. This is the main difference between the continental podsols and atlantic podsols, which in many cases result from degradation of other soils (DUCHAUFOUR, 1959).

When drainage is good the organic substances that descend in the "illuvial" horizon decay rapidly, this horizon is not distinctly richer in organic matter than the eluvial one, and the soil formed is an iron podsol.

On the other hand, when drainage is bad the organic substances that descend in the illuvial horizon accumulate, this horizon is richer in organic matter than the eluvial one, and the soil is a humus podsol. Under intermediary conditions a humus-iron podsol is formed. With very bad drainage the soil is gley podsol or gleyed podsol. When the parent material is an unconsolidated rock rich in clay and bases, the humus produced is not so acid, weathering takes place under less acid conditions, alumina is not leached away; only iron and clay are eluviated. The "eluvial" horizon is bleached but not very sandy; it is not ashy; the "illuvial" horizon is "textural B"; the soil is grey wooded; many grey wooded soils are neutral in all their horizons, even in the "eluvial" one; some are calcareous in their "illuvial" horizons.

In summary, soil distribution may be outlined as follows.

(*a*) *Good drainage, climate not so cold, materials very poor in clays* (granits, gneiss, sands, etc.): iron podsols.

(*b*) *Poor drainage, cold climate, unweathered materials very poor in clays* (granits, gneiss, sands, etc.): humus podsols.

(*c*) *Conditions intermediate between a and b:* humus-iron podsols.

(*d*) *Bad drainage:* gley podsols, gleyed podsols, peaty or organic podsols, peaty on gley, etc.

(*e*) *Materials rich in clay:* grey-wooded.

Continental podsol regions are shown on the maps by the symbol *CP*. They are encountered in Alaska, Canada, United States and Russia.

Sod podsol regions. In the southern part of the area of continental podsols in Russia, coniferous forest becomes more and more mixed with broad-leaf trees, which produce milder humus; moreover forest has been replaced by crops and prairies, which enrich the surface soil with bases that have been absorbed in the lower horizon; in addition some soils receive manure or have been limed. As a consequence the "eluvial" horizon is richer in organic matter and darker in colour. The soil is sod-podsol or sod-podsolic. Naturally not all the soils of the region have a "dark humic" horizon 10 cm or more deep. Enrichment in humus depends on cropping system and duration; it is easier when climate is less humid, less cold, and soil less poor in bases. The distribution of soils in sod-podsolic regions, may be summarized as follows:

(*a*) *Materials very poor in clays, weatherable minerals and carbonates:* podsols.

(*b*) *Intense humification* (broad-leaf forest, prairie, cropping, manures, liming): sod-podsols.

(*c*) *Materials rich in clays, little humification, higher leaching rainfall:* grey-brown podsolic.

(*d*) *Materials rich in clays, little humification, low leaching rainfall:* grey wooded.

(*e*) *Bad drainage:* humus podsols; *gley podsols,* gleyed podsolic.

Sod-podsol regions are shown in the map by the sign *SD*. They are encountered in Russia and Poland. But certainly there are podsols modified by cropping in many other parts of the world.

Gley-podsol regions. Due to the very cold, almost tundra climate, flat topography, and rather high leaching rainfall, poor drainage conditions prevail. Gley-podsols and gleyed podsols with more or less thick peaty horizons and peaty soils prevail. Gley podsol regions are shown in the maps by the symbol *GP*. They are encountered in Canada and Russia.

Permafrost gley-podsol regions. In Siberia leaching rainfall is lower but permafrost shallow; gleization is chiefly due to permafrost. Peaty podsols and gley podsols prevail. These regions are shown in the map by the symbol *PGP*. They are encountered in Siberia.   '

Permafrost podsol regions. In these regions permafrost is encountered at shallow depth, but due to hilly relief and to lower leaching rainfall, waterlogging is less frequent. Gley podsols and gleisols are less frequent. For the same reasons (permafrost and low leaching rainfall) iron eluviation is slow and is counteracted by capillary ascension of ferrous iron. This is why the "eluvial" horizon is yellow (straw coloured) instead of "bleached"; the soil is yellow podsol. Naturally not all soils are yellow

podsols. Those with bad drainage are gley-podsols or gleyed podsols, with a more or less thick peaty horizon; they may be merely peaty. In materials very poor in iron, podsols with "ashy" horizons are formed. Because of the low leaching rainfall base status is rather good and sod podsols abound; some soils are solonized (podsol-planosols). Permafrost continental podsol regions are shown in the maps by the symbol *PP*. They are encountered in Siberia.

Podsol-ando regions. In New Zealand volcanic materials abound; many soils are ando, brun acide, planosols. Podsol-ando regions are shown in the maps by the symbol *P-A*. They are encountered in New Zealand, but certainly there are other parts of the world that have also such regions.

(Temperate) broad-leaf forest regions

When vegetation is broad-leaf forest, and humus forest mull, podzolization is moderate, and podsolic soils are formed. When the materials are rich in clay and bases, or leaching rainfall low, or soil young, podsolization is only incipient and the soil is brunisolic. Therefore under broad-leaf forest two broad soil regions exist.

(*1*) *Grey-brown podsolic regions.* Moderate podsolization.

(*2*) *Brunisolic regions.* Podsolization only incipient (that does not mean necessarily that it will be aggravated with time).

Grey-brown podsolic regions. Due to higher leaching rainfall, older soils, or poorer in lime materials, grey-brown podsolic prevails. The distribution of soils may be outlined as follows: (*a*) medium conditions of parent material and leaching rainfall: grey-brown podsolic; (*b*) young soil from basic rocks and moderate leaching rainfall: braunerde; rendzinas in extreme cases; (*c*) well-drained soil from coarse materials: brun acide; in extreme cases podsols; (*d*) poorly drained soils: gleyed podsolic, fragipan podsols, low humic gley and othe rgleisols; in extreme cases organic soils; (*e*) transition to the chernozemic region: prairie lessivé, planosolic parachernozem.

Grey-brown podsolic regions are shown in the maps by the symbol *GB*. They are encountered in Canada, United States, England, France, Germany, Austria and Yugoslavia.

Brunisolic regions. Due to lower leaching rainfall, more calcareous or richer in bases materials and younger soils, braunerde prevails. The distribution of soils may be outlined as follows: (*a*) medium conditions of leaching rainfall, parent materials and soil age: braunerde; (*b*) higher leaching rainfall and/or coarser materials: brun acide, in extreme cases podsols; (*c*) young soils from soft limestone: rendzinas; (*d*) older soils from calcareous materials: calcareous braunerde; (*e*) poorly drained soils:

gleyed podsolic, ground-water rendzinas, etc.; (*f*) transition to chernozemic: chernozemic braunerde, degraded chernozems.

Brunisolic regions are shown on the maps by the symbol *B*. They are encountered in Chile, Spain, France, Germany, The Netherlands, Luxemburg, Belgium, Switzerland, Rumania, Poland, Czechoslovakia, Russia, China and New Zealand.

Brunisolic–alluvial regions. In certain brunisolic regions (northern Italy for instance) alluvial soils are very common. These soils are usually well provided with bases, they practically have not suffered any podsolization; drainage is sometimes impeded, but it is improved artificially. These regions are shown in the maps by the symbol *B–Al*.

Brunisolic–ando regions. In certain brunisolic regions (New Zealand for instance) volcanic materials abound; and soils referred to as brown are probably recent brown.

CINNAMONIC REGIONS

The fundamental characteristic of these regions is rubification; soil is dried thoroughly from time to time, and iron is de-hydrated irreversibly. Either because the humid season is cold (winter) or because leaching rainfall is low, weathering is siallitic and the soils formed are rich in 2:1 clays. Two cases may be distinguished.

(*1*) Leaching rainfall is high or moderate but falls in winter. The amount of iron released is considerable and soil acquires red colours. It is the case of moist or moderately dry mediterranean climates.

(*2*) Leaching rainfall is very low. Not much iron is released and soil is less red.

Therefore we have two groups of cinnamonic regions.

(*a*) *Mediterranean cinnamonic* with moderate to high leaching rainfall in winter, and redder soils.

(*b*) *Arid cinnamonic* with very low leaching rainfall and less red soils.

Mediterranean cinnamonic regions

Besides their climatic similarity the mediterranean regions show also similarities from geological and geomorphological points of view; soils are young, calcareous materials abound, volcanic materials are frequent; the only mediterranean region that forms part of an old block is Western Australia. Moreover due to topography mediterranean lowlands suffer or have suffered in the past from bad drainage; this favoured the formation of saline, solonetzic and planosolic soils.

Soil distribution may be outlined as follows:

(*a*) Soft limestones and marls: rendzinas and calcareous braunerde;

(*b*) Hard limestone: terra rossa; in many cases terra rossa is not autochtonous but transported.

(*c*) Clayey materials: dark clays, rendzina clays or chernozemic rendzina clays, according to lime content of parent material and moisture conditions; they are often saline or alkaline.

(*d*) Basic consolidated rocks (basalt, etc.), high leaching rainfall, good drainage: red cinnamon.

(*e*) Siliceous materials poor in weatherable minerals, low leaching rainfall: arid brown.

(*f*) Conditions intermediate between *d* and *e*: reddish brown, pale cinnamon.

(*g*) Volcanic materials: ando; usually they are classified in the fore-mentioned groups, but they have a higher cation exchange capacity, thicker humic horizons, greater depth and higher fertility (Thera Island in Greece, Italian soils near Rome, Naples or Catania, many Californian soils, etc.); many of the soils classified as "brun acide", "prairie" and "non-calcic brown" are ando.

(*h*) Kaolinitic (desilicated) materials (some parts of Australia): palaeokaolisols, palaeolateritic.

(*i*) Old soils sometimes from kaolinitic (desilicated) materials, podsolizing vegetation, drainage often poor (some parts of Australia): podsolic, lateritic podsolic.

(*j*) Materials rich in weatherable Na-silicates, rather poor drainage; slightly lessivé brown, planosols, magnesium-solonetz, solonetz, according to drainage conditions, parent material, etc.; many of them cinnamonic.

(*k*) Young alluvial: recent alluvial.

(*l*) Acid materials, podsolizing (tannin rich) vegetation, high leaching rainfall, old soils: brun acide, non-calcic brown.

(*m*) "Exsudational" moisture regime and calcareous ground water: meadow rendzinas or rendzina clays, often with "calcareous accumulations" or "petro-calcic" pans, sometimes saline or alkaline (tirs).

(*n*) Areas of accumulation of salty waters: solonchacks, solonetz, planosols.

(*o*) Rather humid climate with cool summers (California coast near San Francisco): prairie and other chernozemic soils.

Mediterranean cinnamonic regions are encountered in the United States, Portugal, Spain, France, Italy, Greece, Turkey, Cyprus, Lebanon, Syria, Libya, Tunisia, Algeria, Morocco, South Africa. They are shown on the maps by the symbol *MC*.

Mediterranean cinnamonic–ando regions

In certain regions of Italy and many other countries —although that does not appear in the maps—volcanic materials abound and ando soils are common; many recent brown brun acide, dark clays, and grey–brown podsolic are ando or related to them. Mediterranean cinnamonic–ando regions are shown on the maps by the symbol *MC–A*. They are encountered in Italy, but certainly there are similar regions in other parts of the world.

Arid cinnamonic regions

In these regions leaching rainfall (*Ln*) is lower than in the mediterranean cinnamonic regions, less iron is released by weathering and soils are less red. Reddish brown and pale cinnamon soils abound; some of them are so grey that they are called grey cinnamon in Russia. Soil distribution may be outlined as follows.

(*a*) Non-calcareous materials, free drainage, young soils: reddish brown and arid brown.

(*b*) Calcareous materials, good drainage: rendzinas relatively poor in humus, many with "calcareous accumulations" at some depth.

(*c*) Older soils from non-calcareous materials, rich in weatherable Na-silicates: planosols, solonetz.

(*d*) Clayey materials: dark clays, rendzina clays.

(*e*) Poor drainage: planosol, solonetz, solonchacks, meadow rendzinas with calcareous accumulations (tirs), dark clays, often chernozemic or rendzina.

Arid cinnamonic regions are shown on the maps by the symbol *AC*. Some cinnamonic regions cannot be termed neither mediterranean, nor arid, they are just cinnamonic (*C*); they abound in tropical climates, although the scale of the maps of this book seldom permits their separation. Arid cinnamonic and cinnamonic regions are encountered in the United States, Venezuela, Brasil, Paraguay, Argentina, Spain, Turkey, Russia, China, India, Morocco, Mozambique, Madagascar, Rhodesia, Bechuanaland, South Africa and Australia.

CHERNOZEMIC REGIONS

General considerations

The characteristic pedogenetic process of these regions is humification; abundant and highly polymerized humus is synthesized. Vegetation is grassland, soils well provided with calcium, and a deep dark neutral humic horizon is formed. Climax vegetation is pure grassland, when leaching rainfall is low (Ln less than 20% of annual potential evapotranspiration), spring is not dry (rainfall covers more than 50% of potential evapotranspiration), and average daily maximum of the coldest month is below 21°C (PAPADAKIS, 1961). Such conditions are encountered in southeastern Russia, the Danubian Basin, the prairie region of Canada and the United States, and the Argentine Pampa. Soils formed from materials rich in lime are calcareous chernozemic; the humic horizon rests directly on a "calcareous" (ca) horizon, or the two horizons overlap. These soils are richer in organic matter than para-chernozems, and they seldom show signs of clay eluviation; according to thickness and colour of the "humic" horizon, they are divided into chernozems,

kastanozems, and chernozemic brown; but it may be better to specify the thickness of the humic horizon. When the material is marl rich in clay, the soil formed is rendzina chernozemic clay or chernozemic clay.

In many calcareous chernozemic soils a more or less soft calcareous accumulation is formed at some depth; a "gypsic" horizon often underlies the calcareous accumulation, and the higher limit of de-salinization is often high. Calcareous accumulations are less frequent when leaching rainfall is rather high and in para-chernozems.

Soils formed from non-calcareous materials are para-chernozems; the "humic" horizon does not rest directly on a "calcareous" horizon. Para-chernozems are poorer in organic matter, and they often show signs of clay eluviation. Such eluviation is especially important when parent material is rich in easily weatherable Na-silicates, for instance volcanic (PAPADAKIS, 1963b). Poor drainage conditions slow the elimination of Na and favour formation of planosolic para-chernozems, planosols, or solonetz; they also favour gleying and formation of humic glei. Poor drainage conditions also produce "petro-calcic" pans.

Ground waters of chernozemic regions are charged with $CaCO_3$, and salts. In areas with an "exsudational" moisture regime carbonates and salts accumulate; saline subgroups with "calcareous accumulations" are formed. In general the topo-sequence of chernozemic regions goes from chernozemic soils in well drained sites, to chernozemic and plano solic humic-gley, planosols, solonetz and solonchacks in depressions.

When the climate is dry and, from time to time, soil dries out thoroughly to considerable depth, iron sequioxides are irreversibly dehydrated and reddish para-chernozems are formed. When summer is cool (extreme south of Argentina, Patagonia, and very high, rather dry, mountains) and parent materials are siliceous, the decay of organic matter is slow and soil contains a great quantity of undecayed roots; soils are grassland rankers; their "humic" horizons are less than 25 cm thick and they contain a considerable amount of undecayed roots. Three groups of chernozemic regions can thus be distinguished:

(*1*) *Chernozemic sensu stricto.*

(*2*) *Cinnamonic chernozemic*: rubification is common; many soils have "cinnamonic" ("rubified") horizons; reddish chernozemic abound; many soils are "cinnamonic" or "non-calcic brown".

(*3*) *Sub-antarctic chernozemic*: organic matter decomposition is low; soil contains a great quantity of undecayed roots; the humic horizon is thinner; many soils are "cinnamonic-rankers".

The distribution of soils in these regions may be outlined as follows.

(*a*) *Calcareous parent materials and good drainage:* calcareous chernozemic (chernozems, kastanozems, and chernozemic brown) and rendzina chernozems (effervescence with HCl from the surface).

(*b*) *Materials very rich in 2:1 clays:* chernozemic

clays and chernozemic rendzina clays; they may be gleyed.

(*c*) *"Exsudational" moisture regime and/or poor drainage:* meadow chernozems, humic-gley more or less lessivé, saline, alkaline or gypsum; many have "calcareous accumulations" (cac) or "petro-calcic" pans (tosca) at some depth.

(*d*) *Non-calcareous materials rich in weatherable Na-silicates:* slightly planosolic and planosolic para-chernozems, planosols, magnesium solonetz and solonetz, according to drainage conditions.

(*e*) *Non-calcareous materials, poor in weatherable minerals:* sandy para-chernozems; when drainage is poor, planosols and solonetz may be formed.

(*f*) *Transition to forest:* prairie, prairie lessivé and chernozemic braunerde.

(Sensu stricto) chernozemic regions

Region characteristics and soil distribution correspond to the pattern described above. However, there are some differences. In the chernozemic regions of Russia calcareous materials dominate and leaching rainfall is low; the soils are calcareous chernozemic. In North America materials are less calcareous; para-chernozems prevail and planosols abound. In Argentina a great part of the materials are of volcanic origin; soils are para-chernozems with high cation exchange capacity; many are planosolic; planosols abound. Chernozemic regions are shown on the maps by the symbol *Ch*. They are encountered in Canada, United States, Argentina, Uruguay, Austria, Hungary, Yugoslavia, Bulgaria, Rumania, Czechoslovakia, Russia.

Cinnamonic chernozemic regions

Due to a hot and dry summer rubification takes place; cinnamonic (reddish) chernozems are common; some soils are reddish brown or non-calcic brown. These regions are shown on the maps by the symbol *CCh*. They are encountered in the United States where grassland vegetation extends to regions with hot summers.

Sub-antarctic chernozemic regions

Due to a cool summer and siliceous parent materials organic matter decay is slow, the humus formed is mild, but undecayed roots abound. Grassland rankers prevail. Sub-antarctic chernozemic regions are shown on the maps by the symbol *Ch'*. They are encountered in the southern corner of Patagonia (Argentina).

DESERTS

The concept of desert varies from one scientist to another. For pedologists desert soils are formed under an annual rainfall of 10 inches (250 mm) sometimes 20 inches in the United States, and naturally much more in the tropics. The pedologic concept of desert is something broader than that of our climatic classification (PAPADAKIS, 1961).

In the desert, soil is very little protected by vegetation and both aeolian and hydraulic erosion are very severe. Concerning hydraulic erosion one must remember that rains are rare but heavy and in a barren soil they cause much erosion. Soil is not formed on slopes, because the products of weathering, usually sands, are immediately removed by rain and wind, and accumulate in the lowlands, forming dunes. Continuous movement impedes the establishment of vegetation in these dunes and soil formation cannot advance. Some salts, calcium carbonate and clay are produced; that is why the water-table, which is sometimes encountered under these dunes, is usually saline. The principal processes of soil formation in the desert may be outlined as follows.

(*1*) In tablelands the gravel formed by weathering remains on soil surface; little by little a desert pavement is formed. Under this desert pavement weathering proceeds. Calcium–magnesium and sodium carbonates are released, and due to virtual absence of leaching these substances are not leached away. The soil formed is usually rich in Ca, more or less alkaline, clay eluviation takes place and a planosolic soil is formed (planosolic red desert). But the eluvial horizon is only a few centimeters thick (for instance, 5–10 cm for a rainfall between 100 and 200 mm annually). When the parent material contains sufficient Ca and Mg, carbonates accumulate at little depth (e.g., 25 cm).

(*2*) The waters carried by the torrents of the desert do not go to the sea. They flood some lowlands, evaporate, and the silt, clay and salts they contain accumulate. In this way a more or less clayey and saline alkaline soil (desert clays, takyr) is formed. Many of these soils are so saline–alkaline that they are barren of vegetation.

Many of the soils encountered in the desert are allochthone (alluvial soils formed by materials brought from other regions by the rivers traversing the desert). These soils also have the tendency to become saline–alkaline, because water evaporates and the salts it contains accumulate. Near the river, soil is leached from time to time, and soils are in general good. But far from the river no leaching takes place and saline–alkaline soils are formed. In any case the water-table of the alluvial soils of the desert is usually more or less saline or alkaline. It is only near the river that a non-saline water-table is likely to be encountered. Consequently one may distinguish in the desert the following regions:

(*a*) *Hamada* (stony desert): lithosols prevail; rocks are often covered by a varnish.

(*b*) *Erg* (sandy desert): desert sands prevail.

(*c*) *Tablelands:* soil is protected by a desert pavement. Under this pavement a planosolic red desert soil is often formed. Volcanic ashes, received from time to time, greatly facilitate this process, because they weather rapidly and provide the necessary clay and Na bicarbonates. But when the whole parent material is volcanic ash or other soft volcanic material a desert pavement cannot be formed. The soil is a desert crust (soil cemented by the products of weathering); cementation is loose but sufficient to impede wind erosion and permit the process to continue.

(*d*) *Desert clay plains:* soils are rich in clay and deep; but usually they are too saline–alkaline; many are for this reason barren of vegetation; or they have crusts formed by the accumulation of soluble salts, gypsum or lime at the surface.

(*e*) *Alluvial soils of allochthonous origin:* they vary greatly in texture. Those distant from the river have often the characteristics of desert clay plains. A saline–alkaline water-table is very common, except near the river.

Except for those formed in the lowlands and inundated or subirrigated, desert soils are usually poor in organic matter; they lack "humic" horizons and their colours are usually light. In the desert soil dries thoroughly. This is why soils formed in the desert from consolidated rocks suffer rubification and are reddish. But soils formed from materials poor in weatherable minerals have the colour of the parent material.

Deserts are shown on the maps by the symbol *D*. They are encountered in the United States, Mexico, Venezuela, Peru, Ecuador, Bolivia, Chile, Argentina, Russia, China, India, Pakistan, Afghanistan, Iran, Iraq, Syria, Jordan, U.A.R., Libya, Tunisia, Morocco, Spanish Sahara, Senegal, Mauritania, Mali, Niger, Nigeria, Chad, Cameroons, Sudan, Ethiopia, Somalia, Kenya, Angola, Bechuanaland, South Africa, Southwest Africa, and Australia.

KAOLINITIC REGIONS (HUMID TROPICS)

General considerations. Typical regions

When leaching is intense and temperatures are high, weathering of silicates produces 1:1 clays and free iron. The soils formed are kaolisols or kaolinitic podsolic (laterite, etc.). The amount of rainfall and temperatures that are required for such "allitic" weathering depend on parent material; it is less in the case of easily weatherable rocks (granite, gneiss, etc.). It is often admitted that 2:1 clays are transformed into 1:1 clays by leaching, but we have no direct evidence of this process. Usually the kaolinite of tropical soils has been produced directly by weathering. When drainage is poor, the silica and bases released by weathering are not leached away, and 2:1 clays are formed. This is especially true in the case of rocks rich in easily weatherable silicates and is the reason why soils formed under conditions of poor drainage are often rich in 2:1 clays. However, in many cases the materials accumulated in poorly drained lowlands were desilicated before their transport. On the other hand,

river and ground-waters are often rich in silica and bases and a re-silication of kaolinite may take place.

Because of high leaching rainfall seasonal water-logging is very common in the tropics, even in semi-arid areas; this results in the formation of concretionary horizons and laterite (see Chapter 1). As has been stated before such horizons are frequent when the parent material is poor in easily weatherable silicates, and they are rare in the opposite case (basalts, etc.). Poor drainage favours their formation. Sometimes the iron required for this process is provided by ground-waters and in this case an iron pan (laterite) is usually formed.

The lower horizons of kaolisols are usually poor in bases. The top soil may be "eutrophic" (tropical ferruginous and terres de barre) or "dystrophic" (krasnozems, ferralitic, red–yellow podsolic). Soil distribution in São Paulo (Brasil) and West Africa shows that regions with high leaching rainfall have dystrophic soils, and those with lower leaching rainfall eutrophic. The distribution of soils in kaolinitic regions may be outlined as follows:

(*a*) High leaching rainfall, forest vegetation, old consolidated rocks rich in weatherable minerals, free drainage: ferralitic.

(*b*) Same conditions as *a*, but consolidated rocks poor in weatherable minerals: acid ferruginous.

(*c*) Same conditions as *a*, but sandy unconsolidated materials: ferralitic, less red than in *a*.

(*d*) Moderate leaching rainfall, savanna vegetation, consolidated rocks rich in weatherable silicates: terres de barre.

(*e*) Same conditions as *d*, but consolidated rocks poor in weatherable silicates: tropical ferruginous.

(*f*) Same conditions as *d*, but younger unconsolidated rocks: terres de barre.

(*g*) High leaching rainfall, cold winters, old consolidated rocks, poor in weatherable minerals: red–yellow podsolic, hardly kaolinitic.

(*h*) Same as *g*, but old consolidated rocks rich in weatherable minerals: krasnozems.

(*i*) Poor drainage, consolidated rocks poor in weatherable minerals: lateritic podsolic and ground-water laterite.

(*j*) Poor drainage, consolidated rocks rich in weatherable minerals, low or moderate annual rainfall that does not cover entirely annual potential evapotranspiration: dark clays.

(*k*) Same as *j*, transported material rich in 2:1 clays: dark clays or recent alluvial according to clay content.

(*l*) Same as *j*, but transported material rich in 1:1 clays, conditions not favouring re-silication: gleyed kaolisols, ground-water laterite.

(*m*) Young soils from consolidated rocks rich in weatherable minerals, moderate leaching rainfall: red cinnamon.

(*n*) Young soils from calcareous materials: terra rossa.

(*o*) Areas in which salts accumulate, moderate leaching rainfall: solonchacks, solonetz, planosols.

(*p*) Coast dunes, fresh water-table: sandy acid soils ("coconut sands") interspersed with humus podsols; these soils are so sandy that it is difficult to diagnose whether they have "kaolinitic horizons" or not; for the same reason the nature of their clays is of little importance.

(*q*) Saline swamps with mangrove vegetation: cat clays and other more or less saline soils.

(*r*) Recent volcanic materials: ando soils, many of them acid (ando not lessivé, slightly lessivé or lessivé, positive ando, hydrol ando, etc.).

For a detailed study of soil distribution and pedogenesis in kaolinitic regions, see MOHR and VAN BAREN (1954).

Kaolinitic regions are shown in the maps by the symbol *K*. They are encountered in Puerto Rico, Hawaii, Colombia, Venezuela, Ecuador, Guianas, Brasil, Paraguay, Sudan, Kenya, Rwanda–Burundi, Uganda, Tanzania, Mozambique, South Africa, Rhodesia, Bechuanaland, Swaziland, Madagascar, Malawi, Zambia, Congo (both), Angola, Gabon, Spanish Guinea, Cameroons, Central Africa, Chad, Niger, Dahomey, Togo, Ghana, Upper Volta, Ivory Coast, Liberia, Sierra Leone, Guinea, Portuguese Guinea, Senegal, Gambia, Ceylon.

Subtropical kaolinitic regions

In the transition zone from a temperate humid climate (grey–brown podsolic regions) to a humic tropical climate (kaolinitic regions), more especially when parent materials are poor in easily weatherable minerals, the two processes, podsolization and kaolinization, take place at the same time, and a dystrophic kaolinitic podsolic soil (red–yellow podsolic) is formed. The best-known subtropical kaolinitic region is that of southeastern United States. Podsolization has been favoured by parent materials poor in weatherable minerals, and relatively long, humid winters. Podsolization is shown clearly, by free iron content, which increases various times from the eluvial to the illuvial horizon, by the low saturation in bases of the eluvial horizon and the whole profile. It may be that at the beginning the "illuvial" horizon was rich in amorphous clays, but intense leaching and time produced its crystallization into 1:1 clays.

Other subtropical kaolinitic regions have not been identified, but that may be due to insufficient research. It may be also that a monsoon climate with summer rainfall and dry winters is not so favourable for podsolization. In São Paulo (Brasil) red–yellow podsolic abounds in the moister part of the state; because of latitude and altitude climate is not so warm and that favours podsolization. Subtropical kaolinitic regions are shown on the maps by the symbol *SK*.

Young kaolinitic regions

In the West Indies and the Yucatan peninsula of Mexico, soils are young and calcareous materials abound. Many soils have high cation exchange capacity and are more fertile than in other kaolinitic regions. Similar regions certainly exist in other parts of the world. These regions are shown in the maps by the symbol *YK*. They are encountered in Mexico, Cuba, Jamaica, Trinidad.

Kaolinitic–ando regions

In many parts of the humid tropics, Indonesia for instance, volcanic materials abound and ando soils are common. The extent of ando soils is probably much greater than shown in the maps, because the distinction between ando and kaolinitic soils is based on characters shown by laboratory analysis, and emphasis is usually given to morphologic features. Ando, by their high cation exchange capacity and high fertility, are the antitheses of kaolinitic soils, so that kaolinitic–ando regions should be considered separately.

Kaolinitic–ando regions are shown on the maps by the symbols *K–A* or *A–K*. They are encountered in Hawaii, Mexico, Central America, Ecuador, Chile, Ethiopia, Somalia, Kenya, Uganda, Tanzania, Cameroons, Indonesia, Sarawak, and Malaysia, but probably there are several similar regions in other parts of the world.

Kaolinitic–alluvial regions

In some kaolinitic regions, northern India for instance, alluvial soils formed with materials that came from non-kaolinitic regions abound. These soils are completely different both from a pedologic and from a crop-ecologic point of view. That is why these regions are separated. Kaolinitic–alluvial regions are shown on the maps by the symbol *K–Al*. They are encountered in Colombia, India, Nepal, East Pakistan, China, Burma, Thailand, Cambodia, Laos, and Vietnam.

MOUNTAINS

General considerations

As has been said before mountains differ considerably in climate and parent material. Their common characteristics are: (*a*) high frequency of lithosols and rock outcrops; (*b*) extremely free drainage for some soils; (*c*) possibility for others to receive in their upper horizons materials that have been leached from the eluvial horizons of other soils situated higher in the slope; (*d*) exsudational soil moisture, when the slope changes abruptly (ruptures de pente); (*e*) possibility to have, at small distances, different climates and consequently different soil regions, which cannot be mapped separately in small-scale maps. According to climate we may distinguish the following cases: (*1*) high latitude humid mountains; (*2*) mediterranean mountains; (*3*) dry mountains; (*4*) dry mediterranean mountains; (*5*) tropical mountains.

High latitude humid mountains

Above a certain altitude, vegetation of these mountains is coniferous forest. At still higher altitude it is replaced by grasses, forbs and cushion plants, but due to low temperatures and high humidity this vegetation also is podsolizing. However, erosion rejuvenates the soils; new material is deposited on their surface; the higher horizons receive seepage waters from other soils and such waters contain iron, bases, silica, etc. All these processes counteract leaching and podsolization, and podsolic soils are not so frequent, as it might be expected from climate and vegetation; many soils are only braunerde, brun acide or brown podsolic. Highly humic or organic rankers abound; the soil is shallow, but the humic horizon extends to bed-rock. The distribution of soils in the alpine regions may be outlined as follows:

(*a*) *Lower altitudes, deep soils, free drainage:* braunerde, brun acide, brown podsolic, all usually rich in humus.

(*b*) *Steep slopes:* lithosols, rankers, ranker podsols.

(*c*) *Depressions with poor drainage:* peats, organic soils, gleisols.

(*d*) *High altitudes, not very steep slopes:* shallow braunerde or brun acide, alpine humus nano-podsols, lithosols; some of them are polygonal.

(*e*) *Calcareous materials:* proto-rendzinas, often organic, forest rendzinas, calcareous braunerde. Due to the rapid solution of hard limestone by organic acids, the distinction between hard and solft limestone fades in this region (except where summer is dry).

(*f*) *Very cold parts with tundra vegetation:* lithosols, tundra nano-podsols.

High latitude humid mountains are shown on the maps by the symbol *M*. They are encountered in Alaska, Canada, United States, Argentina, Russia, Norway, Sweden, Germany, Czechoslovakia, Poland, Hungary, Rumania, Austria, Yugoslavia, Bulgaria, Italy, Switzerland, France, Spain, Portugal, China, Australia, and New Zealand.

High latitude humid mountains–ando. In the Cordillera of Canada, United States, Chile and New Zealand, volcanic materials abound and many soils are ando. Due to the resistance of ando soils to podsolization, their high humus content, and their fertility, the panorama of soils becomes considerably different. In Chile and New Zealand high humus content is especially favoured by the long growing season of these climates, which results in a high index of annual growth. High latitude humid mountains–ando regions are shown on the maps by the symbol *M–A*.

Mediterranean mountains

The peculiarity of mediterranean mountains is that summer is dry; even at high altitude drought interferes with raw humus accumulation and podsolization is moderate. At lower altitude some irreversible de-hydration of iron sesquioxides takes place and this fact impedes iron eluviation; a certain rubification takes place. For all these reasons even under coniferous forest podsols are seldom encountered. For the same reasons organic soils are uncommon in the mediterranean climates.

The usual pattern of vegetation in the mediterranean mountains is: latifoliates at low altitude, coniferous at higher altitude, then grasses and forbs. Under latifoliates braunerde is common but many soils are slightly "rubified" ("cinnamonic braunerde", brun méditerranéen of French authors). Lithosols are very frequent because mountains are young and erosion is very severe in the Mediterranean. Due to abundance of calcareous materials, terra rossa and rendzina are common. The only podsolic soil that is encountered is grey–brown podsolic; but very often it is an intergrade to non-calcic brown. Mediterranean mountains are shown in the maps by the symbol *M'*. They are encountered in Canada, United States, Portugal, Spain, France, Italy, Yugoslavia, Bulgaria, Greece, Turkey, Cyprus, Syria, Lebanon, Algeria, and Morocco.

Mediterranean mountains–ando. In many mediterranean mountains (Italy, Chile, etc.) volcanic materials abound. Many soils are ando, although that is not always shown in the maps. It seems that under such conditions many soils are, or resemble, brun acide and dark clays. It seems that volcanic materials favour also the formation of prairie soils (perhaps chernozemic ando). Prairie formation is also favoured by the cool summers of some mountains of western United States and Chile.

Dry mountains

Soil dries out thoroughly in dry mountains, "rubification" takes place, and cinnamonic soils abound. However, when summer is cool and not very dry, grassland vegetation prevails and chernozemic soils are common. Naturally lithosolic soils abound. The distribution of soils in dry mountains may be outlined as follows:

(*a*) *Dry, warm summers:* cinnamonic soils, more especially reddish brown.

(*b*) *Cool not very dry summers, grassland vegetation:* chernozemic soils.

(*c*) *More or less steep slopes:* lithosols.

(*d*) *Calcareous materials:* rendzinas, serozems.

Dry mountains are shown on the maps by the symbol *DM*. They are encountered in Bolivia, Argentina, Russia, China, India, Mozambique, Rhodesia and South Africa.

Dry mountains–ando. In the dry mountains of Mexico and elsewhere, volcanic materials abound. Many soils have high cation exchange capacity being ando or chernozemic ando. These soils have a great productivity capacity when given water and fertilizers. Dry mountains–ando are shown on the maps by the symbol *DM–A*. They are encountered in Mexico.

Dry mediterranean mountains

Dry mediterranean mountains have a summer drier than usual; this fact favours "rubification"; cinnamonic soils are more common; the distribution of soils resembles that of mediterranean mountains; but reddish brown are more common. Dry mediterranean mountains are shown on the maps by the symbol *DM'*. They are encoutered in Turkey, Iran, Afghanistan and West Pakistan.

(Humid) tropical mountains

In humid tropical highlands conditions are very favourable for accumulation of organic matter: the "humogenic index" of Pangerango (Indonesia) is 5.1 and of Quito 2.3; by comparison, that of Des Moines, Iowa, in the chernozemic region of the United States is 2.5, and that of Rosario, in the chernozemic region of Argentina 1.9 (PAPADAKIS, 1961). As a consequence, soils are very rich in organic matter, and they have thick humic horizons. Since temperatures are lower de-silication is less advanced; many soils are in phase of leaching 2 (acidification).

In summary, the most important soils of the tropical highlands are: rubrozems; brun acide; braunerde usually rich in humus; organic or highly humic rankers; podsols often "giant"; lithosols. Where leaching rainfall is low, chernozemic soils or chernozemic kaolisols are encountered. In general soils grade from "kaolinitic" to "illitic" as altitude increases.

(Humid) tropical mountains are shown on the maps by the symbol *TM*. They are encountered in Puerto Rico, Hawaii, Cuba, Colombia, Venezuela, Brasil, Bolivia, Argentina, China, Burma, East Pakistan, India, Nepal, Thailand, Cambodia, Vietnam, Ceylon, Madagascar, Ethiopia, Uganda, Kenya, Mozambique, Zambia, Malawi, Congo, Gabon, Cameroons, Angola, Guinea, Portuguese Guinea, Senegal, Sierra Leone, Togo, Upper Volta, Nigeria, Dahomey, South Africa, Rhodesia, Bechuanaland, and Basutoland.

Tropical mountains–ando. In many tropical mountains, volcanic materials abound, and ando soils are very common; due to allophane and to climate these soils are extremely rich in humus; usually they are acid, but some of them are chernozemic. These are the most fertile soils of the tropics and they support dense populations; population heaps up on these mountains.

Tropical mountains–ando are shown on the maps by the symbol *TM–A*. They are encountered in Hawaii, Central America, Ecuador, Indonesia, Sarawak, Malaysia, Kenya, Uganda, Rwanda–Burundi, Cameroons, and Tanzania, but there are possibly others in other parts of the world, the Andes of Central and South America for instance.

INTERMEDIATE REGIONS

Grey-wooded regions

In these regions, which correspond to *p* and *p'* (PAPADAKIS, 1960a), leaching rainfall is low, but due to extremely severe winters vegetation is more or less open woodland (PAPADAKIS, 1961). Under such conditions raw humus accumulates on the soil surface; soluble organic substances are produced; clay and iron are eluviated from the surface and a "bleached" horizon is formed; but bases, clay, silica and iron accumulate in the "illuvial" horizon, which is "textural", usually "neutral", sometimes "calcic"; the soil is grey-wooded. Naturally materials poor in bases and clay give podsols. Soil distribution in these regions may be summarized as follows:

(a) *Materials rich in clay and bases:* grey-wooded.

(b) *Materials poor in clay and weatherable minerals:* iron and humus podsols.

(c) *Poor drainage:* gley podsols and gleyed podsols with more or less thick peaty or organic horizons, low humic gley.

Some soils have thick "humic" horizons (sod podsols and humic gley). Grey-wooded regions are shown on the maps by the symbol *GW*. They are encountered in Canada, United States, Russia, Poland and Rumania.

Brunisolic–cinnamonic regions

The transition from brunisolic to cinnamonic regions is always gradual. But in certain cases the transition zone is sufficiently wide to justify its consideration as a special broad region. This is the case in northern China. Here "rubification" has probably been accentuated by the replacement of forest by crops (production of slightly polymerized acid humus is less abundant in the case of crops; moreover soil dries out more thoroughly when exposed directly to sun rays). Another intermediate region of this type is encountered in Bulgaria. Brunisolic–cinnamonic regions are shown on the maps by the signs *B–C*.

Brunisolic–arid cinnamonic regions

In some cases we pass from a brunisolic climate to an arid climate and a brunisolic–arid cinnamonic transition region may be formed; some soils are arid brown. Brunisolic–arid cinnamonic regions are shown on the maps by the symbol *B–AC*.

Chernozemic–desert regions

In a cold climate we pass often directly from a chernozemic region to the desert, and a chernozemic–desertic region may be formed. This is observed in the United States and China. Chernozemic–desert regions are shown on the maps by the symbol *Ch.D.*

Mediterranean mountains–dry mountains

Certain mountains of southwestern Asia vary from mediterranean (with rather humid winter) to dry (with rather dry winter). They are shown on the maps by the sign *M'–DM'*.

Mediterranean mountains–chernozemic regions

Some mediterranean mountains in western Africa have a grassland climate and vegetation. Humification is prevalent, but erosion also is important. These regions are shown on the maps by the symbol *M'–Ch*.

TRANSITIONAL KAOLINITIC REGIONS

As has been stated before, until recently clay composition was not taken into account when classifying kaolinitic soils; moreover the methods of determining clay were so deficient that soils with high cation exchange capacity were classified as kaolisols. This obliges one to consider some regions, which in addition are more or less marginal to the kaolinitic ones, as transitional.

Brunisolic–kaolinitic regions

On the southeastern coast of the Caucasus (Russia), and adjacent parts of Turkey soils are acid, yellow or red, the profile is latosolic; they are considered as kaolisols (krasnozems), but it may be their cation exhange capacity is higher; they are perhaps on the limit between brunisolic and kaolisols. These regions are shown on the maps by the symbol *B–K*.

High latitude humid mountains–kaolinitic

An analogous case is that of the Appalachian Mountains in the United States. Many soils are classified as kaolisols (red–yellow podsolic). It may be their cation exchange capacity is higher; they are perhaps on the limit between brunisolic and kaolisols. These regions are shown on the maps by the symbol *M–K*.

Kaolinitic–cinnamonic regions

Analogous cases are encountered in Brasil and Australia. But here the presumptions that soils are kaolisols are stronger. These regions are shown on the maps by the symbol *K–C*.

Tropical mountains–kaolinitic

An analogous case is observed in certain tropical mountains. These transitional regions are shown on the maps by the symbol *TM–K*.

Kaolinitic chernozemic regions

Due to climatic and geologic reasons the chernozemic and kaolinitic regions of Uruguay and Rio Grande do Sul (Brasil) approach one another and an extended transition belt is formed; prairie, humic gley, chernozemic kaolinitic and rubrozems abound. This region is shown on the maps by the symbol *K–Ch*.

PALAEO-KAOLINITIC REGIONS

It seems that in a great part of Australia kaolinitic soils were dominant. Later, the climate became drier and soils have been illuviated, rubified, recalcified, rebasified, perhaps resilicated. Sometimes the laterite has been comminuted and transformed in desert sand. Kaolinitic soils are encountered even in the desert.

Something analogous is observed in Africa, where lateritic crusts are encountered in the Sahara, even in the northern Sahara, but there the problem has been studied less.

Kaolinitic–podsolic regions

Clay illuviation is common and many soils are reported as podsolic, or podsols. However, in many cases cation exchange capacity in the lower horizon is so low in relation to clay content that the soil should be classified as kaolinitic. Moreover it is difficult to answer the question if clay illuviation has been produced by slightly polymerized organic acids or Na. It is to be noted, that Eucalyptus, common in Australia, seems to produce podsolizing humus. These regions are called kaolinitic–podsolic or podsolic–kaolinitic. They are shown on the maps by the symbols *K–P* or *P–K*. ·

Kaolinitic podsolic–cinnamonic regions

This case is analogous to the preceding one, but because soil is dried out frequently rubification has taken place and the region is also cinnamonic. For the same reason solonetz and planosols are more frequent. These regions are shown on the maps by the symbol *K–P–C*.

Mediterranean mountains–kaolinitic

An analogous case is observed in South Africa, but there topography is mountainous, and the actual climate is mediterranean. Many soils seem to be kaolisols. This region is shown on the maps by the symbol *M'–K*.

Chapter 7 | Soils of the World (Country by Country)

INTRODUCTION

In this chapter an outline of soils, country by country, will be given. Only the broad soil regions that are encountered in each country, and their subdivisions will be mentioned. Since these subdivisions are named according to the prevailing or characteristic soils, this outline gives also the principal soils that are encountered in each country. It is not necessary to describe these soils, discuss their distribution and their agricultural potentialities. The reader will find in Chapters 3–5 a description and the diagnostics of the soil mentioned; so that no confusion is possible concerning the exact meaning of each term used. The general pattern of soil distribution has been discussed in Chapter 6, separately for each broad region. So that if the reader is interested, for instance, to be informed on the distribution of soils in the mediterranean cinnamonic regions of Spain, he may consult the corresponding section in Chapter 6. The agricultural potentialities of soils, both by soil and by region, are discussed in Chapter 8. So that if the reader wants to have information for instance on the terra rossas of a mediterranean cinnamonic region, he may consult the sections relative to terra rossa and to mediterranean cinnamonic regions of Chapter 8.

The preparation of a world soil map presents great difficulties. As has been stated before great confusion reigns in soil nomenclature. Different terms are used for the same soil in different countries, and the meaning of each term varies considerably from one to another. Even in the same country there are considerable differences; moreover, the terms used are very loosely defined, the definitions are often meaningless. So that a world soil map is not a mere assemblage of national or regional maps. It is an entirely new map; the areas remain naturally the same with the necessary simplifications implied by the change of scale. But the soil of each area is re-classified according to new concepts, usually very different from those of the authors of the national or regional maps.

Until recently such a work would have been impossible. And world maps that appeared even in the sixties are very deficient. But in recent years a great progress has been achieved. Thanks to international technical assistance—FAO, etc.—soil scientists have the opportunity to work, study and classify soils in countries very different from their country of origin; teams are often formed from scientists of very different countries. All that results into a gradual approach of their pedologic con-

cepts into a certain unification. Moreover the World Soil Resources of FAO, directed by Dr Bramao, began to prepare a world soil map. Maps of western Europe, eastern Europe, and Latin America, have been prepared and presented at the International Congress of Wisconsin, in 1960; a map of Africa has been published by the Commission of Technical Cooperation in Africa. All these maps are prepared by a team, including pedologists from different countries. This method has naturally the disadvantage, that it is difficult for each one to depart from his own concepts, he insists on having the soils of his country classified according to him, and the terms of the legend have not exactly the same meaning everywhere in the map. But by working together and being obliged to prepare a legend acceptable to all of them, by travelling together and seeing various soils, they become more permeable to concepts different from their own, and an approach of opinions takes place little by little. If we compare the actual situation with that 10 years ago, we must recognize that an enormous progress has been achieved towards the unification of our pedologic concepts, and such progress is certainly due to the work done by the World Soil Resources Office and international technical assistance. Moreover, the fore-mentioned office of FAO has a very complete collection of maps, of small and large scale, from every part of the world, with abundant literature, and puts all that at the disposal of the scientists that visit Rome. This collection is maintained up to date; many of the maps included have not yet been published. In the preparation of this chapter we have made large use of all this information and we have to acknowledge it in this introduction.

The maps of Latin America and southeastern Asia by FAO World Resources Office (Dr. Bramao, Dudal and collaborators), that of western Europe by TAVERNIER (1960), that of Africa by D'HOORE (1964), that of eastern Europe, Russia and Asia by the Dokutchayev Institute of Pedology, that of China by V. A. Kovda, those of Australia, United States and Canada by their national soil survey services, and those of many other countries have been of great help to the author.

Naturally we are still far from an entirely satisfactory soil map of the world. That presupposes a certain unification of pedologic concepts all over the world, and soil mapping in the field, according to these unified concepts. The maps of this chapter are only an approximation that may contribute to the unification of our concepts and pave the way for the preparation of more and more per-

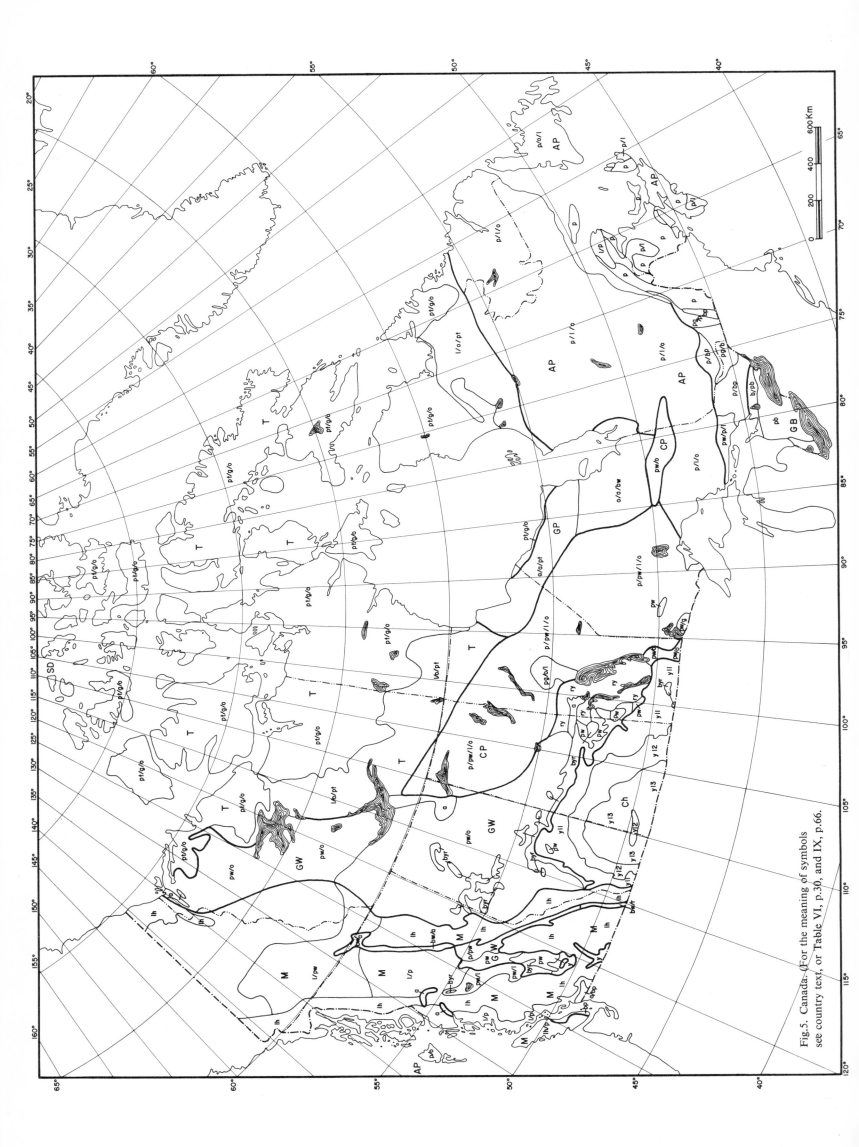

Fig.5. Canada. (For the meaning of symbols see country text, or Table VI, p.30, and IX, p.66.

fect maps later. Since a world map gives a new interpretation of national or continental maps, it may be a useful complement of these maps, in addition to its scientific interest and for using the experience acquired in other parts of the world.

NORTHERN AND CENTRAL AMERICA

Canada

Canada (see Fig.5) includes a variety of broad soil regions: (*1*) subglacial desert (*SD*) and (*2*) tundra (*T*) in the north; (*3*) atlantic podsol (*AP*), (*4*) continental podsol (*CP*), (*5*) gley podsol (*GP*), and (*6*) grey–brown podsolic (*GB*) in the east; (*7*) grey-wooded (*GW*) and (*8*) chernozemic (*Ch*) in the prairies; (*9*) high latitude humid mountains (*M*) and (*10*) atlantic podsol (*AP*) in the west.

(*1*) In the subglacial desert (*SD*) lithosols (*l*) prevail.

(*2*) In the tundra region (*T*) the principal soils are: *g* (gleisolic); *l* (lithosols); *o* (peaty): *pt* (tundra nano-podsols, together with arctic brown and tundra gleisols).

(*3*) In the atlantic podsol regions (*AP*) of the east and west the principal soils are: *a* (alluvial); *bp* (brown podsolic); *l* (lithosols); *o* (peaty); *p* (podsols); *pb* (grey–brown podsolic); *pg* (gley podsols); *pw* (grey-wooded).

(*4*) In the continental podsol (*CP*) region the principal soils are: *g* (gleisolic); *l* (lithosols); *o* (peaty); *p* (podsols); *pw* (grey-wooded).

(*5*) In the gley-podsol (*GP*) region the principal soils are: *a* (alluvial); *bw* (grey-wooded); *o* (peaty); *pt* (tundra nano-podsols), all more or less gleyed.

(*6*) In the grey–brown podsolic region (*GB*) the principal soils are: *b* (braunerde) and *pb* (grey–brown podsolic).

(*7*) In the grey-wooded regions (*GW*) the principal soils are: *a* (alluvial); *byr* (degraded chernozems); *o* (peaty); *p* (podsols); *pw* (grey-wooded); *ry* (chernozemic rendzinas); *y* (chernozemic).

(*8*) In the chernozemic regions (*Ch*) the principal soils are: *byr* (degraded chernozems); *pw* (grey-wooded); y^{11} (black chernozemic); y^{12} (dark brown chernozemic); y^{13} (brown chernozemic).

(*9*) In the high latitude humid mountains (*M*) the principal soils are: *l* (lithosols); *lh* (lithosolic rankers); *p* (podsols); *pw* (grey-wooded).

The mineral fertility of podsols is low, but chernozemic soils are very fertile. Grey-wooded soils have an intermediate fertility.

United States

The United States (Fig.6) have a great variety of soil regions: (*1*) atlantic podsol (*AP*), and (*2*) grey–brown podsolic (*GB*) in the northeast; (*3*) continental podsol (*CP*) and (*4*) grey-wooded (*GW*) in the Great Lakes area;

(*5*) chernozemic (*Ch*) and (*6*) cinnamonic (reddish) chernozemic (*CCh*) in the Great Plains; (*7*) chernozemic–desert (*Ch-D*), (*8*) desert (*D*), (*9*) arid cinnamonic (*AC*), and (*10*) arid cinnamonic-brunisolic (*AC-B*) in the west; (*11*) mediterranean-cinnamonic (*MC*) on the Pacific coast; and (*12*) subtropical kaolinitic (*SK*) on the southeast. The mountainous areas of the north can be classified as (*13*) high latitude humid mountains (*M*) and the Appalachians as (*14*) high latitude humid mountains–kaolinitic (*M–K*). In the west there are (*15*) dry mountains (*DM*) and (*16*) mediterranean mountains (*M'*). Volcanic materials abound on the Pacific coast and some regions can be termed (*17*) atlantic podsol–ando (*AP–A*) or mediterranean mountains ando (*M'–A*).

(*1*) In the atlantic podsol region (*AP*) the more important soils are: *ba* (acid brown); *g* (gleisolic); *p* (podsols) *pb* (grey–brown podsolic); and *pf* (fragipan podsols). All these soils are, in general, more or less poor and acid. Lime and fertilizers are used. Dairying and fruit production (apples) are the most important industries.

(*2*) In the grey–brown podsolic region (*GB*) the more important soils are: *bpg* (gleisolic podsolized brown); *e* (aeolian (dunes)); *pb* (grey–brown podsolic); *pf* (fragipan podsols); *py* (prairie lessivé); *r* (rendzinas); *y* (chernozemic); *yu* (planosolic para-chernozems). Most of these soils are acid and rather poor. Lime and fertilizers are extensively used; but with their aid very high yields of maize, oats, clovers, etc. are obtained.

(*3*) In the continental podsol (*CP*) region the principal soils are: *e* (aeolian (dunes)); *ew* (wet dunes); *o* (peaty); *p* (podsols); *pf* (fragipan podsols); *pw* (grey-wooded); *pwn* (eutrophic grey-wooded). Most of these soils are acid and poor. Lime and fertilizers are extensively used; but with their aid excellent crops of pasture crops are obtained; and dairy is flourishing.

(*4*) The grey-wooded region (*GW*) is small; the principal soils are: *o* (peaty); *e* (aeolian (dunes)); *pwn* (eutrophic grey-wooded). Soil conditions are probably better than in *CP*, but climate is a little drier.

(*5*) In the chernozemic region (*Ch*) the most important soils are: *b* (braunerde); *e* (aeolian (sands)); *mr* (rendzina clays); *myw* (wet chernozemic clays); *t* (solonetz); *ug* (gleisolic planosols); *y* (chernozemic); *yb* (brunisolic para-chernozem (prairie)); *yr* (chernozems); *yrw* (meadow chernozems); *yu* (planosolic para-chernozems); *yug* (gleisolic planosolic para-chernozems); *yw* (meadow chernozemic). Most of these soils are highly fertile. Where rainfall is good high yields are obtained. Wheat is a very important crop in the dry areas; maize and fodder crops in the moister ones.

(*6*) In the region of reddish chernozemic (*CCh*) the more important soils are: *b* (recent brown); *c* (cinnamonic); *mr* (rendzina clays); *myw* (meadow chernozemic clays); *pc* (non-calcic brown); *ug* (gleisolic planosols); *yb* (brunisolic para-chernozems (prairie)); *yc* (cinnamonic para-chernozems); *yr* (chernozems); *yu* (planosolic para-chernozems). These soils are less fertile than those of the

Ch region, but they produce high yields when rainfall is high or irrigated; the principal dryland crops are wheat and sorghum; cotton is grown extensively.

(7) In the chernozemic–desert region (*Ch–D*) the more important soils are: *b* (recent brown); *mr* (rendzina clays); *r* (rendzinas); *rd* (serozems); *u* (planosols); *uc* (cinnamonic planosols); *y* (chernozemic); *yu* (planosolic para-chernozems). Mineral fertility is probably still higher than in the chernozemic region. Soils are naturally poorer in organic matter. Wheat is one of the most important crops.

(8) In the desert region (*D*) the principal soils are: *b* (recent brown); *d* (arid brown and other raw soils of the desert); *e* (aeolian sands); *rd* (serozems); *stt* (takirs); *t* (solonetz); *ud* (planosolic red desert). Mineral fertility is usually high, and with irrigation these soils give high yields. The main problems are salinity, alkalinity, waterlogging, too low or too high permeability, etc.

(9) In the arid cinnamonic region (*AC*) the more important soils are: *c* (cinnamonic); *cy* (reddish brown); *d* (arid brown); *e* (aeolian sands); *pc* (non-calcic brown); *r* (rendzinas); *rd* (serozems); *t* (solonetz); *uc* (cinnamonic planosols); *yw* (meadow chernozemic). Mineral fertility is usually high in these soils, and given irrigation and fertilizers high yields are obtained.

(10) In the arid cinnamonic–brunisolic region (*AC–B*) the principal soils are: *b* (recent brown); *d* (arid brown); *e* (aeolian sands); *pb* (grey–brown podsolic); *u* (planosols); *v* (ando). The importance of ando soils is probably greater than shown by the maps; many of the other soils are also probably ando of dry climate. The mineral fertility of these soils is usually high; with irrigation and fertilizers, they give high yields; wheat is an important crop.

(11) In the mediterranean–cinnamonic region (*MC*) the more important soils are: *b* (recent brown); *h* (organic); *pc* (non-calcic brown); *pcj* (non-calcic brown with duripan); *t* (solonetz); *y* (chernozemic). Most of these soils are ando; their cation exchange capacity and organic matter content are unusually high. They are very fertile. With irrigation and fertilizers, they give high yields. Vegetables and fruits are the more important crops grown·

(12) In the subtropical kaolinitic (*SK*) region of the southeast the more important soils are: *a* (alluvial); *b* (braunerde); *ba* (brun acide); *bpg* (gleisolic podsolized brown); *e* (aeolian sands); *ew* (wet aeolian sands); *g* (gleisolic); *h* (organic); *kl* (latosolic kaolisols (reddish brown lateritic)); *kpd* (red–yellow podsolic); *kpdg* (gleyed dystrophic kaolinitic podsolic); *kpg* (gleyed kaolinitic podsolic); *kpgd* (dystrophic gleyed kaolinitic podsolic); *m* (dark clays); *myw* (meadow chernozemic clays); *oh* (lenists); *ph* (grey–brown podsolic); *pf* (fragipan podsols); *pg* (glei podsols); *ph* (humus podsols); *yw* (meadow chernozemic). It may be the extension of kaolinitic soils (*k*) is far less than shown in the maps; many of the soils qualified as kaolinitic may have too high cation exchange capacities to be classified as kaolisols; the limit between

GB and *AP* on one hand and *SK* on the other may be at lower latitude than shown in the map; a thorough study of adjusted cation exchange capacity/clay ratio would probably change the classification of many of these soils.

Soils of the *SK* region are usually acid and poor; but with fertilizers high yields are obtained. Cotton, tobacco, maize, citrus, fruits, vegetables and fodder crops are produced.

(13) In the high latitude humid mountains of the north (*M*) the most important soils are: *p* (podsols) and *pb* (grey–brown podsolic). In the northwest there are many ando (*v*), probably much more than shown in the map. Podsols are rather poor; but ando are potentially fertile.

(14) In the *M–K* region of the Appalachians, the more important soils are: *b* (braunerde —forest); *ba* (brun acide); *kpd* (red–yellow podsolic); *pf* (fragipan podsols); *pk* (kaolinitic podsols). It may be kaolinitic soils (*k*) are much less important than shown in the map; many of them may have cation exchange capacities too high to enter in this group.

(15) In the dry mountains (*DM*) the principal soils are: *b* (recent brown); *bl* (lithosolic brown); *d* (arid brown); *l* (lithosols); *pw* (grey-wooded); *pwn* (eutrophic grey-wooded); *rd* (serozems); *u* (planosols); *uc* (cinnamonic planosols); *y* (chernozemic); *yu* (para-chernozem lessivé).

(16) In the mediterranean mountains of the west (*M'*) the most important soils are: *b* (recent brown) and *d* (arid brown). But many of the mediterranean mountains are rich in volcanic materials and belong to the mediterranean mountains–ando (*M'–A*) region. The more important soils of this region are: *b* (recent brown); *mr* (rendzina clays); *pc* (non-calcic brown); *u* (planosols); *v* (ando); *vl* (lithosolic ando); *y* (chernozemic). The importance of ando soils is probably greater than shown in the maps; many of the other soils (non-calcic brown, chernozemic, etc.) are probably ando.

(17) In the atlantic podsol–ando (*AP–A*) region of the northwestern coast the most important soils are: *g* (gleisolic); *h* (organic); *p* (podsols); *v* (ando); *vg* (gleisolic ando). The importance of ando soils is probably greater than shown in the maps.

Alaska (Fig.7.) Alaska includes four broad soil regions: (*1*) tundra (*T*); (*2*) continental podsol (*CP*); (*3*) atlantic podsol (*AP*); and (*4*) high latitude humid mountains (*M'*).

(*1*) In the tundra (*T*) the principal soils are: *g* (gleisolic); *lh* (lithosolic rankers); *o* (peaty); and *pt* (tundra nano-podsols grading from non-podsolized arctic brown to gleisolic).

(*2*) In the continental podsol region (*CP*) the principal soils are: *g* (gleisolic) and *b/pt* (brown-tundra nano-podsols).

(*3*) In the atlantic podsol region (*AP*) the principal soils are *ph* (humus podsols).

(*4*) In the high latitude humid mountains the principal soils are: *bl* (lithosolic brown); *g* (gleisolic); *hw* (wet

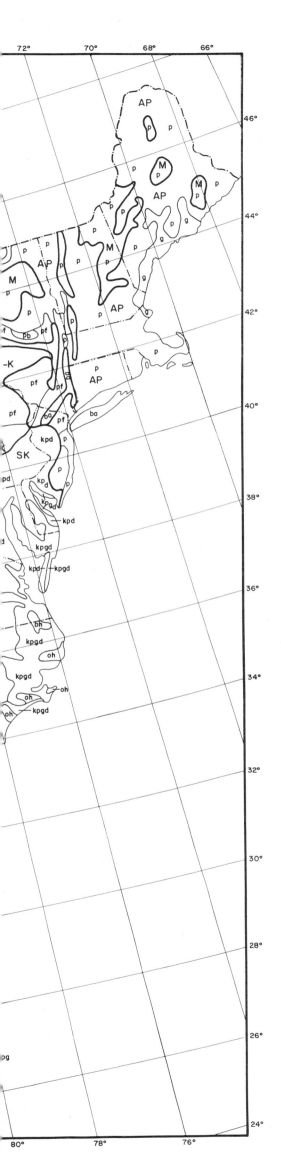

Fig.6 (pp.81–83). The United States of America. Inset left: Hawaiian Islands; inset right: Puerto Rico and St. Croix. (For the meaning of symbols, see country text, or Table VI, p.30, and Table IX, p.66.)

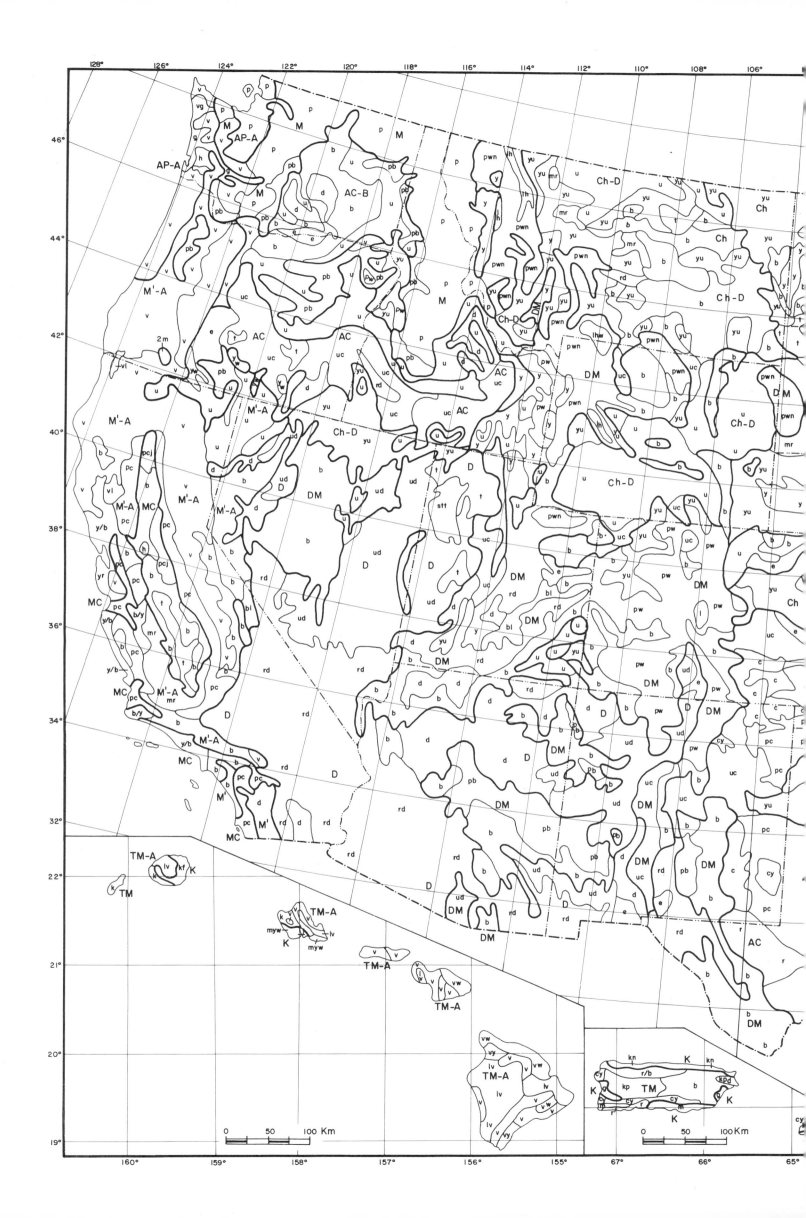

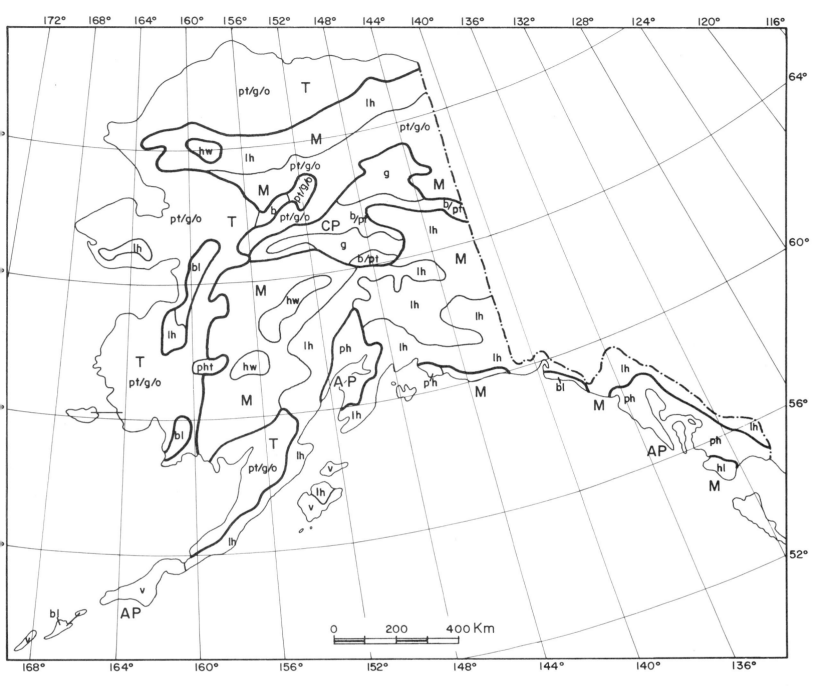

Fig.7. Alaska. (For the meaning of symbols see country text, or Table VI, p.30, and Table IX, p.66.)

organic); *lh* (lithosolic rankers); *o* (peaty); *pt* (tundra nano-podsols including arctic brown and tundra gleisols); *v* (ando); and of course lithosols.

Puerto Rico (Fig.6). In Puerto Rico one may recognize two broad soil regions: (*1*) kaolinitic (*K*), and (*2*) tropical mountains (*TM*).

(*1*) In the kaolinitic region (*K*) the principal soils are: *g* (gleisolic); *kn* (eutrophic kaolisols); *kpd* (red–yellow podsolic); and *m* (dark clays).

(*2*) In the tropical mountains the more important soils are: *b* (recent brown); *cy* (reddish brown); *kp* (kaolinitic podsols); *m* (dark clays); *r* (rendzinas).

It is not yet known to what extent these soils are

really kaolinitic. It may be that the extension of these soils (*k*) is less than shown in the maps. Moreover it is not yet known to what extent these soils are eutrophic or dystrophic; it may be many of the soils classified as red–yellow podsolic (*kdp*) are not dystrophic and those classified as kaolinitic podsolic (*kp*) are eutrophic.

Hawaii Islands (Fig.6). In the Hawaii Islands one may recognize four broad soil regions: (*1*) kaolinitic (*K*); (*2*) kaolinitic–ando (*K–A*); (*3*) tropical mountains (*TM*); (*4*) tropical mountains–ando (*TM–A*).

(*1*) In the kaolinitic region *K*, there are many concretionary kaolisols (*kf*).

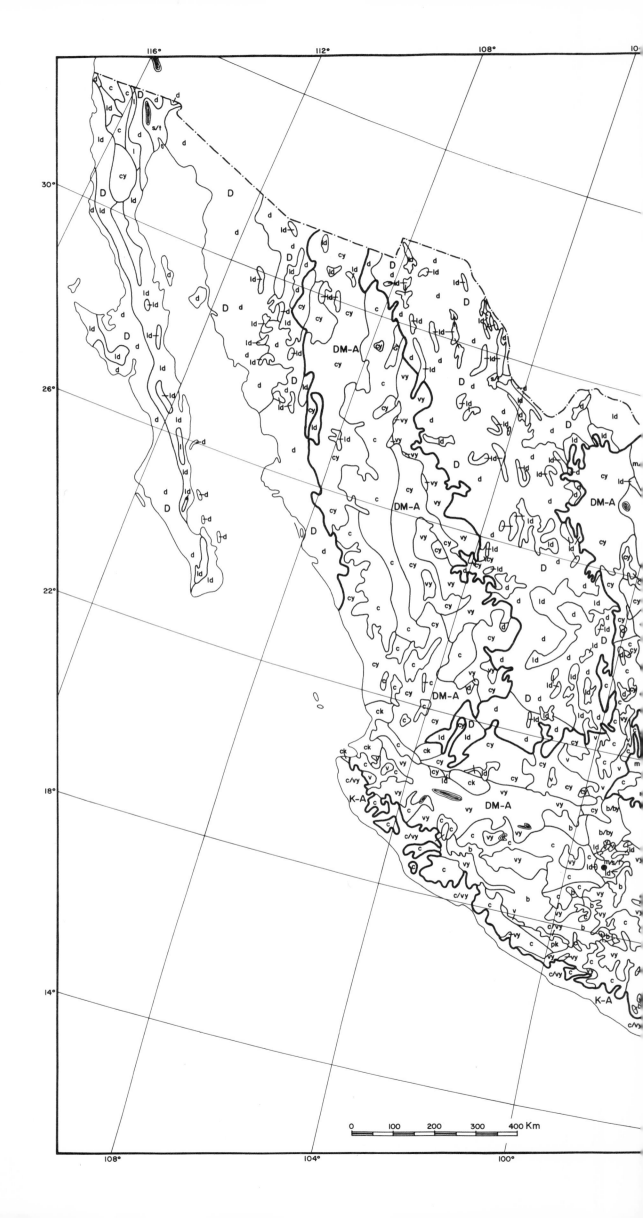

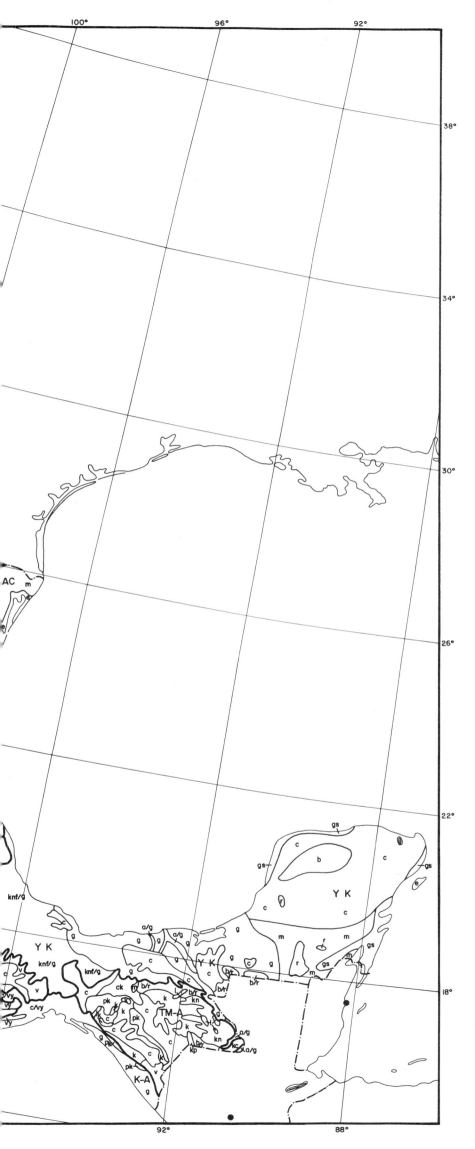

Fig.8 (pp.88, 89). Mexico. (For the meaning of symbols see country text, or Table VI, p.30, and Table IX, p.66.)

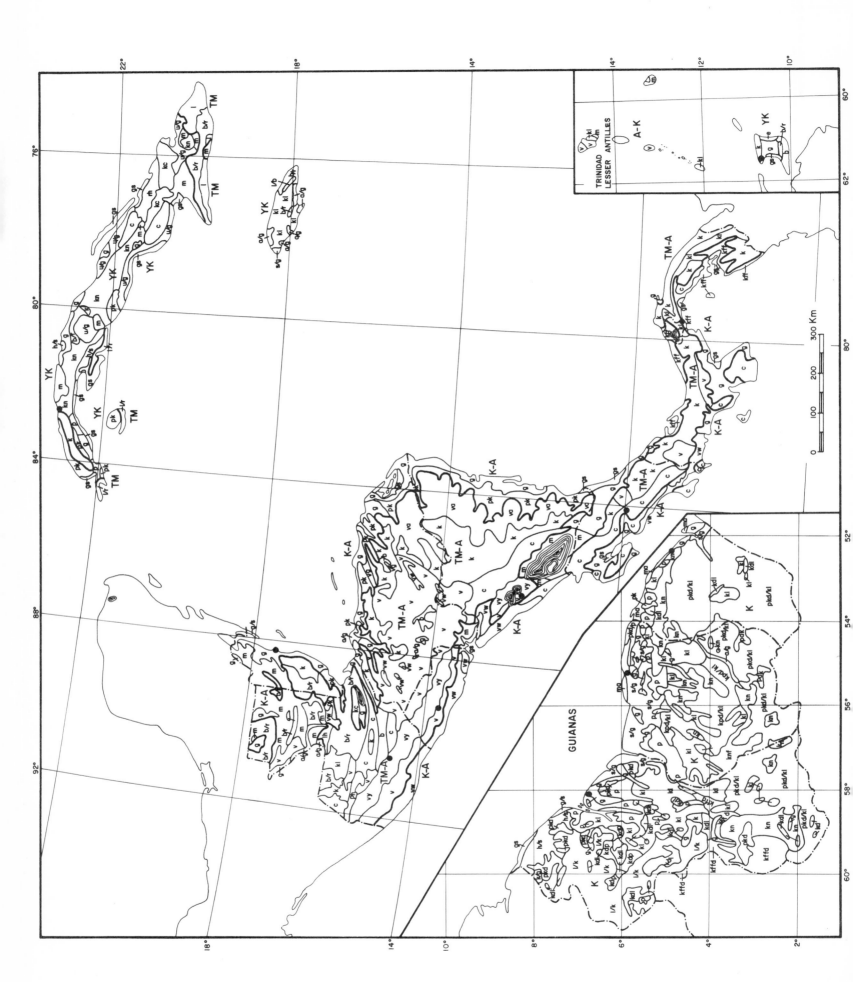

90

(2) In the kaolisol–ando region (*K–A*), meadow chernozemic clays (*myw*).

(3) In tropical mountains (*TM*), kaolisols (*k*).

(4) The most important region is certainly that of tropical mountains–ando (*TM–A*). Its most important soils are: *k* (kaolisols); *lv* (lithosolic ando); *v* (ando); *vw* (hydrol ando) and *vy* (chernozemic ando). All these soils have high cation exchange capacities, high moisture tension, and are very fertile. With fertilizers and irrigation, where it is needed, they give the highest yields of the world in sugar cane and other products (bananas, etc.).

Mexico

Mexico (Fig.8) includes many broad soil regions: (*1*) *AC* (arid cinnamonic) and (*2*) *D* (desert) in the north; (*3*) *K–A* (kaolinitic–ando) and (*4*) *YK* (young kaolinitic) in the south; (*5*) *DM–A* (dry mountains–ando) and (*6*) *TM–A* (tropical mountains–ando) in the mountains.

(*1*) In the arid cinnamonic (*AC*) region the principal soils are: *c* (cinnamonic); *e* (aeolian sands); *g* (gleisolic); *m* (dark clays); *u* (planosols).

(*2*) In the desert (*D*) region the principal soils are: *c* (cinnamonic); *cy* (reddish brown); *d* (arid brown); *l* (lithosols); *ld* (desert lithosols); *s* (saline); *t* (solonetz).

(*3*) In the kaolinitic–ando (*K–A*) region of the southern coast the principal soils are: *c* (cinnamonic); *ck* (cinnamonic–kaolinitic intergrades); *g* (gleisolic); *pk* (kaolinitic podsolic); *vy* (chernozemic ando).

(*4*) In the young kaolinitic (*YK*) region of Yucatan, etc. the principal soils are: *a* (alluvial); *b* (braunerde); *c* (cinnamonic); *e* (aeolian); *g* (gleisolic); *gs* (saline gley (mangrove)); *knf* (tropical ferruginous); *m* (dark clays); *r* (rendzinas).

(*5*) In the dry mountains–ando (*DM–A*) region of the north the principal soils are: *b* (recent brown); *by* (chernozemic braunerde (prairie)); *c* (cinnamonic); *cy* (reddish brown); *ck* (cinnamonic–kaolisol intergrades); *d* (arid brown); *ld* (desert lithosols); *m* (dark clays); *pk* (kaolinitic podsolic); *s* (saline); *t* (solonetz); *v* (ando).

(*6*) In the tropical mountains–ando (*TM–A*) region the principal soils are: *b* (braunerde); *c* (cinnamonic); *ck* (cinnamonic–kaolisol intergrades); *k* (kaolisols); *kc* (kaolisol–cinnamonic intergrades); *kn* (eutrophic kaolisols); *kp* (kaolinitic podsolic); *l* (lithosols); *pk* (kaolinitic podsolic); *r* (rendzinas).

Due to the abundance of volcanic materials and to the young age of soils, the mineral fertility of Mexican soils is satisfactory even in the humid and warm coasts. Naturally productivity is impaired by steep slopes and shallowness of many soils, drought, waterlogging, etc.

Central America

Guatemala (Fig.9). This includes two broad soil regions: (*1*) *K–A* (kaolinitic–ando) and (*2*) *TM–A* (tropical mountains–ando).

(*1*) In the kaolinitic–ando (*K–A*) region the principal soils are: *a* (alluvial); *b* (braunerde); *g* (gleisolic); *kc* (kaolisol–cinnamonic intergrades); *m* (dark clays); *pk* (kaolinitic podsolic); *r* (rendzinas); *v* (ando); *vw* (wet ando).

(*2*) In the tropical mountains–ando (*TM–A*) region the principal soils are: *b* (braunerde); *c* (cinnamonic); *k* (kaolisols); *kl* (latosolic kaolisols); *m* (dark clays); *ph* (humus podsols); *r* (rendzinas); *v* (ando); *vy* (chernozemic ando).

British Honduras (Fig.9). Here there are also two broad regions: (*1*) *K–A* (kaolinitic–ando) and (*2*) *TM–A* (tropical mountains–ando).

(*1*) The principal soils are: *g* (gleisolic); *m* (dark clays); *pk* (kaolinitic podsolic); *s* (saline).

(*2*) The principal soils are: *b* (braunerde); *k* (kaolisols); *r* (rendzinas).

Honduras (Fig.9). This consists of two broad regions:

(*1*) *K–A* (kaolinitic–ando) and (*2*) *TM–A* (tropical mountains–ando).

(*1*) The principal soils are: *a* (alluvial); *g* (gleisolic); *gs* (saline gley (mangrove)); *k* (kaolisols); *m* (dark clays); *pk* (kaolinitic podsolic); *vg* (gleisolic ando); *vw* (wet ando).

(*2*) The principal soils are: *a* (alluvial); *b* (braunerde); *g* (gleisolic); *k* (kaolisols); *v* (ando): *vg* (gleisolic ando); *vw* (wet ando).

San Salvador (Fig.9). Here there are also two broad soil regions: *K–A* (kaolinitic–ando) and *TM–A* (tropical mountains-ando). In the former the principal soils are: *gs* (saline gley (mangrove)); and *vw* (wet ando). In the latter: *v* (ando), and *vy* (chernozemic ando).

Nicaragua (Fig.9). Nicaragua consists of two broad soil regions: (*1*) *K–A* (kaolinitic-ando) and (*2*) *TM–A* (tropical mountains–ando).

(*1*) The principal soils are: *c* (cinnamonic); *g* (gleisolic); *gs* (saline gley (mangrove)); *m* (dark clays); *pk* (kaolinitic podsolic); *vw* (wet ando).

(*2*) The principal soils are: *c* (cinnamonic); *k* (kaolisols); *v* (ando); *va* (acid ando); *vw* (wet ando).

Costa Rica (Fig.9). Two broad soil regions exist: (*1*) *K–A* (kaolinitic–ando) and (*2*) *TM–A* (tropical mountains–ando).

(*1*) The principal soils are: *c* (cinnamonic); *g* (gleisolic); *gs* (saline gley (mangrove)); *m* (dark clays); *vw* (wet ando).

(*2*) The principal soils are: *c* (cinnamonic); *k* (kaolisols); *v* (ando).

Panama (Fig.9) also includes two broad soil regions: (*1*) *K–A* (kaolinitic–ando) and (*2*) *TM–A* (tropical mountains–ando).

(1) The principal soils are: *c* (cinnamonic); *g* (gleisolic); *gs* (saline gley (mangrove)); *kff* (ground-water laterite); *vw* (wet ando).

(2) The principal soils are: *c* (cinnamonic); *k* (kaolisols); *kl* (latosolic kaolisols); *v* (ando).

Due to the abundance of volcanic materials, mineral fertility is usually good in Central America. Naturally many soils are too shallow or too steep, and other waterlogged.

West Indies

Cuba (Fig.9). In Cuba two broad soil regions exist: *(1) YK* (young kaolinitic) and *(2) TM* (tropical mountains).

(1) In the young kaolinitic (*YK*) region the principal soils are: *b* (braunerde); *c* (cinnamonic); *g* (gleisolic); *gs* (saline glei (mangrove)); *h* (organic); *k* (kaolisols); *kc* (kaolisol–cinnamonic intergrades); *kn* (eutrophic kaolisols); *m* (dark clays); *p* (podsols); *pk* (kaolinitic podsolic); *r* (rendzinas); *s* (saline); *u* (planosols).

(2) In the tropical mountains region (*TM*) the principal soils are: *b* (braunerde); *k* (kaolisols); *l* (lithosols); *r* (rendzinas).

Jamaica (Fig.9). Jamaica consists of only one broad soil region: *YK* (young kaolinitic). The principal soils are: *a* (alluvial); *b* (braunerde); *g* (gleisolic); *kl* (latosolic kaolisols); *r* (rendzinas); *s* (saline).

Lesser Antilles (Fig.9). These are included in the ando kaolinitic (*K–A*) region; the principal soils are: *kl* (latosolic kaolisols); *m* (dark clays); *v* (ando).

Trinidad (Fig.9). Only one broad region exists in Trinidad: *YK* (young kaolinitic). The principal soils are: *b* (braunerde); *e* (aeolian sands); *g* (gleisolic); *gs* (saline glei (mangrove)); *k* (kaolisols); *r* (rendzinas).

In general, the West Indies have young soils; in some islands volcanic materials abound, therefore mineral fertility is usually satisfactory. This is one of the reasons these islands maintain a dense population.

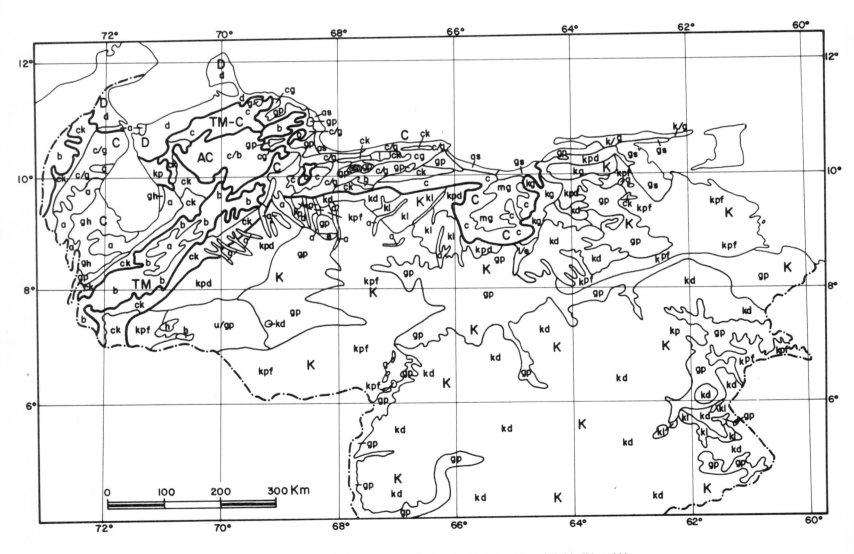

Fig.10. Venezuela. (For the meaning of symbols see country text, or Table VI, p.30 and Table IX, p.66.)

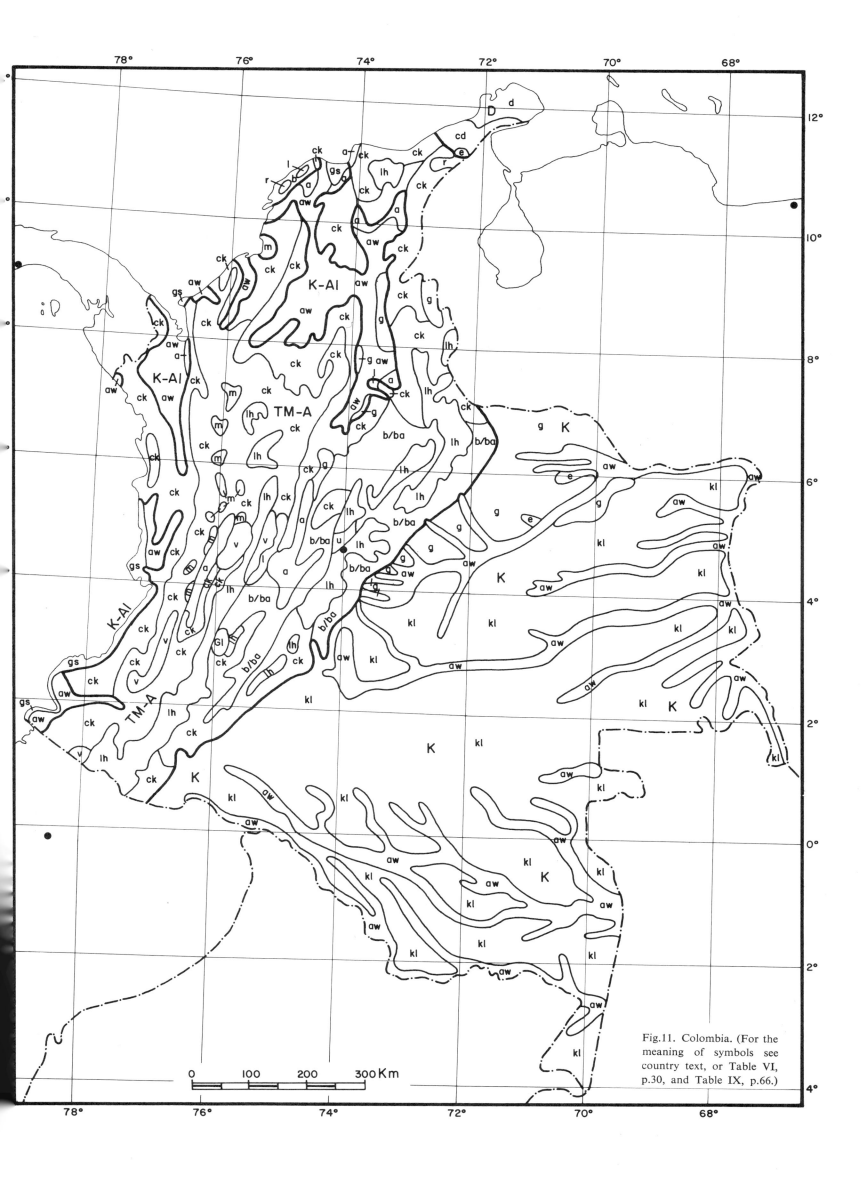

Fig.11. Colombia. (For the meaning of symbols see country text, or Table VI, p.30, and Table IX, p.66.)

SOUTH AMERICA

Venezuela and Guianas

Venezuela (Fig.10). Venezuela consists of five broad soil regions: (*1*) desert (*D*); (*2*) arid cinnamonic (*AC*); (*3*) cinnamonic (*C*); (*4*) kaolinitic (*K*); and (*5*) tropical mountains (*TM*).

(*1*) In the desert (*D*) region the principal soils are: *a* (alluvial) and *d* (arid brown).

(*2*) In the acid cinnamonic (*AC*) region the principal soils are: *b* (recent brown); and *c* (cinnamonic).

(*3*) In the cinnamonic region (*C*) the principal soils are: *a* (alluvial); *as* (saline alluvial); *c* (cinnamonic); *ck* (cinnamonic–kaolisol intergrades); *g* (gleisolic); *gh* (humic gleisolic); *gp* (low humic gley); *mg* (gleyed clays); *gs* (saline gleisols (mangrove)).

(*4*) In the kaolinitic region (*K*) the principal soils are: *a* (alluvial); *e* (aeolian sands); *g* (gleisolic); *gp* (low humic gley); *kd* (dystrophic kaolisols); *kg* (gleisolic kaolisols); *kl* (latosolic kaolisols); *kp* (kaolinitic podsolic); *kpd* (red–yellow podsolic); *kpf* (lateritic podsolic); *l* (lithosols); *gs* (saline gleisols (mangrove)); *u* (planosols).

(*5*) In the tropical mountains (*TM*) the principal soils are: *a* (alluvial); *b* (recent brown).

British Guiana (Fig.9). This country falls in the kaolinitic region (*K*). The principal soils are: *g* (gleisolic); *gs* (saline gley (mangrove)); *h* (organic); *k* (kaolisols); *kd* (dystrophic kaolisols); *kdl* (krasnozems); *kff* (groundwater laterite); *kffd* (dystrophic ground-water laterite); *kl* (latosolic kaolisols); *kn* (eutrophic kaolisols); *l* (lithosols); *p* (podsols); *pkd* (red–yellow podsolic); *s* (saline).

Surinam (Fig.9). Surinam occurs in the kaolinitic region (*K*). The principal soils are: *a* (alluvial); *g* (gleisolic); *kl* (latosolic kaolisols); *kn* (eutrophic kaolisols); *knf* (tropical ferruginous); *kpd* (red–yellow podsolic); *p* (podsols); *pdk* (red–yellow podsolic); *pkd* (red–yellow podsolic); *ma* (acid sulphate clays); *s* (saline).

French Guiana (Fig.9). This country is included in the kaolinitic region (*K*). The principal soils are: *g* (gleisolic); *kdl* (krasnozems); *kl* (latosolic kaolisols); *kn* (eutrophic kaolisols); *ma* (acid sulphate clays); *p* (podsols); *pkd* (red–yellow podsolic); *s* (saline).

Colombia (Fig.11)

Colombia includes four broad soil regions: (*1*) kaolinitic (*K*); (*2*) kaolinitic–alluvial (*K–Al*); (*3*) tropical mountains–ando (*TM–A*); and (*4*) desert (*D*).

(*1*) In the kaolinitic region (*K*) the principal soils are: *aw* (wet alluvial); *e* (aeolian sands); *g* (gleisolic); and *kl* (latosolic kaolisols).

(*2*) In the kaolinitic–alluvial (*K–Al*) region, the principal soils are: *a* (alluvial); *aw* (wet alluvial); *g* (gleisolic); *gs* (saline gleisolic); *m* (dark clays).

(*3*) In the tropical mountains–ando (*TM–A*) region, the principal soils are: *a* (alluvial); *b* (recent brown); *ba* (acid brown); *ck* (cinnamonic–kaolisol intergrades); *g* (gleisolic); *l* (lithosols); *lh* (lithosolic rankers); *m* (dark clays); *r* (rendzinas); *u* (planosols) and *v* (ando).

(*4*) In the small desert (*D*) region the principal soils are: *cd* (reddish desert); *d* (arid brown); and *e* (aeolian sands).

Except for the kaolinitic region of the Amazonas and Llanos the soils of Colombia are seldom kaolinitic; ando is common, perhaps much more than shown in the map. That is why the mineral fertility of the soils of Colombia in the principal agricultural region is good.

Ecuador

Ecuador (Fig.12). Ecuador consists of three broad soil regions: (*1*) *D* (desert); (*2*) *A–K* (ando–kaolinitic) and *K* (kaolinitic); and (*3*) *TM–A* (tropical mountains–ando).

(*1*) In the desert (*D*) region the principal soils are (*d*) arid brown, and other soils of the desert.

(*2*) In the ando–kaolinitic (*A–K*) region the principal soils are: *aw* (wet alluvial); *gs* (saline gley (mangrove)); *kc* (kaolisol–cinnamonic intergrades); *m* (dark clays); *r* (rendzinas). In the kaolinitic (*K*) region the principal soils are *kl* (latosolic kaolisols).

(*3*) In the tropical mountains–ando (*TM–A*) region the principal soils are: *a* (alluvial); *b* (braunerde); *by* (chernozemic brown (prairie)); *c* (cinnamonic); *ck* (cinnamonic-kaolisol intergrades); *cy* (reddish brown); *d* (arid brown); *g* (gleisolic); *gs* (saline gley (mangrove)); *l* (lithosols); *ll* (rock outcrops); *m* (dark clays); *v* (ando); *vy* (chernozemic ando).

Excepting the Amazonian lowlands, soils are young in Ecuador; volcanic materials abound. That is why the mineral fertility is usually satisfactory. Naturally many soils are shallow, steep, suffer from drought or waterlogging. The extent of ando soils is greater than shown on Fig.12.

Peru (Fig.13)

Peru consists of three broad soil regions: (*1*) desert (*D*) on the coast; (*2*) tropical mountains (*TM*) in the "sierra" (mountains); and (*3*) kaolinitic (*K*) in the Amazonas.

(*1*) In the desert (*D*) region the principal soils are: *a* (alluvial); *cd* (red desert); *d* (arid brown); and *l* (lithosols). Many of them are of volcanic origin and should be classified as ando (of the desert region); desert crusts, that slack in water and are formed from the products of the rapid meteorization of volcanic ashes are a characteristic of these ando soils. Sand dunes abound.

(*2*) In the tropical mountains (*TM*) the more important soils are: *a* (alluvial); *b* (recent brown); *by* (chernozemic brown (prairie)); *kc* (kaolisol-cinnamonic intergrades); *l* (lithosols); *m* (dark clays); *o* (peaty).

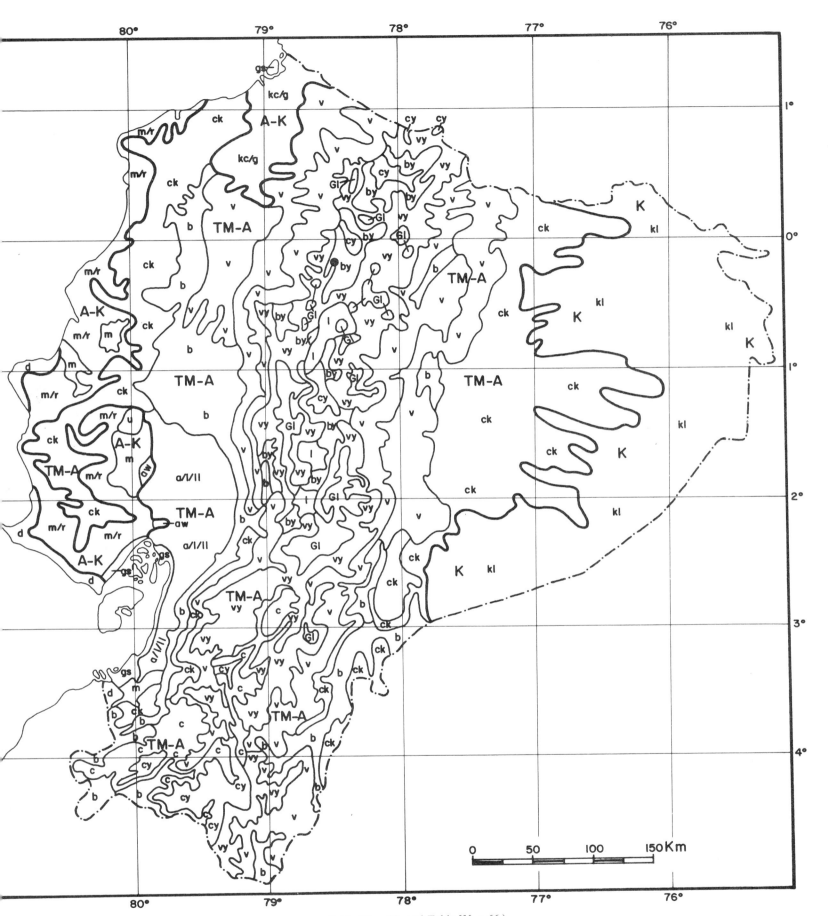

Fig.12. Ecuador. (For the meaning of symbols see country text, or Table VI, p.30, and Table IX, p.66.)

95

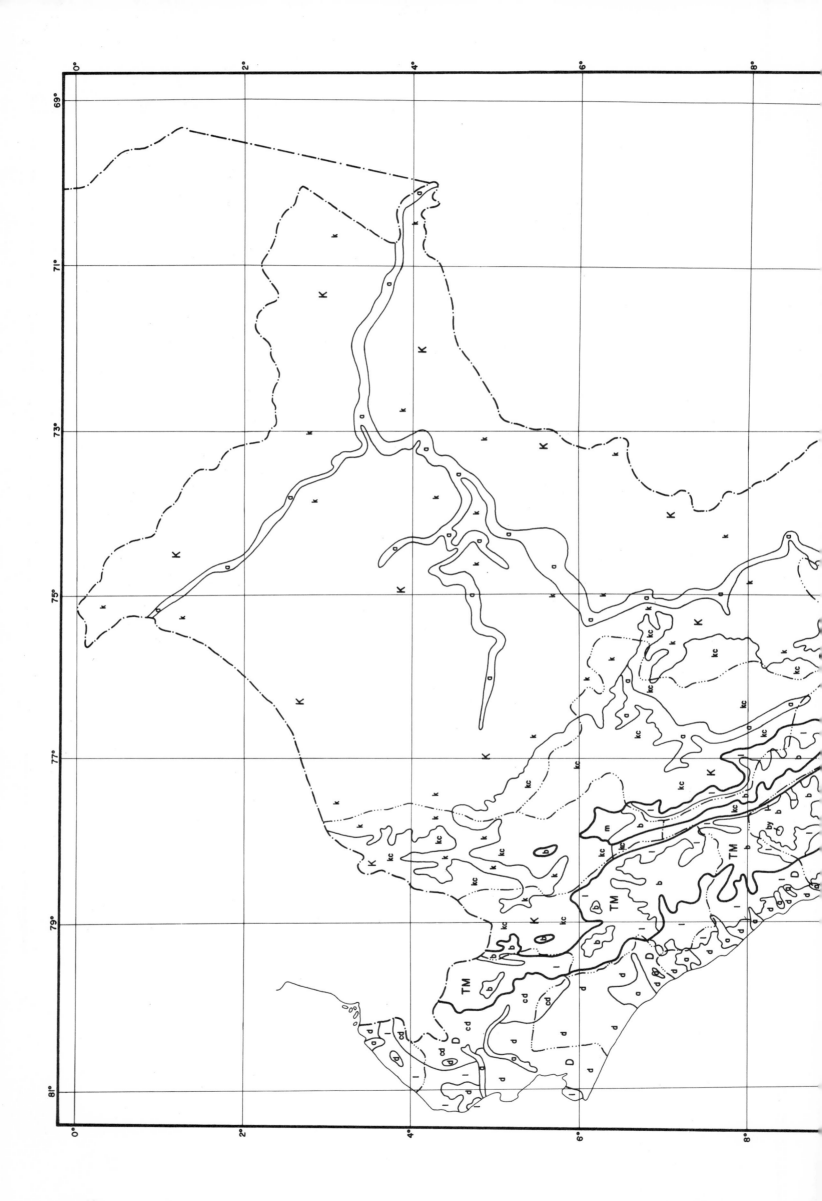

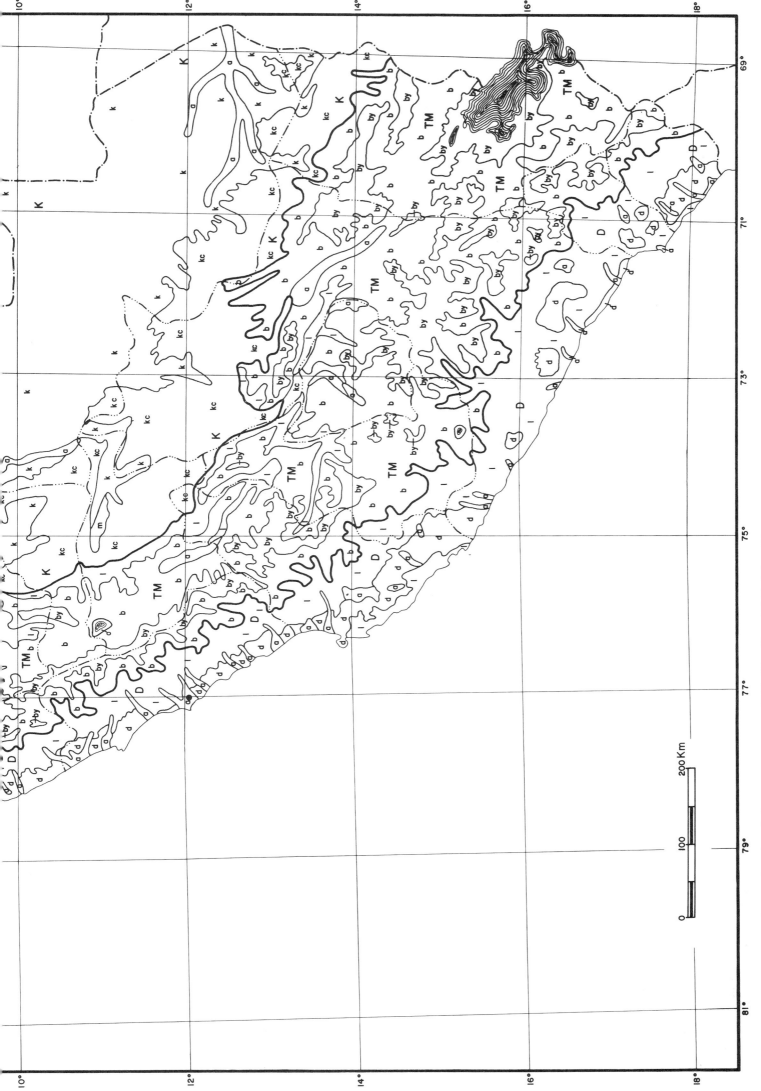

Fig. 13 (pp.96, 97). Peru. (For the meaning of symbols see country text, or Table VI, p.30, and Table IX, p.66.)

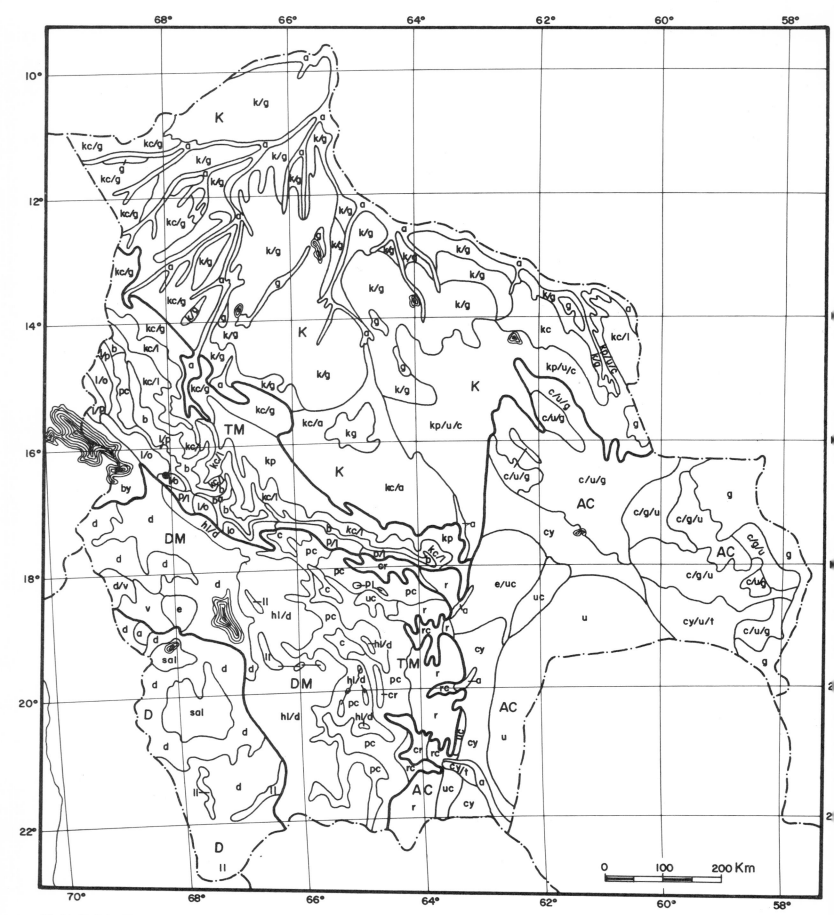

Fig.14. Bolivia. (For the meaning of symbols see country text, or Table VI, p.30, and Table IX, p.66.)

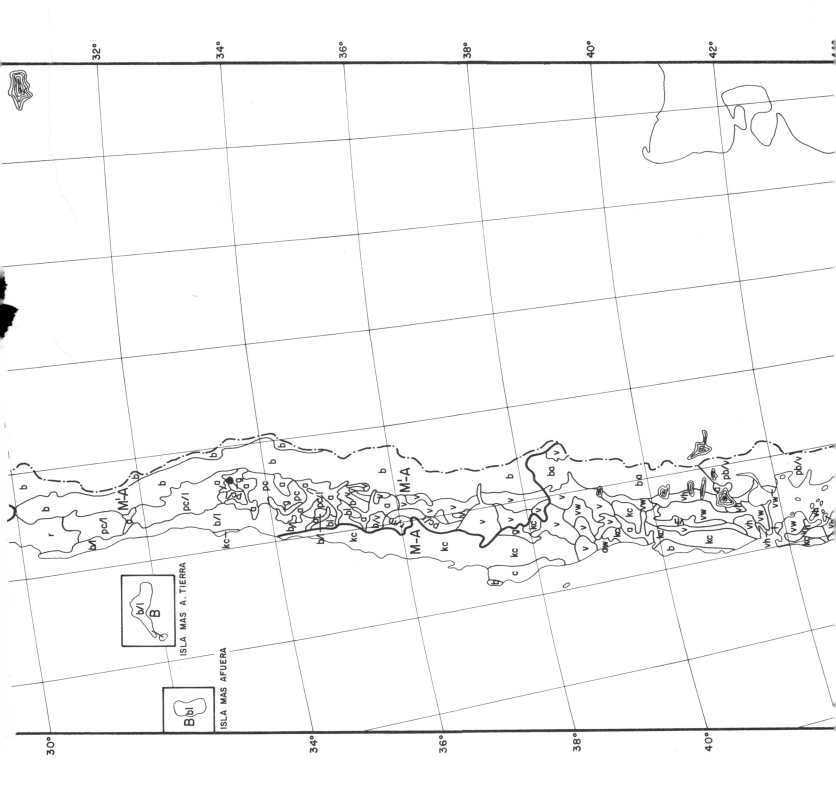

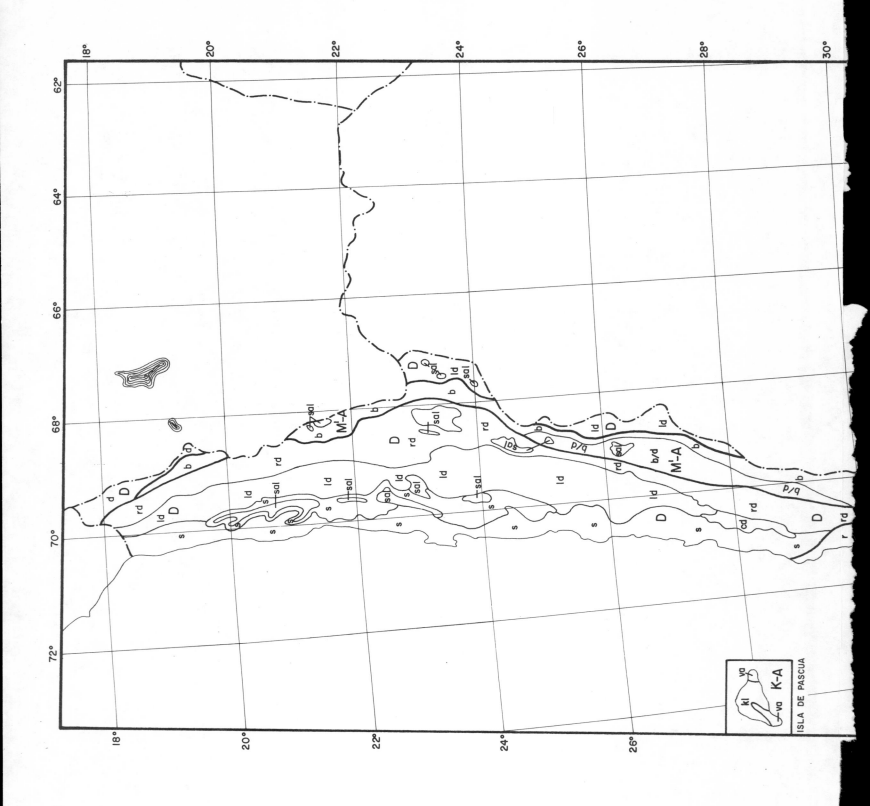

ISLA DE PASCUA

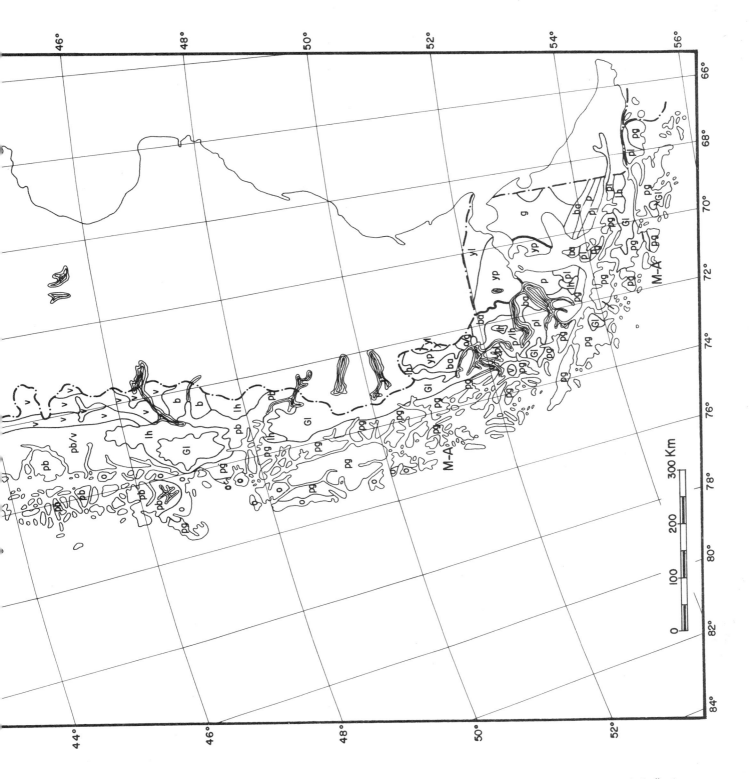

Fig. 15 (pp.99–101). Chile. (For the meaning of symbols see country text, or Table VI, p.30, and Table IX, p.66.)

(3) In the kaolinitic *(K)* region the principal soils are: *a* (alluvial); *k* (kaolisols); *kc* (kaolisol–cinnamonic intergrades) and *m* (dark clays).

Because of the Andes, Peru has very few kaolinitic soils, except in the Amazonas, the soils of which are more or less unknown and which is almost void of population. Even there alluvial soils abound. That is why the soil factor is favourable in Peru.

Bolivia (Fig.14)

Bolivia includes five broad soil regions: *(1)* desert *(D)*; *(2)* dry mountains *(DM)*; *(3)* arid cinnamonic *(AC)*; *(4)* tropical mountains *(TM)*; *(5)* kaolinitic *(K)*.

(1) In the desert region *(D)* of the southwestern high plateau the principal soils are: *d* (arid brown and other soils of the desert); *ll* (rock outcrops); there are many "salares" (salt accumulations).

(2) In the dry mountains *(DM)* of the rest of the high plateau the principal soils are: *c* (cinnamonic); *d* (arid brown); *e* (aeolian sands); *hl* (lithosolic rankers); *ll* (rock outcrops); *pc* (non-calcic brown); *rc* (cinnamonic rendzinas); *v* (ando).

(3) In the tropical mountains *(TM)* the principal soils are: *b* (recent brown); *ba* (brun acide); *by* (chernozemic brown (prairie)); *g* (gleisolic); *kc* (kaolisol–cinnamonic intergrades); *kp* (kaolinitic podsolic); *l* (lithosols); *o* (peaty); *p* (podsols); and *pc* (non-calcic brown).

(4) In the arid cinnamonic region *(AC)* of the southwest (Chaco, etc.) the principal soils are: *a* (alluvial); *c* (cinnamonic); *cr* (terra rossa); *cy* (reddish brown); *e* (aeolian sands); *r* (rendzinas); *rc* (cinnamonic rendzinas); *t* (solonetz); *u* (planosols); *pc* (non-calcic brown).

(5) In the kaolinitic region *(K)* of the Amazonas, etc. the principal soils are: *a* (alluvial); *c* (cinnamonic); *g* (gleisolic); *k* (kaolisols); *kc* (kaolisol–cinnamonic intergrades); *kp* (kaolinitic podsolic); *l* (lithosols); *u* (planosols).

It may be there are many ando soils in Bolivia. Due to the influence of the Andes, there are no kaolinitic soils in Bolivia, except in the Amazonas.

Chile (Fig.15)

In Chile we may recognize three broad soil regions: *(1)* desert *(D)* in the north; *(2)* mountainous with mediterranean climate and many volcanic materials *(M'–A)* in middle Chile; *(3)* mountainous with humid cool climate and many volcanic materials *(M–A)* in the south; there is also a brunisolic *(B)* region in the Juan Fernandez Islands, and a kaolinitic–ando *(K–A)* region in the Pascua Islands.

(1) In the desert *(D)* region the most important soils are *cd* (reddish desert); *d* (arid brown and other raw soils of the desert); *ld* (desert lithosols); *rd* (serozems); *s* (saline). Volcanic materials abound and many soils are ando. Many of them have a desert crust (the surface horizon is loosely cemented with products of weathering as soluble salts, gypsum, carbonates, silica; cementation is sufficient to impede erosion, but the crust slacks in water). Other soils are covered by a desert pavement. There are many "salares" (plains covered with salts).

(2) In the mountainous region of central Chile *(M'–A)* the principal soils are: *a* (alluvial); *b* (recent brown); *bl* (lithosolic recent brown); *d* (arid brown); *g* (gleisolic); *kc* (kaolisol–cinnamonic intergrades); *l* (lithosols); *m* (dark clays); *o* (peaty); *pc* (non-calcic brown); *r* (rendzinas); *v* (ando). Ando soils are probably much more common than suggested by the maps; many dark clays, non-calcic brown, and soils resembling kaolisols are probably of volcanic origin and have clay mineralogical composition that would justify their classification as ando.

(3) In the mountainous region of south Chile *(M–A)* the principal soils are: *a* (alluvial); *aw* (wet alluvial); *b* (recent brown); *ba* (brun acide); *c* (cinnamonic); *g* (gleisolic); *kc* (kaolinitic–cinnamonic intergrades); *l* (lithosols); *lh* (lithosolic–alpine rankers); *o* (peaty); *p* (podsols); *pb* (grey–brown podsolic); *pg* (gley podsols); *pl* (lithosolic podsols); *v* (ando); *vh* (humus ando); *vw* (wet ando); *yl* (grassland rankers); *yp* (prairie lessivé). The importance of ando soils is still more important than shown in the map. Brun acide, and soils resembling kaolisols, are probably ando; it is the same for many recent brown; podsolization is weak in southern Chile, probably due to the volcanic origin of the materials.

On the Juan Fernandez Islands *(B* region) the principal soils are: *b* (recent brown); *bl* (lithosolic brown) and *l* (lithosols).

On the Pascua Islands *(K–A* region) the principal soils are *kl* (latosolic kaolisols) and *va* (acid ando). But it is possible that all soils there are ando.

The predominance of volcanic materials and the mountainous relief of Chile result in soils of very high cation exchange capacity, potentially very fertile. However, ando soils fix phosphorus and many soils need phosphorus fertilizers. Some soils of the south have been intensely leached because of the high leaching rainfall. Naturally nitrogen fertilizers give everywhere good results. As the country is mountainous, stoniness, shallowness and steep slopes are often limiting factors. Some soils of the south are waterlogged; some of the soils of the north saline.

Argentina (Fig.16)

Argentina includes a variety of broad soil regions: *(1)* chernozemic *(Ch)* in the pampean region; *(2)* desert *(D)* in the west; *(3)* arid cinnamonic *(AC)*, and arid cinnamonic–gleisolic *(AC–G)* in the north; *(4)* kaolinitic–cinnamonic *(K–C)* in the northeast; *(5)* sub-antartic chernozemic *(Ch')* and *(6)* atlantic podsolic *(AP)* in the extreme south; *(7)* sub-glacial desert–tundra *(SD–T)* in the Antarctic; *(8)* dry mountains *(DM)*, *(9)* tropical mountains *(TM)* and *(10)* high latitude humid mountains *(M)* in the cordillera, etc.

(1) In the chernozemic *(Ch)* region the principal

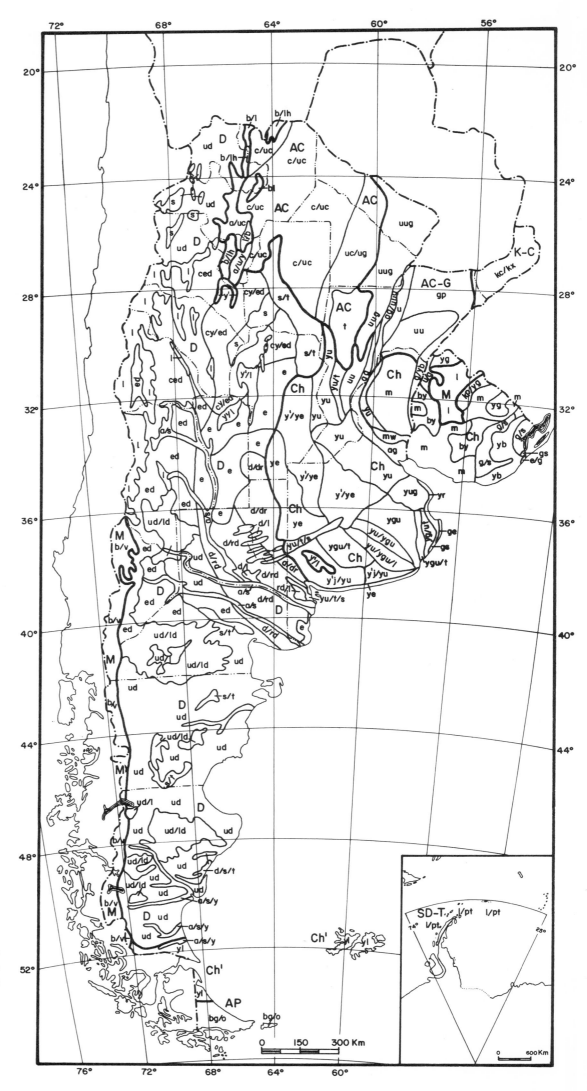

Fig.16. Argentina, Uruguay. (For the meaning of symbols see country text, or Table VI, p.30, and Table IX, p.66.)

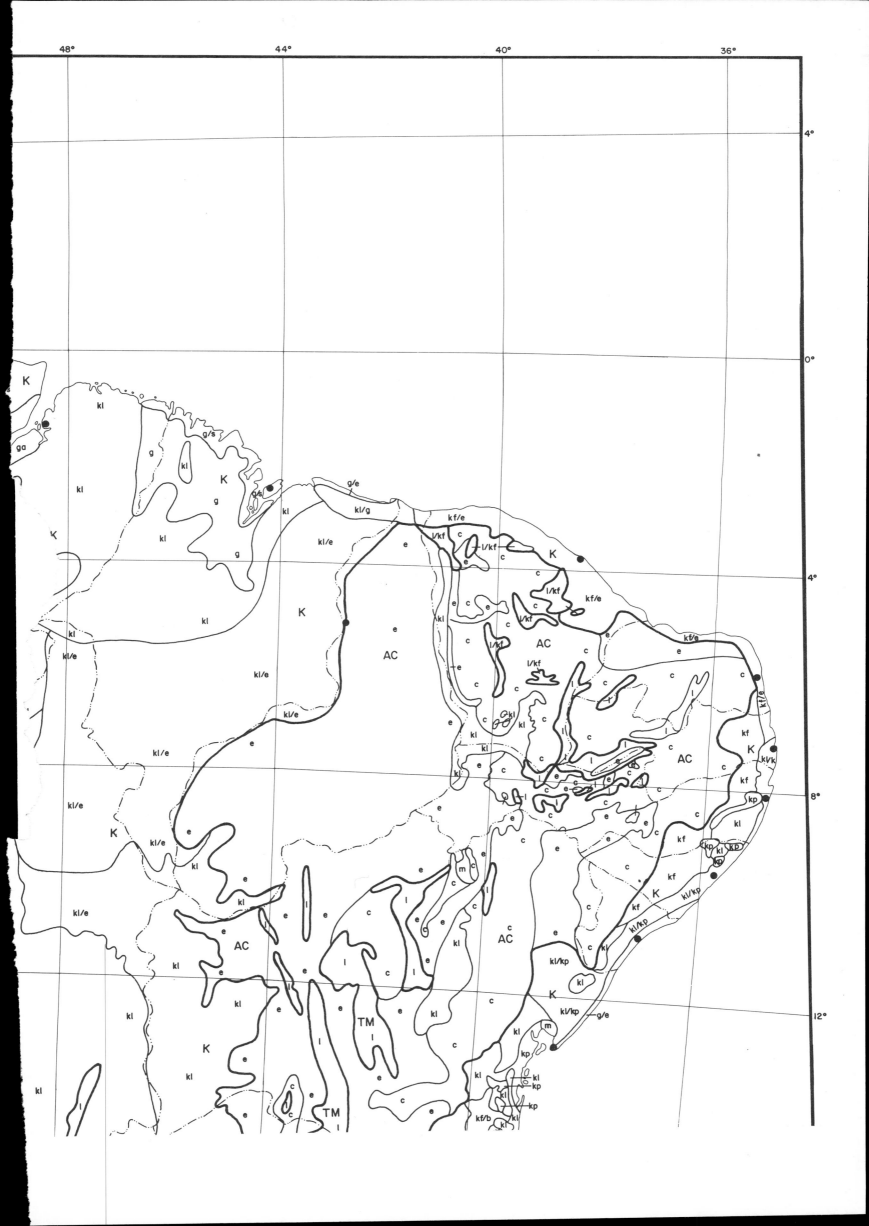

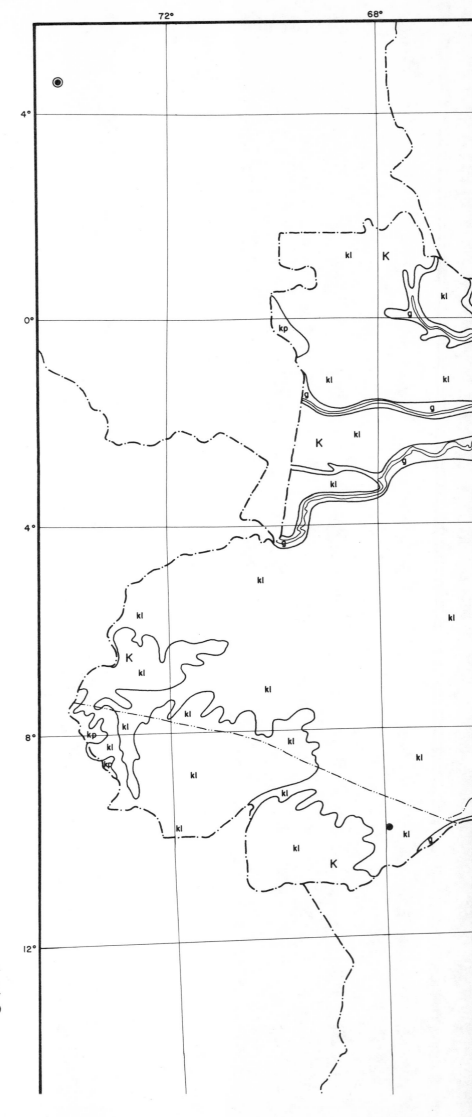

Fig.17A (pp.110–112). Northern
half of Brazil. (For the meaning
of symbols see country text, or
Table VI, p.30, and Table IX, p.66.)

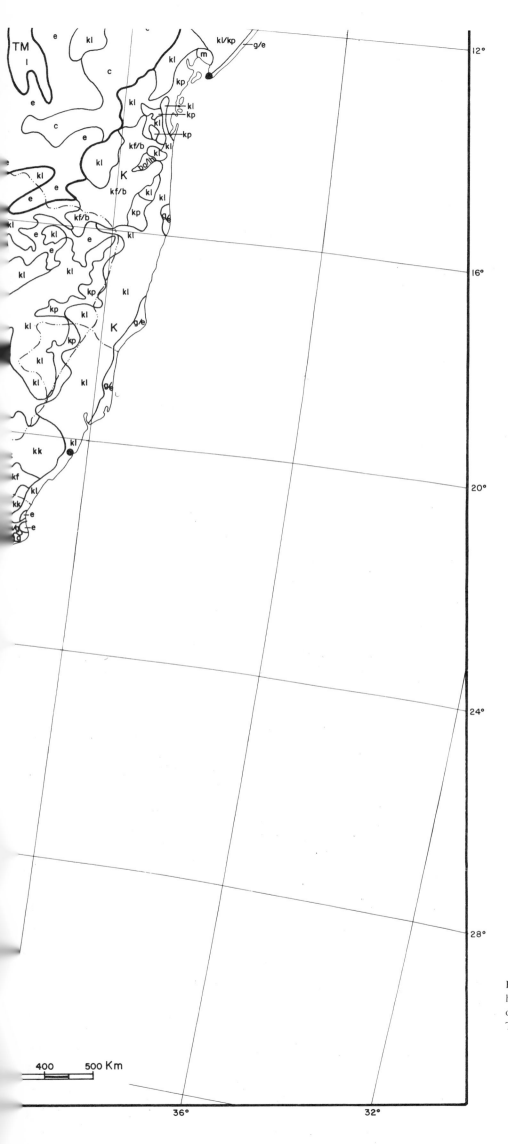

Fig.17B (pp.113–115). Southern
half of Brazil. (For the meaning
of symbols see country text, or
Table VI, p.30, and Table IX, p.66.)

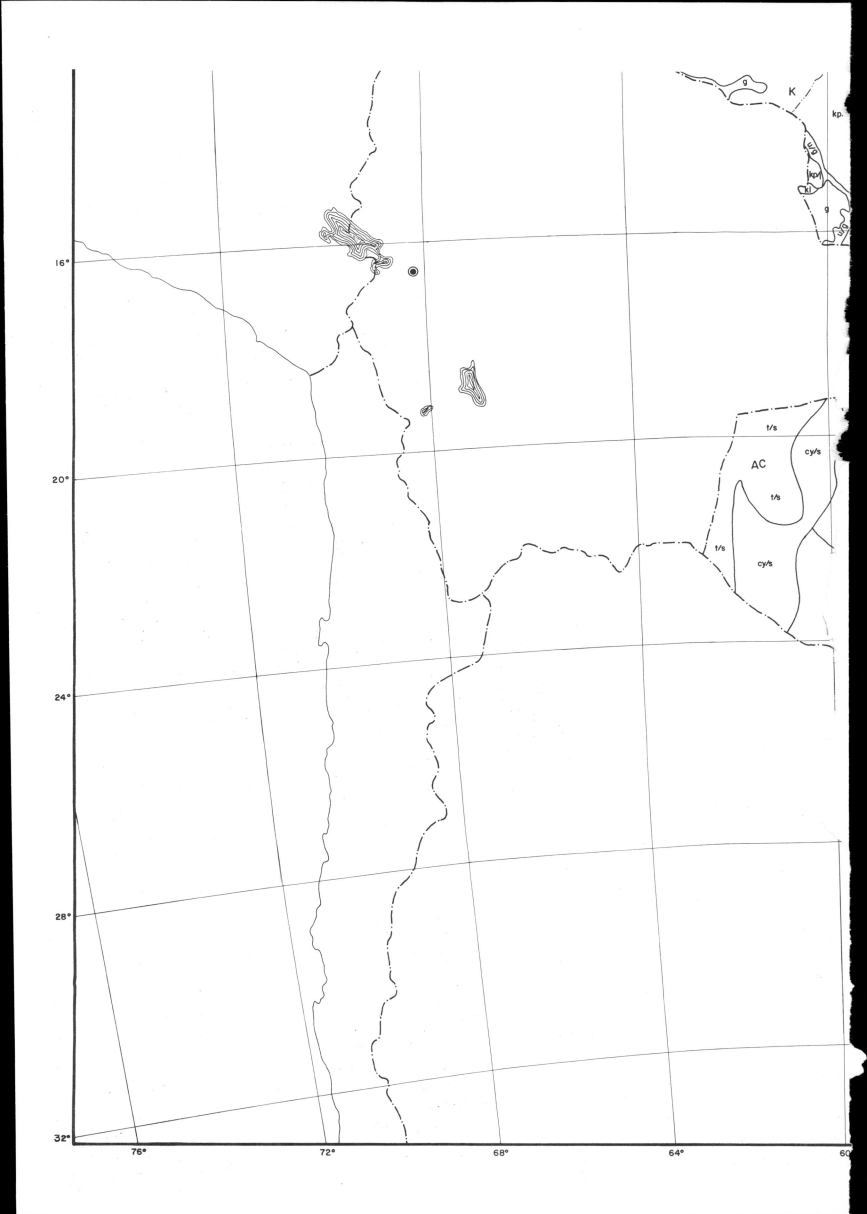

soils are: *ag* (gleisolic alluvial); *g* (gleisolic); *gs* (saline gleisols); *l* (lithosols); *m* (dark clays); *mw* (wet dark clays); *s* (saline); *t* (solonetz); *y'* (para-chernozems); *yb* (brunisolic para-chernozems (prairie)); *ye* (aeolian para-chernozems); *ygu* (planosolic gleisolic para-chernozems); *y'j* (petro-calcic para-chernozems); *yu* (planosolic para-chernozems); *yug* (gleisolic planosolic para-chernozems).

(2) In the desert region (*D*) the principal soils are: *a* (alluvial); *ced* (reddish desert sands); *cy* (reddish brown); *d* (arid brown and other raw soils of the desert); *dr* (serozems); *e* (aeolian sands); *ed* (desert sands); *l* (lithosols); *ld* (desert lithosols); *s* (saline); *t* (solonetz); *ud* (planosolic red desert); *y* (chernozemic); *y'* (para-chernozems); *yu* (planosolic para-chernozems).

(3) In the arid-cinnamonic (*AC*) region the principal soils are: *a* (alluvial); *c* (cinnamonic); *t* (solonetz); *u* (planosols); *pc* (non-calcic brown); *ug* (gleisolic planosols); *y* (chernozemic).

(3) In the arid-cinnamonic–gleisolic (*AC–G*) region the principal soils are: *ag* (gleisolic alluvial); *gp* (low humic gley); *u* (planosols); *uu* (clay-pan planosols); *uug* (gleisolic clay-pan planosols).

(4) In the kaolinitic–cinnamonic (*K–C*) region the principal soils are: *kc* (kaolisol–cinnamonic intergrades); *kx* (terra roxa).

(5) In the sub-antarctic chernozemic region (*Ch'*) the principal soils are: *a* (alluvial); *s* (saline); *t* (solonetz); *y* (chernozemic); *yl* (grassland rankers).

(6) In the atlantic podsol region (*AP*) the principal soils are: *o* (peaty); *bg* (gleisolic brunisolic more or less podzolized).

(7) In the sub-glacial desert–tundra (*SD–T*) region the principal soils are: *l* (lithosols); and *pt* (tundra nano-podsols).

(8) In the dry mountains (*DM*) the principal soils are: *a* (alluvial); *b* (recent brown); *l* (lithosols); *pc* (non-calcic brown); *y'* (para-chernozems).

(9) In the tropical mountains (*TM*) the principal soils are: *b* (recent brown); *bl* (lithosolic brown); *lh* (lithosolic rankers).

(10) In the high latitude mountains (*M*) the principal soils are lithosols (*l*).

In Argentina a great part of the parent materials, from which the soils have been formed, is volcanic ash coming from the cordillera. Many of the fore-mentioned soils are ando. Their cation exchange capacity is unusually high, compared to their clay content. And this is one of the causes of their high fertility. A re-classification of the soils, to take account of their volcanic character (high cation exchange capacity and moisture tension) is needed.

Uruguay (Fig.16)

Uruguay includes one only broad soil region, which is chernozemic (*Ch*); although it has some low mountains (*M*). In the chernozemic region (*Ch*) the principal soils are: *by* (chernozemic brown (prairie)); *e* (aeolian sands); *g* (gleisolic); *gs* (saline gleisolic); *kp* (kaolinitic podsolic); *m* (dark clays); *yb* (prairie); *yg* (humic gley); *u* (planosols).

In the high latitude humid mountains (*M*) soils are chiefly lithosols.

The chernozemic region of Uruguay is the transition to the chernozemic–kaolinitic region of Brasil; that is why few soils are typical chernozemic.

Paraguay (Fig.17)

Paraguay includes two broad soil regions: (*1*) arid cinnamonic (*AC*) in the west; (*2*) kaolinitic (*K*) in the east.

(*1*) In the arid cinnamonic (*AC*) region, the principal soils are: *a* (alluvial); *cy* (reddish brown); *g* (gleisolic); *kd* (dystrophic kaolisols); *kp* (kaolinitic podsolic); *m* (dark clays); *s* (saline); t (solonetz); *u* (planosols).

(2) In the kaolinitic region (*K*) the principal soils are: *c* (cinnamonic); *kl* (latosolic kaolisol); *kp* (kaolinitic podsolic); *kx* (terra roxa); *rg* (gleisolic rendzinas).

Brazil (Fig.17)

The greater part of Brazil forms a kaolinitic soil region (*K*); but in the north we encounter an arid cinnamonic region (*AC*); and in the south kaolinitic–cinnamonic (*K–C*) and kaolinitic–chernozemic (*K–Ch*); moreover there are some tropical mountains (*TM*).

(*1*) In the kaolinitic (*K*) region the principal soils are: *b* (recent brown); *ba* (acid brown); *c* (cinnamonic); *e* (aeolian sands); *g* (gleisolic); *ga* (alluvial gleisolic); *gs* (saline gleisols–mangrove); *k* (kaolisols); *kc* (kaolinitic cinnamonic intergrades); *kdl* (latosolic dystrophic kaolisols); *kdp* (red–yellow podsolic); *kf* (concretionary kaolisols); *kg* (gleisolic kaolisols); *kh* (rubrozems); *kk* (ferralitic); *kl* (latosolic kaolisols); *kn* (eutrophic kaolisols); *kp* (kaolinitic podsolic); *kpf* (lateritic podsolic); *kx* (terra roxa); *l* (lithosols); *lh* (lithosolic rankers); *m* (dark clays); *s* (saline); *u* (planosols). Many of the kaolisols are dystrophic and consequently of low mineral fertility; but others are eutrophic. Except for the states of São Paulo and Rio de Janeiro for which analytical data abound, the author has not been able to classify the soils of other states concerning this important character.

(2) In the arid cinnamonic region (*AC*) of the north the principal soils are: *c* (cinnamonic); *e* (aeolian sands); *kl* (latosolic kaolisols); *m* (dark clays).

(3) In the kaolinitic–cinnamonic region (*K–C*) of the south, the principal soils are: *kdl* (latosolic dystrophic kaolisols); *kl* (latosolic kaolisols); *kp* (kaolinitic podsolic) and *kx* (terra roxa). Cation exchange capacity of these soils seems to be high, placing them on the limit with cinnamonic soils; it may be some of them are cinnamonic.

(4) In the kaolinitic–chernozemic region (*K–Ch*) the principal soils are: *e* (aeolian sands); *g* (gleisolic);

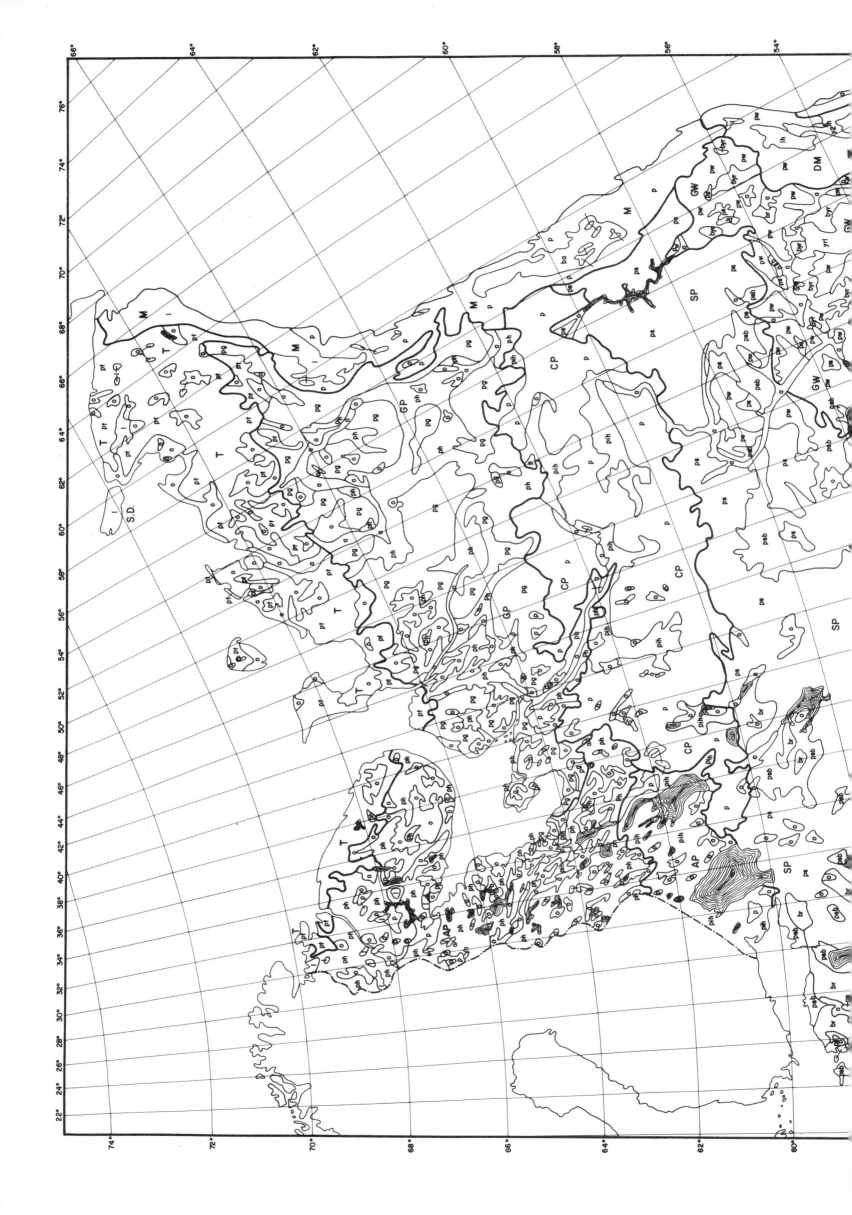

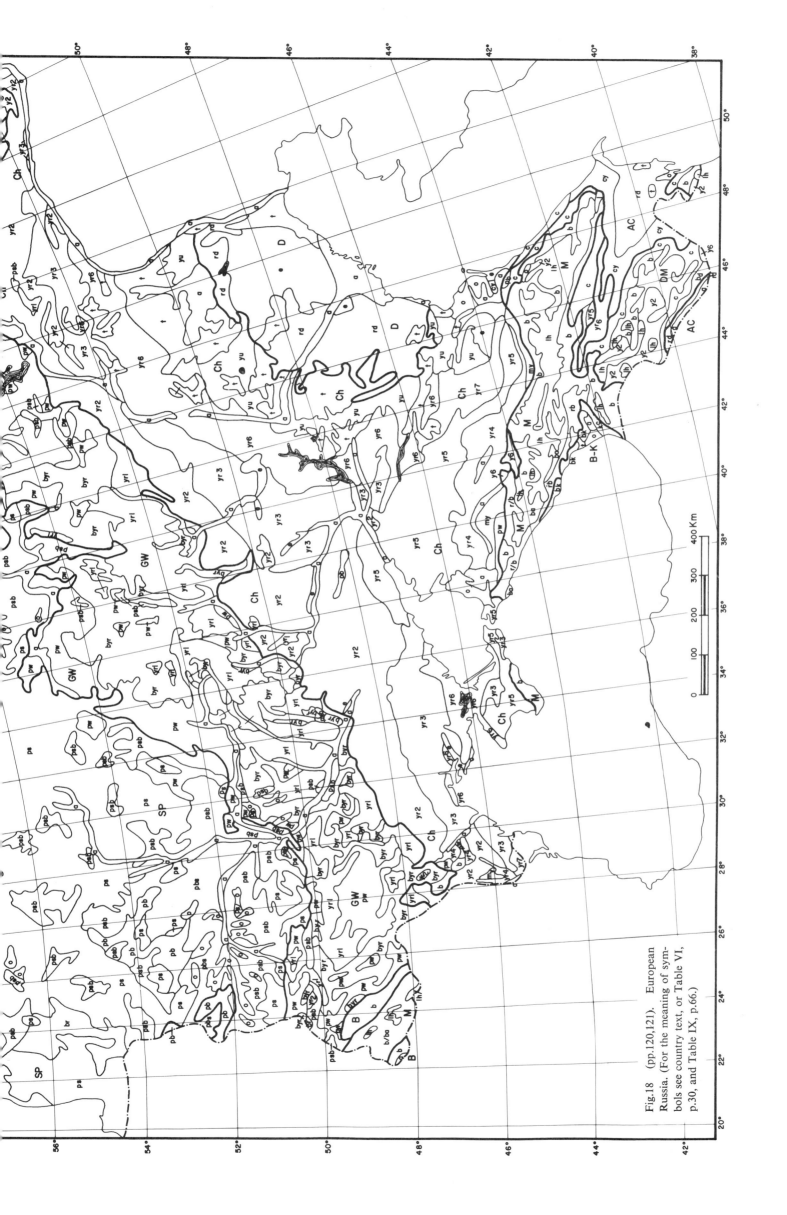

Fig.18 (pp.120,121). European Russia. (For the meaning of symbols see country text, or Table VI, p.30, and Table IX, p.66.)

400 Km

0   100   200   300

kc (kaolinitic–cinnamonic intergrades); *kdp* (red–yellow podsolic); *kh* (rubrozems); *kp* (kaolinitic podsolic); *kx* (terra roxa); *ky* (chernozemic kaolisols); *m* (dark clays); *s* (saline); *u* (planosols); *yb* (brunisolic para-chernozems (prairie); *yg* (humic glei).

(*5*) In tropical mountains (*TM*) the principal soils are: *ba* (brun acide); *e* (aeolian sands); *kf* (concretionary kaolisols); *kh* (rubrozems); *kx* (terra roxa); *l* (lithosols); *lh* (lithosolic rankers); *p* (podsols).

EUROPE

European Russia

European Russia (Fig.18) includes a great variety of broad soil regions: (*1*) sub-glacial desert (*SD*), (*2*) tundra (*T*), (*3*) atlantic podsol (*AP*), (*4*) gley podsol (*GP*), (*5*) continental podsol (*CP*), (*6*) sod podsol (*SD*) and (*7*) grey-wooded (*GW*) in the north; (*8*) chernozemic (*Ch*), (*9*) arid cinnamonic (*AC*) and (*10*) desert (*D*) in the south; (*11*) brunisolic (*B*) in the southwest; and (*12*) brunisolic–kaolinitic (*B–K*) on the southeastern coast of the Black Sea. There are also some mountainous regions: (*13*) high latitude humid mountains (*M*) and (*14*) dry mountains (*DM*).

(*1*) In the sub-glacial desert (*SD*) soils are usually lithosols (*l*).

(*2*) In the tundra region (*T*) the principal soils are: *a* (alluvial); *l* (lithosols); *o* (peaty); *ph* (humus podsols); *pt* (tundra nano-podsols including arctic brown and tundra gleisols).

(*3*) Two atlantic podsol (*AP*) regions can be distinguished: in the northern one the principal soils are: *l* (lithosols); *ph* (humus podsols); *pih* (humus–iron podsols). In the southern region the principal soils are: *o* (peaty); *p* (podsols); *pb* (grey–brown podsolic); and *pih* (humus–iron podsols).

(*4*) In the gley podsol (*GP*) region the principal soils are: *a* (alluvial); *o* (peaty); *p* (podsols); *pg* (gley podsols); *ph* (humus podsols); *pt* (tundra nano-podsols).

(*5*) In the continental podsol (*CP*) region the principal soils are: *o* (peaty); *p* (podsols); and *pih* (humus–iron podsols).

(*6*) In the sod podsol (*SP*) region the principal soils are: *a* (alluvial); *br* (calcareous braunerde); *o* (peaty); *pb* (grey–brown podsolic); *pbs* (grey–brown podsolic–sod podsol intergrades); *ps* (sod podsols); *psb* (grey–brown podsolic–sod podsol intergrades); *pw* (grey-wooded).

(*7*) In the grey-wooded (*GW*) region the principal soils are: *a* (alluvial); *b* (braunerde); *br* (calcareous braunerde); *byr* (degraded chernozems); *ps* (sod podsols); *psb* (grey–brown podsolic–sod podsol intergrades); *pw* (grey-wooded); *yr1* (typical chernozems); *yr2* (ordinary chernozems).

(*8*) In the chernozemic (*Ch*) region the principal soils are: *a* (alluvial); *b* (braunerde); *byr* (degraded chernozems); *cy* (reddish brown); *e* (aeolian sands); *my* (chernozemic clays); *o* (peaty); *pb* (grey–brown podsolic);

psb (grey–brown podsolic–sod podsol intergrades); *pw* (grey-wooded); *t* (solonetz); *yr1* (typical chernozems); *yr2* (ordinary chernozems); *yr3* (southern chernozems); *yr4* (chernozems and meadow chernozems); *yr5* (chernozems with promycelium of carbonates near the surface); *yr6* (dark chestnut); *yr7* (chestnut); *yu* (leached parachernozems). The soils of this region have usually been formed from calcareous materials, and have suffered little leaching; that is why the humic horizon is immediately underlain or overlapped with a calcareous horizon; they are calcareous chernozems (*yr*), and their fertility is very high.

(*9*) In the arid cinnamonic (*AC*) region the principal soils are: *a* (alluvial); *c* (cinnamonic); *cy* (reddish brown); *o* (peaty); *rd* (serozems); *t* (solonetz).

(*10*) In the desert (*D*) region the principal soils are: *a* (alluvial); *e* (aeolian sands); *rd* (serozems); *t* (solonetz).

(*11*) In the brunisolic (*B*) region of the southwest the principal soils are: alluvial (*a*), and braunerde (*b*).

(*12*) In the brunisolic–kaolinitic region (*B–K*) of the Black Sea coast of the southern Caucasus the principal soils are: *a* (alluvial); *bk* (jeltozems); *kc* (kaolisol–cinnamonic intergrades); *o* (peaty). It is not known to what extent the soils of this region are kaolinitic; it may be that their cation exchange capacities, compared to their clay content, are too high for their classification as kaolisols.

(*13*) In the high latitude humid mountains (*M*) the principal soils are: *b* (recent brown); *ba* (brun acide); *l* (lithosols); *lh* (lithosolic rankers); *p* (podsols); *r* (rendzinas); *y2* (mountain chernozems); *y6* (mountain chestnut).

(*14*) In the dry mountains (*DM*) the principal soils are: *b* (braunerde); *bd* (arid brown); *c* (cinnamonic); *l* (lithosols); *lh* (lithosolic rankers); *pw* (grey-wooded); *y2* (chernozems); *y6* (chestnut); *yr5* (chernozems with promycelium of carbonates near the surface).

The soils of northern Russia (regions *SD*, *T*, *AP*, *GP*, *CP*) are usually poor; but with liming and fertilizers they can give high yields, where climate permits it. Those of southern Russia regions (*Ch*, *AC*, *D*) are fertile, especially those of the chernozemic (*Ch*) region, which are among the most fertile of the world. Soils of the sod podsolic (*SP*) and grey-wooded (*GW*) regions are intermediary.

Scandinavian countries

Finland (Fig.19). Finland is included in only one broad soil region: atlantic podsol (*AP*); however, there are some high latitude humid mountains (*M*).

The principal soils of the atlantic podsol region (*AP*) are: *a* (alluvial); *e* (aeolian sands); *o* (peaty); *p* (podsols); *pb* (grey–brown podsolic). In the mountains the principal soils are lithosols (*l*) and rankers (*lh*).

Sweden (Fig.19). Sweden consists of two broad soil regions: (*1*) atlantic podsol (*AP*), and (*2*) high latitude humic mountains (*M*).

Fig.19. Scandinavia. (For the meaning of symbols see country text, or Table VI, p.30, and Table IX, p.66.)

(*1*) In the atlantic podsol region (*AP*) the principal soils are: *o* (peaty); *p* (podsols); *pb* (grey–brown podsolic).

(*2*) In the high latitude humid mountains (*M*) the principal soils are: *lt* (tundra lithosols); *o* (peaty); *p* (podsols); of course lithosols abound.

Norway (Fig.19). Norway consists of two broad soil regions: (*1*) atlantic podsol (*AP*), and (*2*) high latitude humid mountains (*M*).

(*1*) The principal soils of the atlantic podsol region (*AP*) are: *o* (peaty); *p* (podsols); *pb* (grey–brown podsolic).

(*2*) The principal soils of high latitude humic mountains (*M*) are: *lt* (tundra lithosols); *o* (peaty); *p* (podsols); of course lithosols abound.

Denmark (Fig.19). Denmark consists of only one broad soil region, atlantic podsol (*AP*). The principal soils are: *a* (alluvial); *o* (peaty); *p* (podsols); *pb* (grey–brown podsolic).

Great Britain and Ireland (Fig.20)

The United Kingdom. Two broad soil regions are found here: (*1*) atlantic podsol (*AP*), and (*2*) grey–brown podsolic (*GB*); the mountains can be included in the atlantic podsol region.

(*1*) In the atlantic podsol region (*AP*) the principal soils are: *a* (alluvial); *ba* (brun acide); *l* (lithosols); *o* (peaty); *p* (podsols); *pb* (grey–brown podsolic); *ph* (humus podsols).

(*2*) In the grey–brown podsolic region (*GB*) the

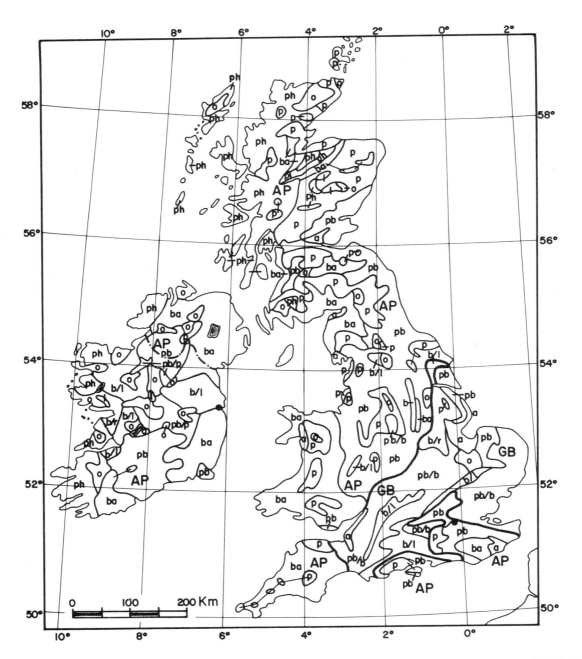

Fig.20. Great Britain and Ireland. (For the meaning of symbols see country text, or Table VI, p.30, and Table IX, p.66.)

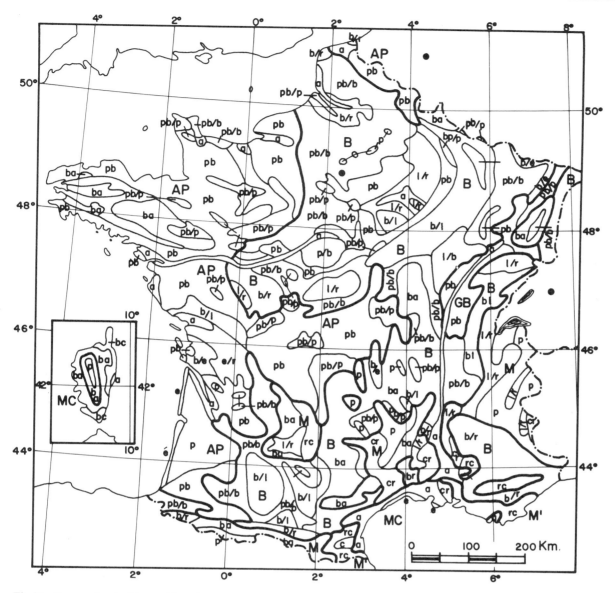

Fig.21. France. Inset: Corse. (For the meaning of symbols see country text, or Table VI, p.30, and Table IX, p.66.)

principal soils are: *a* (alluvial); *b* (braunerde); *l* (lithosols); *p* (podsols); *pb* (grey–brown podsolic); *r* (rendzinas).

Ireland. Ireland has only one broad soil region: atlantic podsol (*AP*). The principal soils are: *b* (braunerde); *ba* (brun acide); *l* (lithosols); *o* (peaty); *p* (podsols); *pb* (grey–brown podsolic); *ph* (humus podsols).

France (Fig.21)

France consists of six broad soil regions: (*1*) atlantic podsol (*AP*); (*2*) grey–brown podsolic (*GB*); (*3*) brunisolic (*B*); (*4*) mediterranean cinnamonic (*MC*); (*5*) high latitude humid mountains (*M*), and mediterranean mountains (*M'*).

(*1*) In the atlantic podsol region (*AP*) the principal soils are: *a* (alluvial); *b* (braunerde); *ba* (brun acide); *e* (aeolian sands); *l* (lithosols); *p* (podsols); *pb* (grey–brown podsolic).

(*2*) In the grey–brown podsolic region (*GB*) the principal soils are: *a* (alluvial); *p* (podsols); *pb* (grey–brown podsolic).

(*3*) In the brunisolic (*B*) region the principal soils are: *a* (alluvial); *b* (braunerde); *ba* (brun acide); *cr* (terra rossa); *l* (lithosols); *pb* (grey–brown podsolic), and *r* (rendzinas).

(*4*) In the mediterranean cinnamonic region (*MC*) the principal soils are: *a* (alluvial); *ba* (brun acide); *bc* (cinnamonic brown—brun méditerranéen): *cr* (terra rossa); and *e* (aeolian sands).

(*5*) In high latitude humid mountains (*M*) the principal soils are: *b* (braunerde); *ba* (brun acide); *l* (lithosols); *p* (podsols); *pb* (grey–brown podsolic); *r* (rendzinas); *rc* (terra rossa).

(*6*) In mediterranean mountains (*M'*) the principal soils are: *b* (braunerde); *ba* (brun acide); *c* (cinnamonic); *p* (podsols); *r* (rendzinas); *rc* (terra rossa); and of course lithosols.

These broad regions are intermingled and cannot

be satisfactorily shown in a small-scale map. But the brunisolic regions (*B*) have fertile soils and occupy extensive areas in France. That is why agricultural production in France has been always high and there was no need to force it.

Western central Europe (Fig.22)

The Netherlands. Two broad soil regions exist here: (*1*) atlantic podsols (*AP*); and (*2*) brunisolic (*B*).

(*1*) In the atlantic podsol region (*AP*) the principal soils are: *a* (alluvial); *o* (peaty); *p* (podsols); and *pb* (grey–brown podsolic).

(*2*) In the brunisolic region (*B*) the principal soils are *pb* (grey–brown podsolic).

The original soils of The Netherlands were mostly podsols (*p*) and consequently poor; however, the Dutch

acquired much land from the sea; moreover by manures, fertilizers, lime, sown prairies of grasses and legumes they increased the fertility of their podsols; and now they have very fertile land. This land is man-made both qualitatively and quantitatively. It is the best example of man's influence on soils.

Belgium. There are also two broad soil regions in evidence here: (*1*) atlantic podsol (*AP*); and (*2*) brunisolic (*B*).

(*1*) In the atlantic podsol (*AP*) region the principal soils are: *a* (alluvial); *p* (podsols); and *pb* (grey–brown podsolic).

(*2*) In the brunisolic region (*B*) the principal soils are: *b* (braunerde); *ba* (brun acide); *pb* (grey–brown podsolic).

The brunisolic region has fairly fertile soils. In the atlantic podsol region soils have been greatly improved by

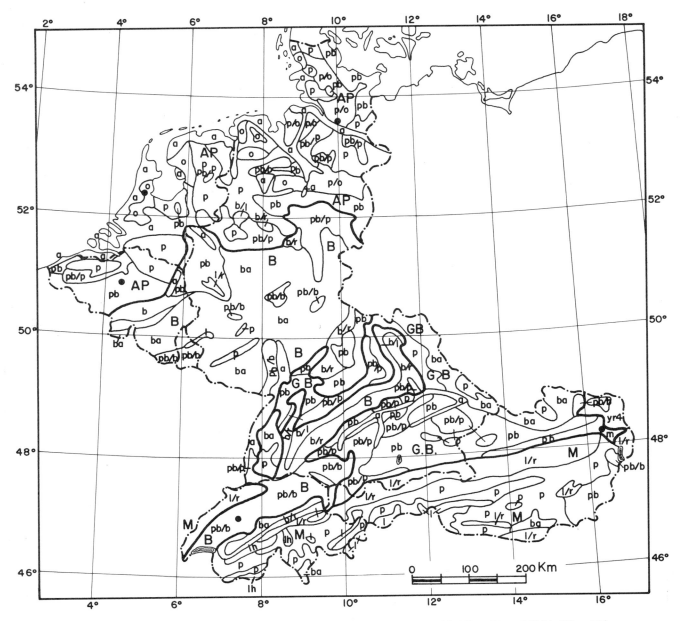

Fig.22. Western central Europe. (For the meaning of symbols see country text, or Table VI, p.30, and Table IX, p.66.)

liming, fertilizers, manuring ánd rotation with grasses and legumes; now they are very fertile; the podsol (*p*) soils are the less fertile.

Luxembourg. Only one broad soil region exists: brunisolic (*B*). The principal soils are: *b* (braunerde); *l* (lithosols); *pb* (grey–brown podsolic).

Because of the high frequency of brunisolic soils, Luxemburg may be considered as having fairly good land.

West Germany. West Germany consists of four broad soil regions: (*1*) atlantic podsol (*AP*); (*2*) grey–brown podsolic (*GB*); (*3*) brunisolic (*B*); and (*4*) high latitude humid mountains (*M*).

(*1*) In the atlantic podsol region (*AP*) the principal soils are: *a* (alluvial); *o* (peaty); *p* (podsols); *pb* (grey–brown podsolic).

(*2*) In the grey–brown podsolic region (*GB*) the principal soils are: *a* (alluvial); *ba* (brun acide); *p* (podsols); *pb* (grey–brown podsolic).

(*3*) In the brunisolic region (*B*) the principal soils are: *a* (alluvial); *b* (braunerde); *ba* (brun acide); *l* (lithosols); *p* (podsols); *pb* (grey–brown podsolic).

(*4*) In high latitude humid mountains (*M*) the principal soils are: *ba* (brun acide); *l* (lithosols); *p* (podsols); *r* (rendzinas).

The broad regions are intermingled and cannot be shown satisfactorily on a small-scale map. The brunisolic regions of Germany have good soils; and the podsols of the atlantic podsol region, which are chiefly sandy, have been considerably improved by introducing sown pastures of legumes and grasses, manuring, fertilizing and liming; they are now good soils.

Switzerland. There exist two broad soil regions: (*1*) brunisolic (*B*); and (*2*) high latitude humid mountains (*M*).

(*1*) In the brunisolic region (*B*) the principal soils are: *b* (braunerde); and *pb* (grey–brown podsolic).

(*2*) In high latitude humid mountains (*M*) the principal soils are: *b* (braunerde); *ba* (brun acide); *l* (lithosols); *lh* (lithosolic rankers); *p* (podsols); *r* (rendzinas).

Belonging to a brunisolic region, the plains of Switzerland have good soils. But the country is mountainous. However, the mountains also have soils of high mineral fertility (*r*, *b*, *lh*); naturally many of them are shallow or stony.

Austria. Austria consists of three broad soil regions: (*1*) chernozemic (*Ch*); (*2*) grey–brown podsolic (*GB*); and (*3*) high latitude humid mountains (*M*).

(*1*) In the chernozemic region (*Ch*) the principal soils are: *b* (braunerde); *m* (dark clays); *pb* (grey–brown podsolic); and *yr4* (chernozems and meadow chernozems).

(*2*) In the grey–brown podsolic region (*GB*) the principal soils are: *b* (braunerde); *ba* (brun acide); *p* (podsols); and *pb* (grey–brown podsolic).

(*3*) In high latitude humid mountains (*M*) the principal soils are: *b* (braunerde); *ba* (brun acide); *l* (lithosols);

p (podsols); *pb* (grey–brown podsolic); *r* (rendzinas).

For the high frequency of brunisolic soils and a few chernozemic, mineral fertility is good in Austria; but the country is very mountainous.

Eastern central Europe (Fig.23)

Czechoslovakia. Four broad soil regions exist here: (*1*) atlantic podsol (*AP*); (*2*) brunisolic (*B*); (*3*) chernozemic (*Ch*); and (*4*) high latitude humid mountains (*M*).

(*1*) In the atlantic podsol (*AP*) region the principal soils are: *pb* (grey–brown podsolic).

(*2*) In the brunisolic (*B*) region the principal soils are: *a* (alluvial); *b* (braunerde); *byr* (degraded chernozems); and *pb* (grey–brown podsolic).

(*3*) In the chernozemic region (*Ch*) the principal soils are: *b* (braunerde); *byr* (degraded chernozems); *e* (aeolian sands); and *yr4* (chernozems and meadow chernozems).

(*4*) In the high latitude humid mountains (*M*) the principal soils are: *b* (braunerde); *ba* (brun acide); *p* (podsols); *pb* (grey–brown podsolic); *rb* (brunified rendzinas).

Except for some soils of the atlantic podsol (*AP*) and the mountainous (*M*) region the soils of Czechoslovakia are fairly good.

Eastern Germany. Here there are three broad soil regions: (*1*) atlantic podsol (*AP*); (*2*) brunisolic (*B*); and (*3*) high latitude humid mountains (*M*).

(*1*) In the atlantic podsol region (*AP*) the principal soils are: *a* (alluvial); *b* (braunerde); *byr* (degraded chernozems); *g* (gleisolic); *o* (peaty); *pb* (grey–brown podsolic); *pih* (iron–humus podsols); and *psb* (grey–brown podsolic–sod podsol intergrades).

(*2*) In the brunisolic region (*B*) the principal soils are: *ba* (brun acide); *byr* (degraded chernozems); *br* (calcareous brown); *pb* (grey–brown podsolic).

(*3*) In high latitude humid mountains (*M*) the principal soils are: *ba* (brun acide); *p* (podsols); *pb* (grey–brown podsolic); *rb* (brunified rendzinas); and of course lithosols.

Poland. This country consists of five broad soil regions: (*1*) atlantic podsol (*AP*); (*2*) sod podsol (*SP*); (*3*) brunisolic (*B*); (*4*) grey-wooded (*GW*), and (*5*) high latitude humid mountains (*M*).

(*1*) In the atlantic podsol region (*AP*) the principal soils are: *a* (alluvial); *byr* (degraded chernozems); *o* (peaty); *pb* (grey–brown podsolic); *psb* (grey–brown podsolic-sod podsol intergrades).

(*2*) In the sod podsol region (*SP*) the principal soils are: *a* (alluvial); *psb* (grey–brown podsolic–sod podsol intergrades).

(*3*) In the brunisolic region (*B*) the principal soils are: *a* (alluvial); *b* (braunerde); *br* (calcareous brown); *pb* (grey–brown podsolic).

(*4*) In the grey-wooded region (*GW*) the principal

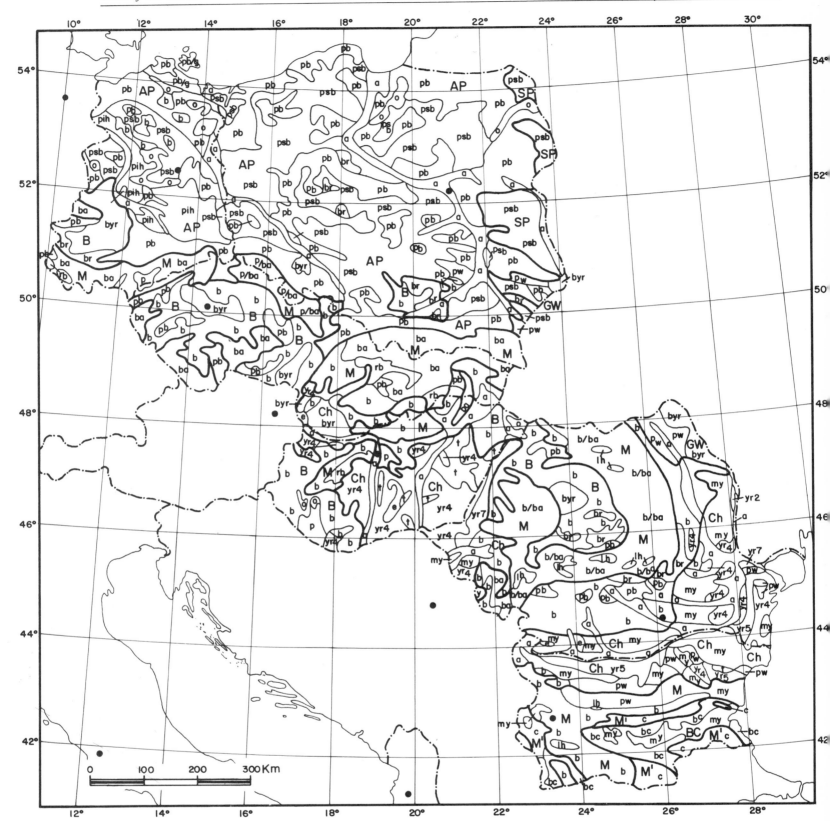

Fig.23. Eastern central Europe (Eastern Germany, Czechoslovakia, Poland, Hungary, Rumania, Bulgaria). (For the meaning of symbols see country text, or Table VI, p.30, and Table IX, p.66.)

soils are: *br* (calcareous brown); *byr* (degraded cherno-zems); *pb* (grey–brown podsolic); *psb* (grey–brown pod-solic–sod podsol intergrades); *pw* (grey-wooded).

(5) In the high latitude humid mountains (*M*) the principal soils are: *ba* (brun acide); *p* (podsols); and of course lithosols.

Hungary. Three broad soil regions are in evidence here: (*1*) chernozemic (*Ch*); (*2*) brunisolic (*B*); and (*3*) high latitude humid mountains (*M*).

(*1*) In the chernozemic region (*Ch*) the principal soils are: *a* (alluvial); *e* (aeolian sands); *t* (solonetz); and *yr4* (chernozems and meadow chernozems).

(*2*) In the brunisolic region (*B*) the principal soils are: *a* (alluvial); *b* (braunerde); *o* (peaty); *p* (podsols); *yr4* (chernozems and meadow chernozems).

(*3*) In the high latitude humid mountains (*M*) the principal soils are: *b* (braunerde); *ba* (brun acide); and *rb* (brunified rendzina).

Because of the high frequency of chernozemic and brunisolic soils, Hungary is a fertile country.

Rumania. Rumania consists of four broad soil regions: (*1*) chernozemic (*Ch*); (*2*) grey-wooded (*GW*); (*3*) brunisolic (*B*); and (*4*) high latitude humid mountains (*M*).

(*1*) In the chernozemic (*Ch*) region the principal soils are: *a* (alluvial); *b* (braunerde); *e* (aeolian sands); *my* (chernozemic clays); *pw* (grey-wooded); *yr2* (ordinary chernozems); *yr4* (chernozems and meadow chernozems); *yr5* (chernozem with CaCO$_3$ promycelium to the surface); *yr7* (dark chestnut).

(*2*) In the grey-wooded (*GW*) region the principal soils are: *a* (alluvial); *byr* (degraded chernozems); *pw* (grey-wooded).

(*3*) In the brunisolic region (*B*) the principal soils are: *b* (braunerde); *br* (calcareous brown); *byr* (degraded chernozems); *pb* (grey–brown podsolic).

(*4*) In high latitude humid mountains (*M*), the principal soils are: *b* (braunerde); *ba* (brun acide); *lh* (lithosolic rankers); and of course lithosols.

Because of the high frequency of chernozemic and brunisolic soils, Rumania may be considered as a country with fertile land.

Bulgaria. Four broad soil regions exist in Bulgaria: (*1*) chernozemic (*Ch*); (*2*) brunisolic–cinnamonic (*B–C*); (*3*) high latitude humid mountains (*M*); and (*4*) mediterranean mountains (*M'*).

(*1*) In the chernozemic region (*Ch*) the principal soils are: *b* (braunerde); *my* (chernozemic clays); *pw* (grey-wooded); and *yr5* (chernozems).

(*2*) In the brunisolic–cinnamonic region (*B–C*) the principal soils are: *bc* (cinnamonic brown); *cy* (reddish brown); and *my* (chernozemic clays).

(*3*) In the high latitude humid mountains (*M*) the principal soils are: *b* (braunerde); *lh* (lithosolic rankers); *my* (chernozemic clays); and *pw* (grey-wooded); of course lithosols abound.

(*4*) In the mediterranean mountains (*M'*) the principal soils are: *bc* (cinnamonic brown); *c* (cinnamonic); *my* (chernozemic clays); of course lithosols abound.

Yugoslavia (Fig.24). Yugoslavia consists of five broad soil regions: (*1*) chernozemic (*Ch*); (*2*) grey–brown podsolic (*GB*); (*3*) mediterranean cinnamonic (*MC*); (*4*) high latitude humid mountains (*M*); and (*5*) mediterranean mountains (*M'*).

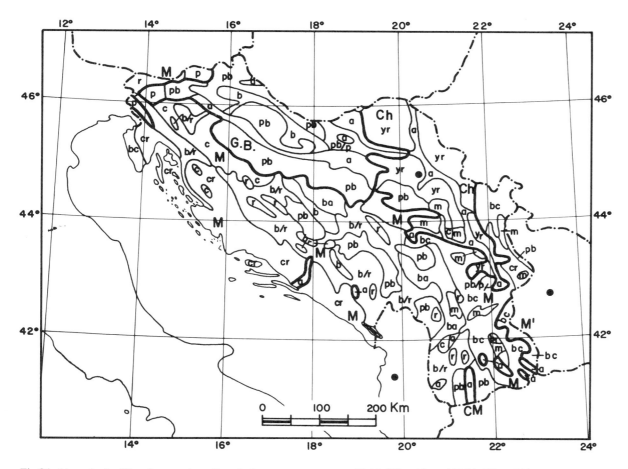

Fig.24. Yugoslavia. (For the meaning of symbols see country text, or Table VI, p.30, and Table IX, p.66.)

(*1*) In the chernozemic region (*Ch*) the principal soils are: *a* (alluvial); *c* (cinnamonic); *m* (dark clays); *pb* (grey–brown podsolic); and *yr* (chernozems).

(*2*) In the grey–brown podsolic region (*GB*) the principal soils are: *a* (alluvial); *b* (braunerde); *p* (podsols); and *pb* (grey–brown podsolic).

(*3*) In the mediterranean cinnamonic region (*MC*), alluvial soils (*a*) abound.

(*4*) In the high latitude humid mountains (*M*) the principal soils are: *a* (alluvial); *b* (braunerde); *ba* (brun acide); *bc* (cinnamonic brown); *c* (cinnamonic); *cr* (terra rossa); *m* (dark clays); *p* (podsols); *pb* (grey–brown podsolic); and *r* (rendzinas).

(*5*) In the mediterranean mountains (*M'*) cinnamonic brown (*bm*) and of course lithosols abound.

The chernozemic and grey–brown podsolic plains of Yugoslavia are fairly fertile. The mountains too, but naturally much land is too steep, too shallow or too stony.

Mediterranean countries

Spain (Fig.25). Spain consists of a variety of broad soil regions: (*1*) atlantic podsol (*AP*); (*2*) brunisolic (*B*); (*3*) mediterranean cinnamonic (*MC*); (*4*) arid cinnamonic (*AC*); (*5*) high latitude humid mountains (*M*); and (*6*) mediterranean mountains (*M'*).

(*1*) In the atlantic podsol region (*AP*) the principal soils are: *ba* (brun acide); *p* (podsols); *pb* (grey–brown podsolic).

(*2*) In the brunisolic region (*B*) the principal soils are: *b* (braunerde); *l* (lithosols); *pb* (grey–brown podsolic); *r* (rendzinas).

(*3*) In the mediterranean cinnamonic region (*MC*) the principal soils are: *a* (alluvial); *bc* (cinnamonic brown); *c* (cinnamonic); *cr* (terra rossa); *cy* (reddish brown); *e* (aeolian sands); *l* (lithosols); *m* (dark clays).

(*4*) In the arid cinnamonic region (*AC*) the principal soils are: *a* (alluvial); *cy* (reddish brown); *cr* (terra rossa); *l* (lithosols); and *rd* (serozems).

(*5*) In the high latitude humid mountains (*M*) the principal soils are: *ba* (brun acide); *p* (podsols); *pb* (grey–brown podsolic); and of course lithosols.

(*6*) In the mediterranean mountains (*M'*) the prin-

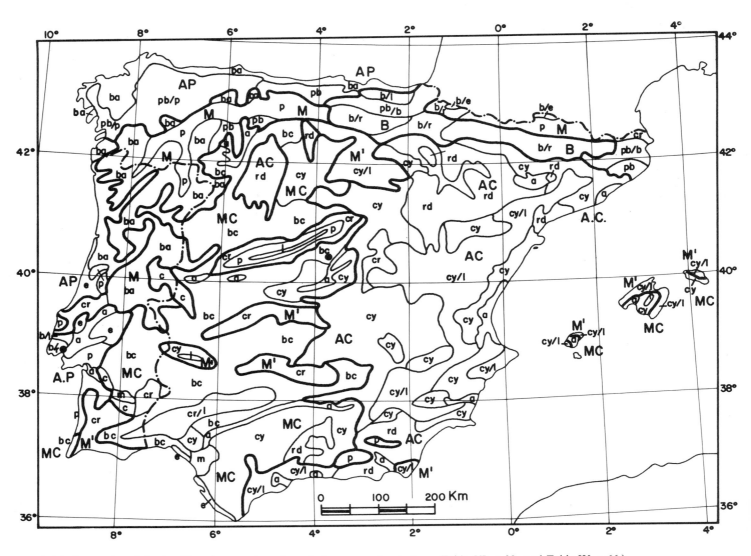

Fig.25. Portugal and Spain. (For the meaning of symbols see country text, or Table VI, p.30, and Table IX, p.66.)

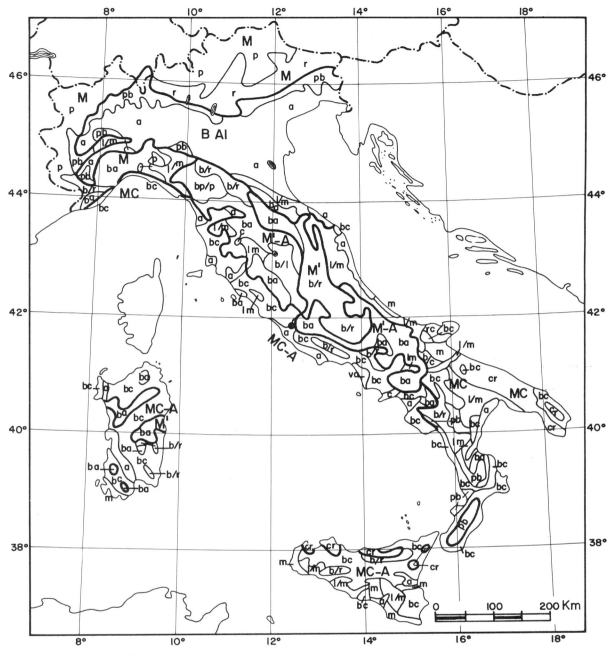

Fig.26. Italy. (For the meaning of symbols see country text, or Table VI, p.30, and Table IX, p.66.)

cipal soils are: *c* (cinnamonic); *cy* (reddish brown); *l* (lithosols); and *p* (podsols).

Portugal (Fig.25). Portugal consists of four broad soil regions: (*1*) atlantic podsol (*AP*); (*2*) mediterranean cinnamonic (*MC*); (*3*) high latitude humid mountains (*M*); and (*4*) mediterranean mountains (*M'*).

(*1*) In the atlantic podsol region (*AP*) the principal soils are: *a* (alluvial); *b* (braunerde); *ba* (brun acide); *bc* (cinnamonic brown); *e* (aeolian sands); *l* (lithosols); *p* (podsols); *r* (rendzinas).

(*2*) In the mediterranean cinnamonic region (*MC*) there are many cinnamonic brown (*bc*) soils.

(*3*) In high latitude humid mountains the principal soils are: *ba* (brun acide); *cr* (terra rossa); *p* (podsols).

(*4*) In mediterranean mountains (*M'*) terra rossa (*cr*) abound.

Italy (Fig.26). Italy is composed of a variety of broad soil regions: (*1*) brunisolic–alluvial (*B–Al*) in the north; (*2*) mediterranean cinnamonic (*MC*), and (*3*) mediterranean cinnamonic–ando (*MC–A*) on the peninsula and islands; (*4*) high latitude humid mountains (*M*) in the north; (*5*) mediterranean mountains (*M'*), and (*6*) mediterranean mountains–ando (*M'–A*) on the peninsula and islands.

(*1*) In the brunisolic–alluvial (*B–Al*) region the principal soils are: *a* (alluvial); *ba* (brun acide); *l* (lithosols); *m* (dark clays); *pb* (grey-brown podsolic). The materials of this region come from geologically young mountains, and

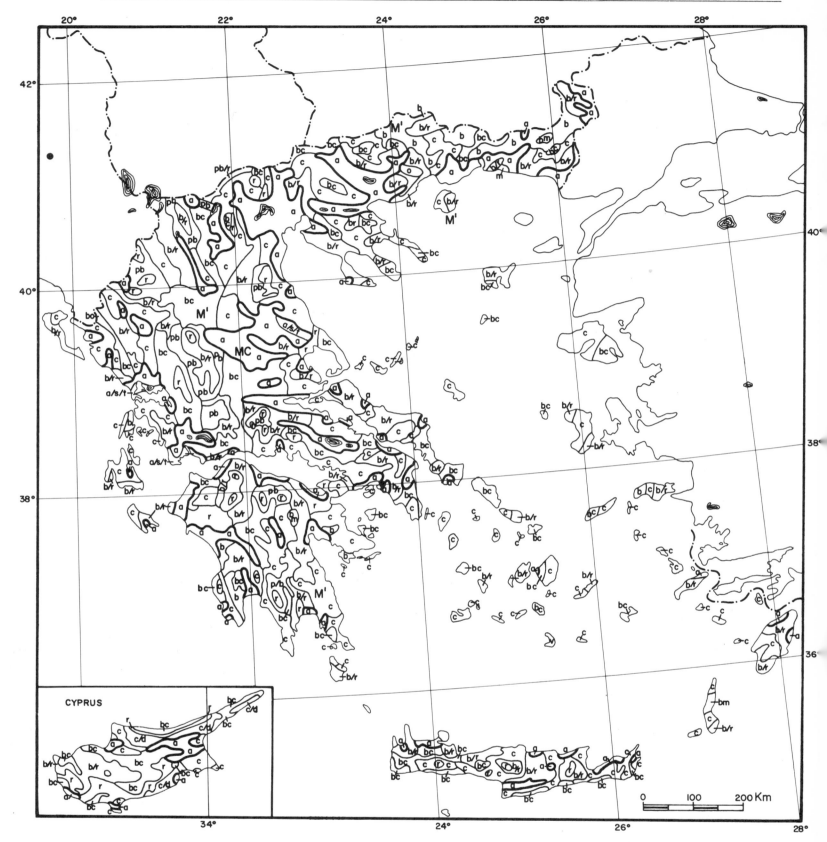

Fig.27. Greece and Cyprus. (For the meaning of symbols see country text, or Table VI, p.30, and Table IX, p.66.)

the region is highly fertile; the Po valley is one of the best agricultural regions of the world.

(2) In the mediterranean cinnamonic region (*MC*) the principal soils are: *a* (alluvial); *b* (recent brown); *ba* (brun acide); *bc* (cinnamonic brown); *cr* (terra rossa); *m* (dark clays); *pb* (grey–brown podsolic); *r* (rendzinas).

(3) In the mediterranean cinnamonic–ando region (*MC–A*) the principal soils are: *a* (alluvial); *b* (recent brown); *ba* (brun acide); *bc* (cinnamonic brown); *cr* (terra rossa); *m* (dark clays); *pb* (grey–brown podsolic); and *r* (rendzinas).

(4) In the high latitude humid mountains (*M*) the

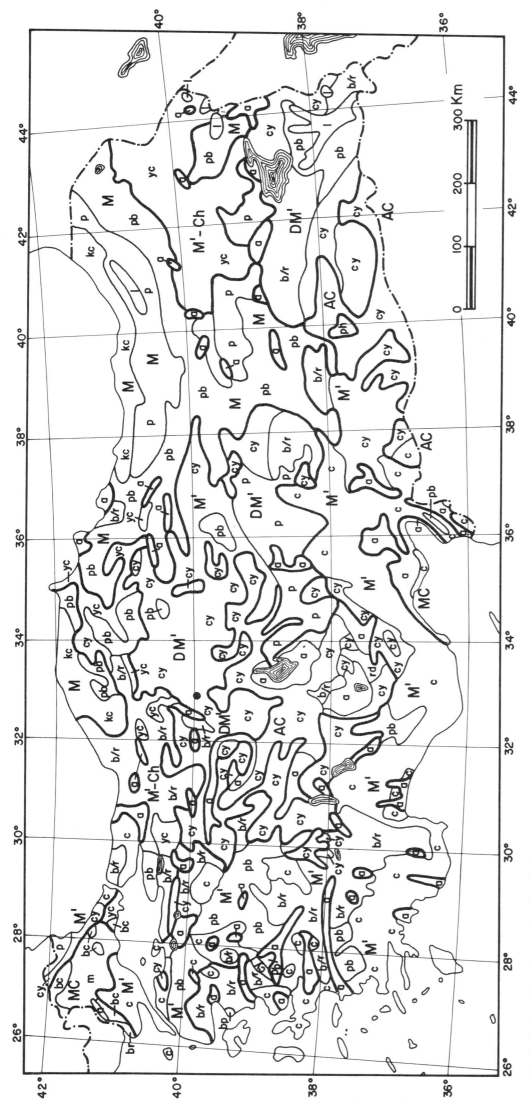

Fig. 28. Turkey. (For the meaning of symbols see country text, or Table VI, p.30, and Table IX, p.66.)

principal soils are: *b* (braunerde); *ba* (brun acide); *l* (lithosols); *p* (podsols); *r* (rendzinas).

(5) In the mediterranean mountains (*M'*) the principal soils are: *b* (recent brown); *ba* (brun acide); *cr* (terra rossa); *pb* (grey–brown podsolic); *r* (rendzinas).

(6) In the mediterranean mountains–ando (*M'–A*) the principal soils are: *b* (recent brown); *ba* (brun acide); *bp* (grey–brown podsolic); *c* (cinnamonic); *l* (lithosols); *m* (dark clays).

Italy is a volcanic country; it still has many active volcanoes. Unfortunately ando soils are not shown on the maps, but they abound, and a great part of the fertility of italian soils is due to this fact. Rome has been able to feed a great population even during times in which very little food was coming from abroad, due to the high fertility of its soils formed principally of volcanic materials; and the case of Naples, of Catania, etc., is analogous. A re-classification of Italian soils on the basis of the relation of their cation exchange capacity and moisture tension to clay content is needed.

Greece (Fig.27). Greece is very mountainous and includes two broad soil regions: (*1*) mediterranean mountains (*M'*) and (*2*) mediterranean cinnamonic (*MC*).

(*1*) In the mediterranean mountains (*M'*) the principal soils are: *a* (alluvial); *b* (recent brown); *bc* (cinnamonic brown); *c* (cinnamonic); *m* (dark clays); *pb* (grey–brown podsolic); *r* (rendzinas); *v* (ando.)

(2) In the plains (*MC*) the principal soils are: *a* (alluvial); *m* (dark clays); *o* (peaty); *s* (saline) and *t* (solonetz).

Since Greek mountains are geologically young, Greek soils have usually good mineral fertility; moreover there are some ando, much more than it is shown on the maps. The soils of the plains (*MC*) are practically all alluvial; many have been formed under conditions of bad drainage and this is the reason of the high frequency of saline, solonetz and planosols (*u* and *uc*) not shown in the map but frequent; now they have been naturally or artificially drained.

Cyprus (Fig.27). Cyprus is composed of two broad soil regions: (*1*) mediterranean mountains (*M'*) and (*2*) mediterranean cinnamonic (*MC*).

(*1*) In the mountains (*M'*) the principal soils are: *b* (recent brown); *bc* (cinnamonic brown); *c* (cinnamonic); *d* (arid brown); *r* (rendzinas).

(2) In the plains (*MC*) alluvial soils prevail.

The mountains of the island are geologically young and that is why soils have usually good mineral fertility.

Turkey (Fig.28). Turkey has a variety of broad soil regions: (*1*) mediterranean cinnamonic (*MC*); (*2*) arid cinnamonic (*AC*); (*3*) mediterranean mountains (*M'*); (*4*) mediterranean mountains–chernozemic (*M'–Ch*); (*5*) dry mediterranean mountains (*DM'*) and (*6*) high latitude humid mountains (*M*).

(*1*) In the mediterranean cinnamonic region (*MC*)

the principal soils are: *a* (alluvial); *bc* (cinnamonic brown); *br* (calcareous brown); and *m* (dark clays).

(2) In the arid cinnamonic (*AC*) region the principal soils are: *a* (alluvial); *b* (recent brown); *cy* (reddish brown); *r* (rendzinas); and *rd* (serozems).

(3) In the mediterranean mountains (*M'*) the principal soils are: *b* (braunerde); *bc* (cinnamonic brown); *c* (cinnamonic); *cy* (reddish brown); *pb* (grey–brown podsolic); *r* (rendzinas); and of course lithosols.

(4) In the mediterranean mountains–chernozemic (*M'–Ch*) region the principal soils are: *b* (braunerde); *bc* (cinnamonic brown); *cy* (reddish brown); *l* (lithosols); *pb* (grey–brown podsolic); *r* (rendzinas); and *yc* (reddish para-chernozems).

(5) In the dry mediterranean mountains (*DM'*) the principal soils are: *b* (braunerde); *cy* (reddish brown); *l* (lithosols); *p* (podsols); *pb* (grey–brown podsolic); *r* (rendzinas).

(6) In the high latitude humid mountains (*M*) the principal soils are: *b* (braunerde); *kc* (kaolisol–cinnamonic intergrades); *l* (lithosols); *p* (podsols); *r* (rendzinas).

Because of the high frequency of brunisolic and cinnamonic soils mineral fertility is usually good in Turkey; however, many soils are poor in organic matter and, as the country is mountainous, the percentage of culturable land is not so high.

AFRICA

North Africa

Morocco (Fig.29). Morocco is composed of four broad soil regions: (*1*) mediterranean cinnamonic (*MC*); (*2*) mediterranean mountains (*M'*); (*3*) arid cinnamonic (*AC*); and (*4*) desert (*D*).

(*1*) In the mediterranean cinnamonic (*MC*) region the principal soils are: *a* (alluvial); *c* (cinnamonic); *d* (arid brown); *mw* (wet clays).

(2) In the mediterranean mountains (*M'*) the principal soils are: *c* (cinnamonic); *cb* (brown cinnamonic); *d* (arid brown); *l* (lithosols); *ll* (rock outcrops); *r* (rendzinas).

(3) In the arid cinnamonic (*AC*) region the principal soils are: *a* (alluvial); *d* (arid brown); *e* (aeolian sands); *mw* (wet clays); *r* (rendzinas).

(4) In the desert region (*D*) the principal soils are: *a* (alluvial); *d* (arid brown and other raw soils of the desert); *e* (aeolian sands); *ed* (desert sands); *kff* (laterites); *l* (lithosols); *ll* (rock outcrops); *r* (rendzinas); *s* (saline); *stt* (takirs).

Moroccan soils are in general young and of good mineral fertility; the presence of laterites in the desert region is due to a humid former climate.

Algeria (Fig.29) includes three broad soil regions: (*1*) mediterranean cinnamonic (*MC*) in the north; (*2*)

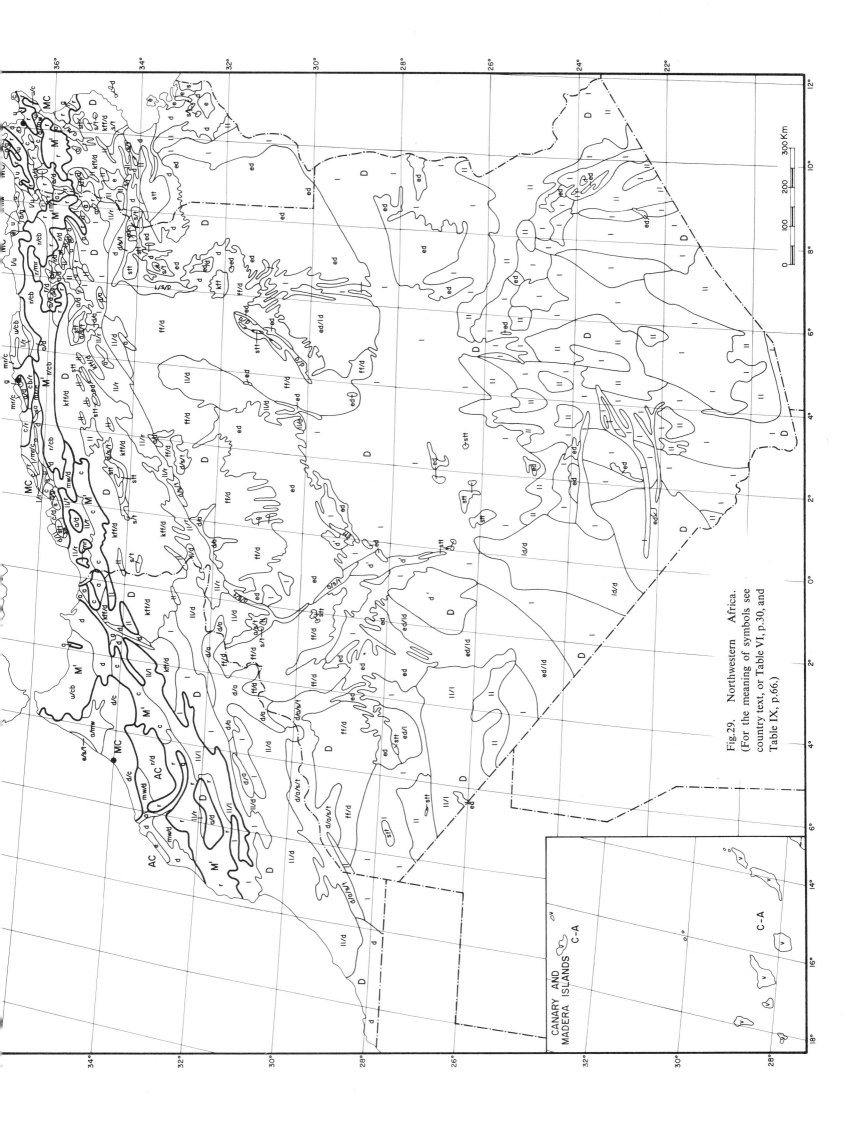

Fig. 29. Northwestern Africa.
(For the meaning of symbols see
country text, or Table VI, p.30, and
Table IX, p.66.)

CANARY AND
MADERA ISLANDS

C-A

desert (*D*) in the south; and (*3*) the mediterranean mountains (*M′*).

(*1*) In the mediterranean cinnamonic region (*MC*) the principal soils are: *a* (alluvial); *c* (cinnamonic); *cb* (brown cinnamonic); *d* (arid brown); *e* (aeolian sands); *g* (gleisolic); *l* (lithosols); *ll* (rock outcrops); *mr* (rendzina clays); *mw* (wet clays); *r* (rendzinas); *s* (saline); *se* (aeolian saline—lunettes); *stt* (takirs); *u* (planosols).

(*2*) In the desert region (*D*) the principal soils are: *a* (alluvial); *d* (arid brown and other raw soils of the desert); *d′* (gypsisols); *ed* (desert sands); *ff* (lateritic rocks); *kff* (laterite); *l* (lithosols); *ld* (desert lithosols); *ll* (rock outcrops); *mr* (rendzina clays); *r* (rendzinas); *s* (saline); *stt* (takirs); *t* (solonetz).

(*3*) In the mediterranean mountains (*M′*) the principal soils are: *c* (cinnamonic); *cb* (brown cinnamonic); *ll* (rock outcrops); *r* (rendzinas).

The presence of lateritic rocks and laterite in the desert is interesting; it is certainly due to a former more humid climate. The soils of the mediterranean regions (*MC* and *M′*) are young and their mineral fertility good.

Tunisia (Fig.29). Tunisia consists of three broad soil regions: (*1*) mediterranean cinnamonic (*MC*); (*2*) mediterranean mountains (*M′*); and (*3*) desert (*D*).

(*1*) In the mediterranean cinnamonic region (*MC*) the principal soils are: *a* (alluvial); *c* (cinnamonic); *d* (arid brown); *g* (gleisolic); *mw* (wet clays); *r* (rendzinas); *u* (planosols).

(*2*) In the mediterranean mountains (*M′*) rendzinas (*r*) prevail.

(*3*) In the desert region (*D*) the principal soils are: *a* (alluvial); *d* (arid brown and other raw soils of the desert); *e* (aeolian sands); *ed* (desert sands); *kff* (laterites); *l* (lithosols); *ll* (rock outcrops); *r* (rendzinas); *s* (saline); *stt* (takirs); *t* (solonetz).

Tunisian soils are usually young and their mineral fertility is good, but some laterites, perhaps due to former more humid climate, are reported in the south.

Canary and Madeira Islands (Fig.29). Here volcanic materials prevail and fertile ando soils abound.

Libya (Fig.30). Libya is composed of two broad soil regions: (*1*) mediterranean–cinnamonic (*MC*); and (*2*) desert (*D*).

(*1*) In the mediterranean–cinnamonic region (*MC*) the principal soils are: *c* (cinnamonic); *d* (arid brown).

(*2*) In the desert region (*D*) the principal soils are: *d* (arid brown); *ed* (desert sands); *ld* (desert lithosols); *ll* (rock outcrops); *s* (saline).

The United Arab Republic (Fig.30). This country consists of only one broad soil region, which is desert (*D*). The principal soils are: *aw* (wet alluvial); *d* (arid brown); *ed* (desert sands); *ll* (rock outcrops); *s* (saline); *stt* (takir); *t* (solonetz).

West Africa (from the Atlantic to Togo)

Mauritania (Fig.31) is included in one only broad soil region, which is desert (*D*). The principal soils are: *a* (alluvial); *aw* (wet alluvial); *d* (arid brown and other raw soils of the desert); *ed* (desert sands); *kff* (laterite); *ld* (desert lithosols); *ll* (rock outcrops); *m* (dark clays); *md* (desert clays); *s* (saline); *stt* (takirs); *t* (solonetz).

In spite of the actual desert climate there are lateritic rocks and laterite soils, formed under more humid conditions.

Spanish Sahara. This has only one broad soil region, which is desert (*D*). The principal soils are: *d* (arid brown and other raw soils of the desert); *ed* (desert sands); *ld* (desert lithosols); *l* (rock outcrops); *stt* (takir).

Senegal (Fig.31). This includes three broad soil regions: (*1*) desert (*D*) in the north; (*2*) kaolinitic (*K*) in the south; and (*3*) tropical mountains (*TM*).

(*1*) In the desert (*D*) region the principal soils are: *a* (alluvial); *aw* (wet alluvial); *d* (arid brown); *kff* (laterites); *ll* (rock outcrops); *s* (saline); *t* (solonetz).

(*2*) In the kaolinitic (*K*) region, the principal soils are: *a* (alluvial); *aw* (wet alluvial); *g* (gleisolic); *gs* (saline gleisols—mangrove); *kff* (laterites); *kk* (ferralitic); *kln* (terres de barre); *kn* (eutrophic kaolisols); *knf* (tropical ferruginous); *ll* (rock outcrops); *m* (dark clays); *mr* (rendzina clays); *s* (saline); *t* (solonetz).

(*3*) In tropical mountains (*TM*), lateritic rocks (*ff*) abound.

Gambia. This consists of only one broad soil region, which is kaolinitic (*K*); the principal soils are: *a* (alluvial); *aw* (wet alluvial); *gs* (saline gleisols—mangrove); *kk* (ferralitic); *knf* (tropical ferruginous); *s* (saline); *t* (solonetz).

Guinea (Fig.31). Guinea consists of two broad soil regions: (*1*) kaolinitic (*K*); and (*2*) tropical mountains (*TM*).

(*1*) In the kaolinitic (*K*) region the principal soils are: *b* (recent brown); *gs* (saline gleisols—mangrove); *ff* (lateritic rock); *kd* (dystrophic kaolisols); *kff* (laterites); *kk* (ferralitic); *kln* (terres de barre); *knf* (tropical ferruginous); *knl* (terres de barre); *l* (lithosols); *ll* (rock outcrops).

(*2*) In the tropical mountains (*TM*) the principal soils are: *ff* (lateritic rocks); *l* (lithosols).

Portuguese Guinea. This country belongs to only one broad soil region, which is kaolinitic (*K*); however, there are some tropical mountains (*TM*).

In the kaolinitic region (*K*) the principal soils are: *gs* (saline gleisols—mangrove); *kff* (laterites); *kk* (ferralitic); *knf* (tropical ferruginous). In tropical mountains (*TM*) the principal soils are: *l* (lithosols); *ff* (lateritic rocks).

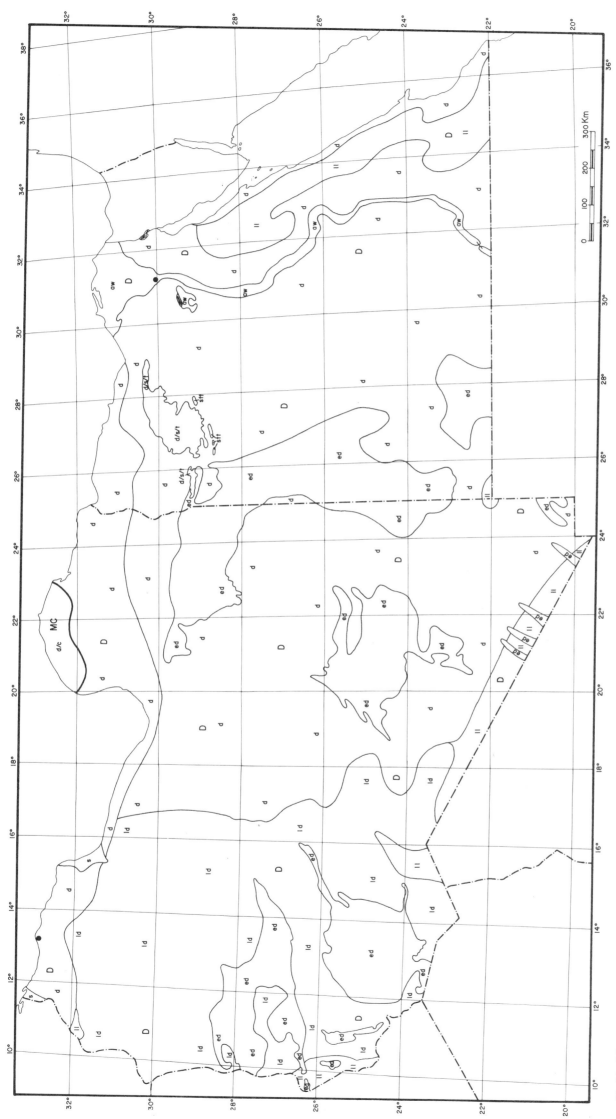

Fig.30. Libya and United Arab Republic. (For the meaning of symbols see country text, or Table VI, p.30, and Table IX, p.66.)

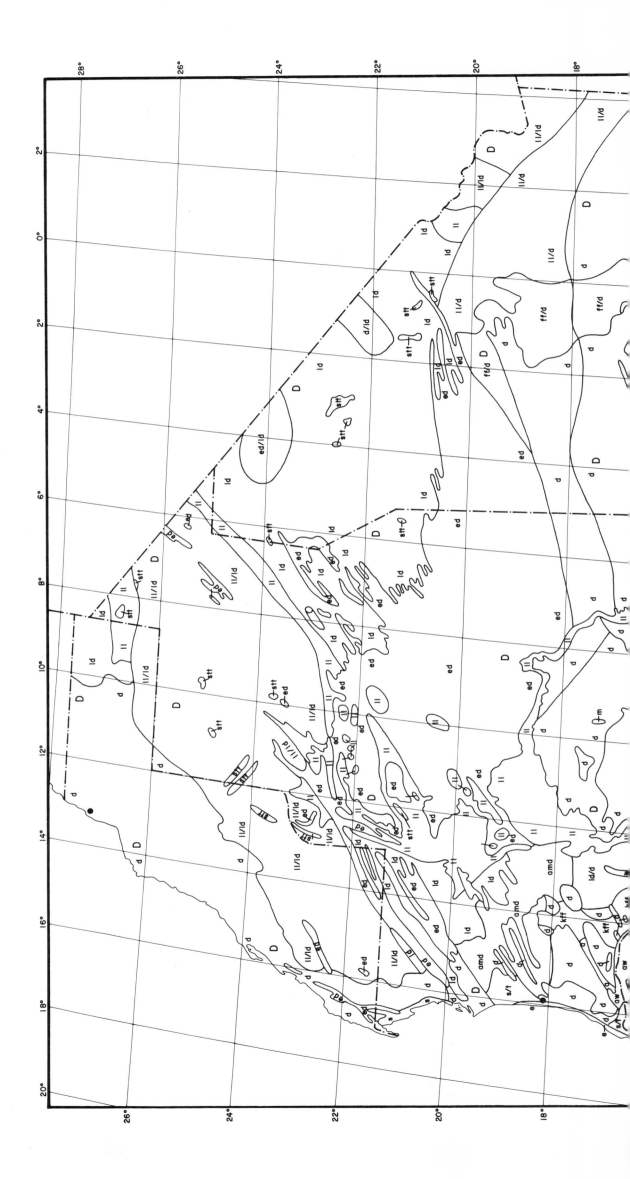

138

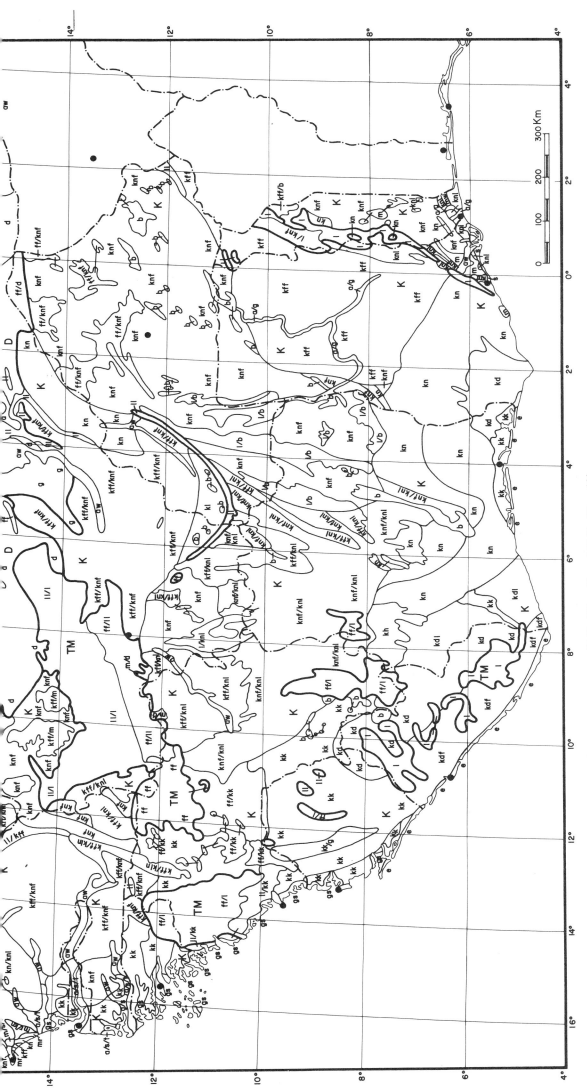

Fig.31 (pp.138, 139). West Africa. (For the meaning of symbols see country text, or Table VI, p.30, and Table IX, p.66.)

Sierra Leone (Fig.31). Sierra Leone belongs to only one broad soil region, which is kaolinitic (*K*); however, there are some tropical mountains (*TM*).

In the kaolinitic (*K*) region the principal soils are: *e* (aeolian sands); *g* (gleisolic); *gs* (saline gleisols—mangrove); *kd* (dystrophic kaolisols); *kk* (ferralitic). In the tropical mountains (*TM*) the principal soils are: *l* (lithosols); *ff* (lateritic rocks).

Mali (Fig.31). Mali can be divided into three broad soil regions: (*1*) desert (*D*) in the north; (*2*) kaolinitic (*K*) in the south; and (*3*) tropical mountains (*TM*).

(*1*) In the desert (*D*) region the principal soils are: *aw* (wet alluvial); *d* (arid brown and other raw soils of the desert); *ed* (desert sands); *ff* (lateritic rocks); *g* (gleisolic); *ld* (desert lithosols); *ll* (rock outcrops); *m* (dark clays).

(*2*) In the kaolinitic (*K*) region the most important soils are: *aw* (wet alluvial); *b* (recent brown); *d* (arid brown); *ff* (lateritic rock); *kff* (laterite); *kn* (eutrophic kaolisols); *knf* (tropical ferruginous); *knl* (terres de barre); *l* (lithosols); *ll* (rock outcrops); *m* (dark clays).

(*3*) In the tropical mountains (*TM*) the principal soils are: *ff* (lateritic rocks); *l* (lithosols); *ll* (rock outcrops).

A great part of the population lives near the rivers on non-kaolinitic soils (*aw, b, d, m*).

Liberia (Fig.31). Liberia falls into only one broad soil region, which is kaolinitic (*K*). The principal soils are: *b* (recent brown); *e* (aeolian sands); *kd* (dystrophic kaolisols); *kdf* (acid ferruginous); *l* (lithosols). The great majority of soils is dystrophic and this fact makes it difficult to grow cocoa; but these dystrophic soils are sufficiently fertile for rubber, oil palm and coconut.

Ivory Coast (Fig.31). The Ivory Coast can be included in one broad soil region, which is kaolinitic (*K*). The principal soils are: *b* (recent brown); *e* (aeolian sands); *ff* (lateritic rocks); *kdf* (acid ferruginous); *kdl* (dystrophic latosolic kaolisols); *kh* (rubrozems); *kk* (ferralitic); *kn* (eutrophic kaolisols); *knf* (tropical ferruginous); *knl* (terres de barre); *l* (lithosols).

The abundance of eutrophic soils (*kn, knf, knl, b*) in southern Ivory Coast (east and central) permitted the development of cocoa production.

Upper Volta (Fig.31). This country can be subdivided into three broad soil regions: (*1*) small desert (*D*) in the north; (*2*) extensive kaolinitic (*K*) in the south; and (*3*) tropical mountains (*TM*).

(*1*) In the desert region (*D*) the most important soils are: *d* (arid brown) and *ff* (lateritic rocks).

(*2*) In the kaolinitic region (*K*) the principal soils are: *aw* (wet alluvial); *b* (recent brown); *ff* (lateritic rocks); *kff* (laterites); *kl* (latosolic kaolisols); *knf* (tropical ferruginous); *knl* (terres de barre); *l* (lithosols).

(*3*) In the tropical mountains (*TM*) rock outcrops (*ll*) abound.

Ghana (Fig.31). This country consists of only one broad soil region, which is kaolinitic (*K*). The principal soils are: *a* (alluvial); *aw* (wet alluvial); *b* (recent brown); *g* (gleisolic); *kd* (dystrophic kaolisols); *kff* (laterite); *kk* (ferralitic); *knf* tropical ferruginous); *kn* (eutrophic kaolisols); *knl* (terres de barre); *l* (lithosols); *m* (dark clays); *s* (saline).

Eutrophic kaolisols (*knf, kn*) are concentrated in the cocoa producing region; this crop cannot be grown in dystrophic soils; and the existence of abundant eutrophic soils under an equatorial moderately humid climate is the reason Ghana became one of the leading cocoa producing countries in the world. Dystrophic kaolisols (*kd*) abound in the southwest. Laterites (*kff*) in the Volta Basin, which is a region very scarcely populated. Non-kaolisols (*g, a, aw, s*) abound in the lower Volta plains.

Togo (Fig.31). Togo consists of two broad soil regions: (*1*) kaolinitic (*K*); and (*2*) tropical mountains (*TM*).

(*1*) In the kaolinitic (*K*) region the principal soils are: *a* (alluvial); *b* (recent brown); *g* (gleisolic); *kff* (laterites); *kn* (eutrophic kaolisols); *knf* (tropical ferruginous); *knl* (terres de barre); *m* (dark clays); *mw* (wet clays).

(*2*) In the tropical mountains (*TM*) the principal soils are: *knf* (tropical ferruginous); *l* (lithosols). Terres de barre (*knl*) are situated near the coast and they sustain a dense population. The eutrophic soils (*knf*, etc.) of the mountains permitted the development of an important cocoa production.

Nigeria, Niger, Dahomey, Cameroons, Chad and Central African Republic

Nigeria (Fig.32). Nigeria falls mainly into a kaolinitic (*K*) region. But there is a small desert region (*D*) near Lake Chad; and some tropical mountains (*TM*).

(*1*) In the kaolinitic region (*K*) the principal soils are: *a* (alluvial); *aw* (wet alluvial); *ag* (gleisolic alluvial); *b* (recent brown); *d* (arid brown); *e* (aeolian sands); *ff* (laterite rocks); *g* (gleisolic); *gs* (saline gleisols—mangrove); *kdl* (dystrophic latosolic kaolisols); *kn* (eutrophic kaolisols); *knf* (tropical ferruginous); *knl* (terres de barre); *l* (lithosols); *m* (dark clays); *mr* (rendzina clays); *mw* (wet clays); *v* (ando); *vh* (humus rich ando); terres de barre are very extensive in southern Nigeria and they sustain a very dense population. Moreover Nigeria has many alluvial soils, some dark clays and a few ando. All that explains the high population density of the country.

(*2*) In the desert region (*D*) the principal soils are: *a* (alluvial); *g* (gleisolic); *m* (dark clays); *mw* (wet clays); *oh* (lenist); *s* (saline); *t* (solonetz).

(*3*) In the tropical mountains (*TM*) the principal soils are: *b* (recent brown); *knf* (tropical ferruginous); *l* (lithosols).

Niger (Fig.32). This country can be divided into two broad soil regions: (*1*) desert (*D*) in the north; (*2*) kaolinitic (*K*) in the south; and (*3*) a small area of some tropical mountains (*TM*).

(*1*) In the desert region (*D*) the principal soils are: *a* (alluvial); *aw* (wet alluvial); *d* (arid brown and other raw soils of the desert); *ed* (desert sands); *ff* (laterite rock); *g* (gleisolic); *kff* (laterite); *ld* (desert lithosols); *ll* (rock outcrops); *md* (desert clays); *r* (rendzinas); *s* (saline).

(*2*) In the kaolinitic (*K*) region the principal soils are: *a* (alluvial); *aw* (wet alluvial); *b* (recent brown); *d* (arid brown); *g* (gleisolic); *ff* (laterite rocks); *kff* (laterite); *kn* (eutrophic kaolisols); *knf* (tropical ferruginous); *knl* (terres de barre); *l* (lithosols); *ll* (rock outcrops); *mw* (wet clays); *r* (rendzinas).

(*3*) In the tropical mountains (*TM*) the most important soils are lithosols (*l*).

Dahomey (Fig.32). This country falls in only one broad soil region, which is kaolinitic (*K*); although there are some tropical mountains (*TM*).

In the kaolinitic (*K*) region the principal soils are: *a* (alluvial); *aw* (wet alluvial); *ff* (laterite rock); *g* (gleisolic); *kff* (laterite); *knf* (tropical ferruginous); *knl* (terres de barre); *l* (lithosols); *m* (dark clays); *mw* (wet clays). Terres de barre (*knl*) of the coast of Dahomey are the best kaolinitic soils; and the greater part of Dahomey population is concentrated on them.

In the tropical mountains (*TM*) the principal soils are lithosols (*l*).

Cameroons (Fig.32). Cameroons can be divided into five broad soil regions: (*1*) desert (*D*) in the extreme north; (*2*) kaolinitic (*K*) in the south; (*3*) kaolinitic–ando (*K–A*) in the west; the mountains can be further subdivided: (*4*) tropical mountains (*TM*), and (*5*) tropical mountains–ando (*TM–A*).

(*1*) In the desert (*D*) region the principal soils are: *m* (dark clays); *mw* (wet clays); *oh* (lenists); *s* (saline); *t* (solonetz).

(*2*) In the kaolinitic region (*K*) the principal soils are: *g* (gleisolic); *kdl* (latosolic dystrophic kaolisols); *kff* (laterites); *kk* (ferralitic); *kl* (latosolic kaolisols); *kn* (eutrophic kaolisols); *knf* (tropical ferruginous); *knl* (terres de barre); *l* (lithosols); *m* (dark clays); *oh* (lenists).

(*3*) In the kaolinitic–ando (*K–A*) region the principal soils are: *v* (ando); and *vh* (humus rich ando).

(*4*) In the tropical mountains–ando (*TM–A*) region the principal soils are: *kh* (rubrozems); *kl* (latosolic kaolisols); *l* (lithosols) and *vl* (lithosolic ando).

(*5*) In the tropical mountains (*TM*) the principal soils are: *mw* (wet clays) and *l* (lithosols).

Cameroons has many ando soils; the maps show only a part of them; a careful study of their cation exchange capacity, moisture tensions, apparent density, etc., is necessary. These volcanic soils are fertile, and on them the greater part of the population is concentrated.

Fernando Póo (Fig.32). This area has soils of volcanic origin (*v*—ando). The high production of the island is to a considerable extent due to this fact.

Chad (Fig.32) includes two broad soil regions: (*1*) desert (*D*) in the north; (*2*) kaolinitic (*K*) in the south; and (*3*) there is a small area of dry mountains (*DM*).

(*1*) In the desert (*D*) region the most important soils are: *a* (alluvial); *d* (arid brown and other raw soils of the desert); *ed* (desert sands); *l* (lithosols); *ld* (desert lithosols); *md* (desert clays); *mw* (wet clays); *oh* (lenists); *s* (saline); *t* (solonetz).

(*2*) In the kaolinitic (*K*) region the principal soils are: *a* (alluvial); *g* (gleisolic); *kff* (laterites); *kl* (latosolic kaolisols); *kn* (eutrophic kaolisols); *ll* (rock outcrops); *m* (dark clays); *mw* (wet clays); *s* (saline); *t* (solonetz).

(*3*) In the tropical mountains (*TM*) the principal soils are: *a* (alluvial); *g* (gleisolic); *l* (lithosols).

The soils of the desert region (*D*) have good mineral fertility. Concerning those of the kaolinitic region little is known concerning their cation exchange capacity/clay ratio and their absorbed cations/clay ratio; that is why their classification is incomplete.

Central African Republic (Fig.32). This area includes one broad soil region, kaolinitic (*K*); and there are also a few tropical mountains (*TM*).

(*1*) In the kaolinitic (*K*) region, the principal soils are: *a* (alluvial); *g* (gleisolic); *kff* (laterite); *kk* (ferralitic); *kl* (latosolic kaolisols); *kn* (eutrophic kaolisols); *knf* (tropical ferruginous); *knl* (terres de barre); *l* (lithosols).

(*2*) In the tropical mountains (*TM*) chiefly lithosols are encountered.

Ethiopia, Somalia, Sudan

Ethiopia (Fig.33). Ethiopia consists of three kinds of broad soil regions: (*1*) desert (*D*) in the north, east and southwest; (*2*) kaolinitic–ando (*K–A*) in the interior high valleys and plateaux; and (*3*) tropical mountains (*TM*).

(*1*) In the desert (*D*) region the principal soils are: *a* (alluvial); *aw* (wet alluvial); *d* (arid brown and other raw soils of the desert); *knf* (tropical ferruginous); *l* (lithosols); *ll* (rock outcrops); *mw* (wet clays); *s* (saline).

(*2*) In the kaolinitic–ando (*K–A*) region the principal soils are: *a* (alluvial); *kh* (rubrozems); *m* (dark clays); *mw* (wet clays); *o* (peaty).

(*3*) In the tropical mountains (*TM*) the principal soils are: *kh* (rubrozems); *l* (lithosols); *m* (dark clays).

The soils of Ethiopia are mostly good; their mineral fertility is high (soils of the desert, dark clays), and/or they are rich in organic matter (rubrozems). Fertilizers can considerably increase their yields when climate is favourable for the crop.

French Somaliland (Fig.33). This country falls into one broad soil region: *D* (desert); the principal soils are: *l*

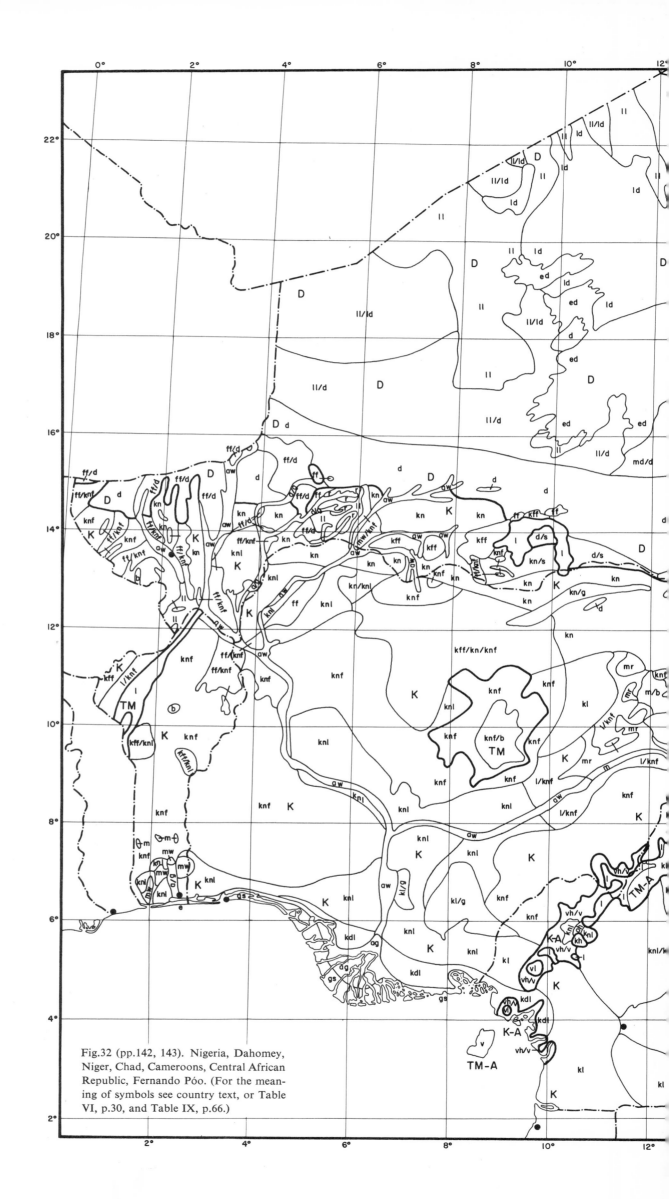

Fig.32 (pp.142, 143). Nigeria, Dahomey, Niger, Chad, Cameroons, Central African Republic, Fernando Póo. (For the meaning of symbols see country text, or Table VI, p.30, and Table IX, p.66.)

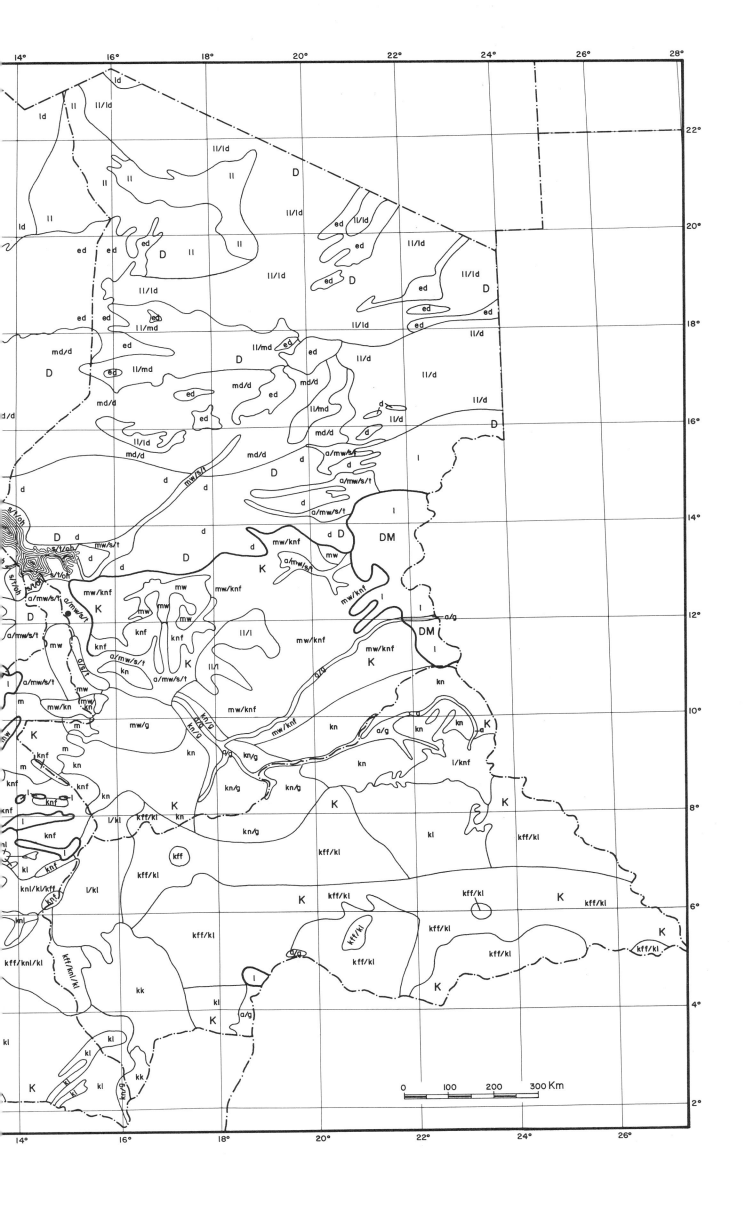

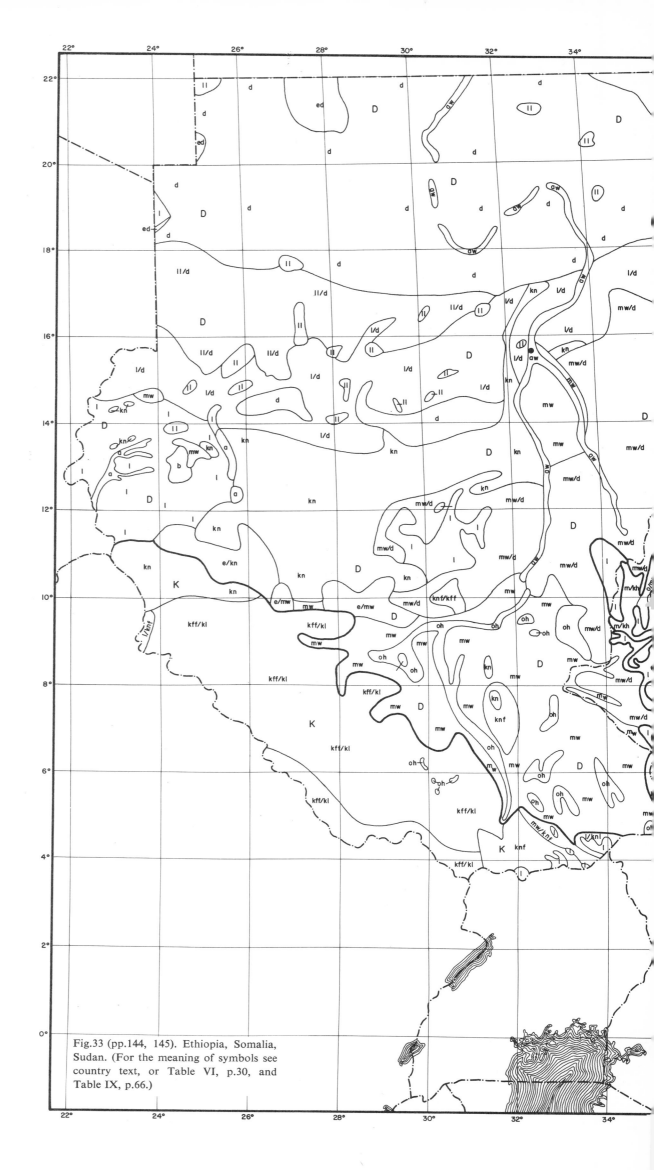

Fig.33 (pp.144, 145). Ethiopia, Somalia, Sudan. (For the meaning of symbols see country text, or Table VI, p.30, and Table IX, p.66.)

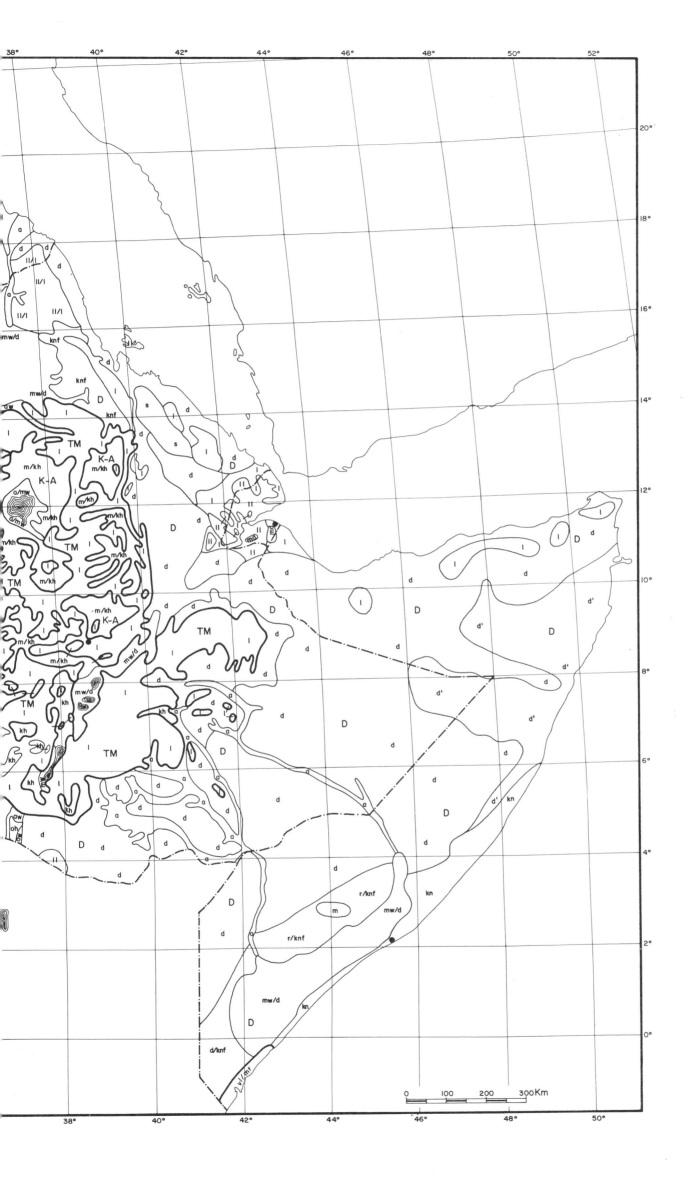

(lithosols); *ll* (rock outcrops); *md* (desert clays); the mineral fertility of all these soils is good.

Somalia (Fig.33). Somalia falls into only one broad soil region: *D* (desert); but on the southern coast the kaolinitic–ando region of Kenya (*K–A*) extends into Somalia.

The principal soils of the desert (*D*) region are: *d* (arid brown and other raw soils of the desert); *d'* (gypsisols); *kn* (eutrophic kaolisols); *knf* (tropical ferruginous); *l* (lithosols); *m* (dark clays); *mw* (wet clays); *r* (rendzinas). The principal soils of the kaolinitic–ando (*K–A*) region are: *v* (ando) and *mr* (rendzina clays).

The mineral fertility of Somalia soils is, in general, good. Naturally fertilizers increase yields considerably when irrigation is carried out or rainfall is sufficient.

Sudan (Fig.33). Sudan consists of two broad soil regions: (*1*) desert (*D*) in the north, and (*2*) kaolinitic (*K*) in the south.

(*1*) The principal soils of the desert (*D*) region are: *a* (alluvial); *aw* (wet alluvial); *b* (recent brown); *d* (arid brown and other raw soils of the desert); *e* (aeolian sands); *ed* (desert sands); *kff* (laterite); *kn* (eutrophic kaolisols); *knf* (tropical ferruginous); *l* (lithosols); *ll* (rock outcrops); *mw* (wet clays); *oh* (lenists); *vl* (lithosolic ando).

(*2*) The principal soils of the kaolinitic region (*K*) are: *kff* (laterite); *kn* (eutrophic kaolisols); *knf* (tropical ferruginous); *knl* (terres de barre); *kl* (latosolic kaolisols); *l* (lithosols); *mw* (wet clays); *oh* (lenists).

Although the climate of Sudan is dry, kaolinitic soils extend into the desert region; and laterites are frequent. The soils of Sudan are not as good as it could be expected from the climate, but from an agronomic point of view the important soils are those near the rivers which can be irrigated or are flooded; and these soils are good (not kaolinitic).

East Africa

Kenya (Fig.34). Kenya includes five broad soil regions: (*1*) desert (*D*); (*2*) kaolinitic (*K*); (*3*) ando–kaolinitic (*A–K*); (*4*) tropical mountains (*TM*), and (*5*) tropical mountains–ando (*TM–A*).

(*1*) In the desert region (*D*) the principal soils are: *a* (alluvial); *d* (arid brown and other raw soils of the desert); *knf* (tropical ferruginous); *ll* (rock outcrops); *s* (saline).

(*2*) In the kaolinitic region (*K*) the principal soils are: *a* (alluvial); *d* (arid brown); *e* (aeolian sands); *g* (gleisolic); *kff* (laterites); *kl* (latosolic kaolisols); *kn* (eutrophic kaolisols); *knf* (tropical ferruginous); *knl* (terres de barre); *l* (lithosols); *mw* (wet clays); *mr* (rendzina clays); *t* (solonetz).

(*3*) In the ando–kaolinitic region (*A–K*) the principal soils are: *a* (alluvial); *g* (gleisolic); *kh* (rubrozems); *knf* (tropical ferruginous); *kng* (gleisolic eutrophic kaolisols); *mw* (wet clays); *ve* (aeolian ando); *vn* (eutrophic ando).

(*4*) In the tropical mountains (*TM*) the principal soils are: *ag* (gleisolic alluvial); *d* (arid brown); *ff* (lateritic rocks); *kh* (rubrozems); *knf* (tropical ferruginous); *l* (lithosols); *lh* (lithosolic rankers); *ll* (rock outcrops); *mw* (wet clays); *ve* (aeolian ando).

(*5*) In the tropical mountains–ando region (*TM–A*) the principal soils are: *knf* (tropical ferruginous); *l* (lithosols); *ll* (rock outcrops); *o* (peaty); *ve* (aeolian ando); *vn* (eutrophic ando).

Due to the presence of many ando soils, many non-kaolinitic, and to the fact that the greater part of kaolinitic soils is eutrophic, the mineral fertility of Kenya soils is in general good.

Uganda (Fig.34). Uganda can be divided into three broad soil regions: (*1*) kaolinitic (*K*); (*2*) tropical mountains (*TM*); and (*3*) tropical mountains–ando (*TM–A*).

(*1*) In the kaolinitic region (*K*) the principal soils are: *a* (alluvial); *d* (arid brown); *g* (gleisolic); *kh* (rubrozems); *kff* (laterites); *kk* (ferralitic); *kl* (lithosolic kaolisols); *kng* (gleisolic eutrophic kaolisols); *knl* (terres de barre); *l* (lithosols); *mw* (wet clays).

(*2*) In the tropical mountains (*TM*) the principal soils are: *ff* (lateritic rocks); *l* (lithosols); *ll* (rock outcrops).

(*3*) In the tropical mountains–ando (*TM–A*) the principal soils are: *kh* (rubrozems); *kl* (latosolic kaolisols); *knl* (terres de barre); *l* (lithosols); *vl* (lithosolic ando); *vn* (eutrophic ando).

Due to the frequency of ando soils, perhaps greater than shown in the map, the mineral fertility of Uganda soils is in general good.

Rwanda and Burundi (Fig.34). This area consists of two broad soil regions: (*1*) kaolinitic (*K*); and (*2*) tropical mountains–ando (*TM–A*).

(*1*) In the kaolinitic region (*K*) the principal soils are: *kff* (laterites); *kh* (rubrozems); *kl* (latosolic kaolisols); *kn* (eutrophic kaolisols); *knl* (terres de barre); *mw* (wet clays).

(*2*) In the tropical mountains–ando region (*TM–A*) the principal soils are: *b* (recent brown); *kh* (rubrozems); *knl* (terres de barre); *l* (lithosols); *mw* (wet clays); *s* (saline); *vl* (lithosolic ando); *vn* (eutrophic ando).

Due to the great frequency of ando soils, mineral fertility is usually good in Rwanda-Burundi soils.

Tanzania (Fig.34). Here four broad soil regions are in evidence: (*1*) kaolinitic (*K*); (*2*) ando–kaolinitic (*K–A*); (*3*) tropical mountains-kaolinitic (*TM–K*); and (*4*) tropical mountains–ando (*TM–A*).

(*1*) In the kaolinitic (*K*) region the principal soils are: *a* (alluvial); *d* (arid brown); *g* (gleisolic); *k* (kaolisols); *kff* (laterites); *kk* (ferralitic); *kl* (latosolic kaolisols); *kn* (eutrophic kaolisols); *knf* (tropical ferruginous); *l* (lithosols); *mr* (rendzina clays); *mw* (wet clays); *oh* (lenists); *r* (rendzinas); *s* (saline); *t* (solonetz); *vn* (eutrophic ando).

(*2*) In the ando–kaolinitic region (*A–K*) the principal soils are: *knf* (tropical ferruginous); *knl* (terres de

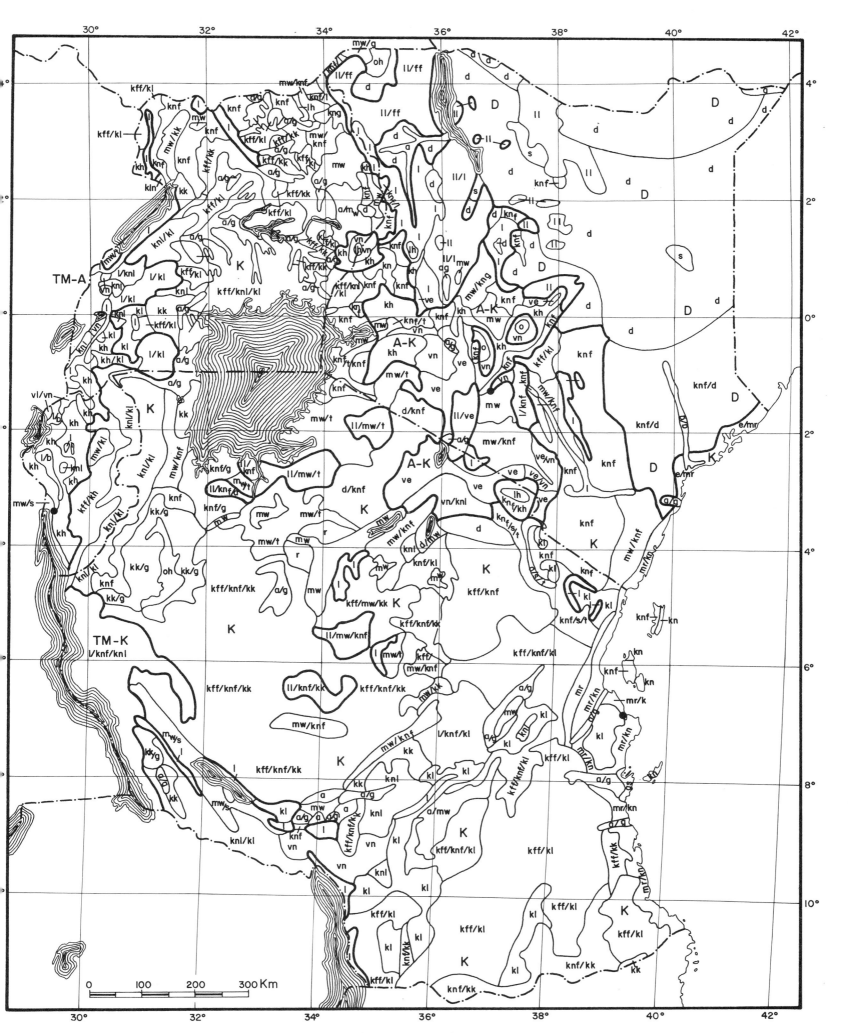

Fig.34. East Africa. (For the meaning of symbols see country text, or Table VI, p.30, and Table IX, p.66.)

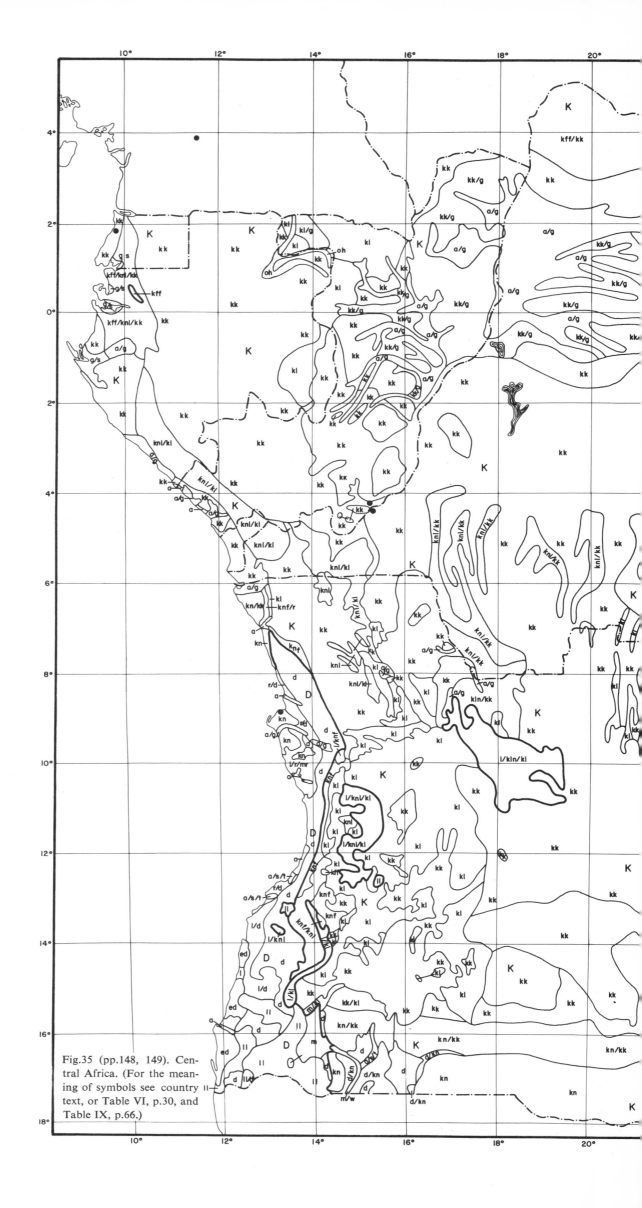

Fig.35 (pp.148, 149). Central Africa. (For the meaning of symbols see country text, or Table VI, p.30, and Table IX, p.66.)

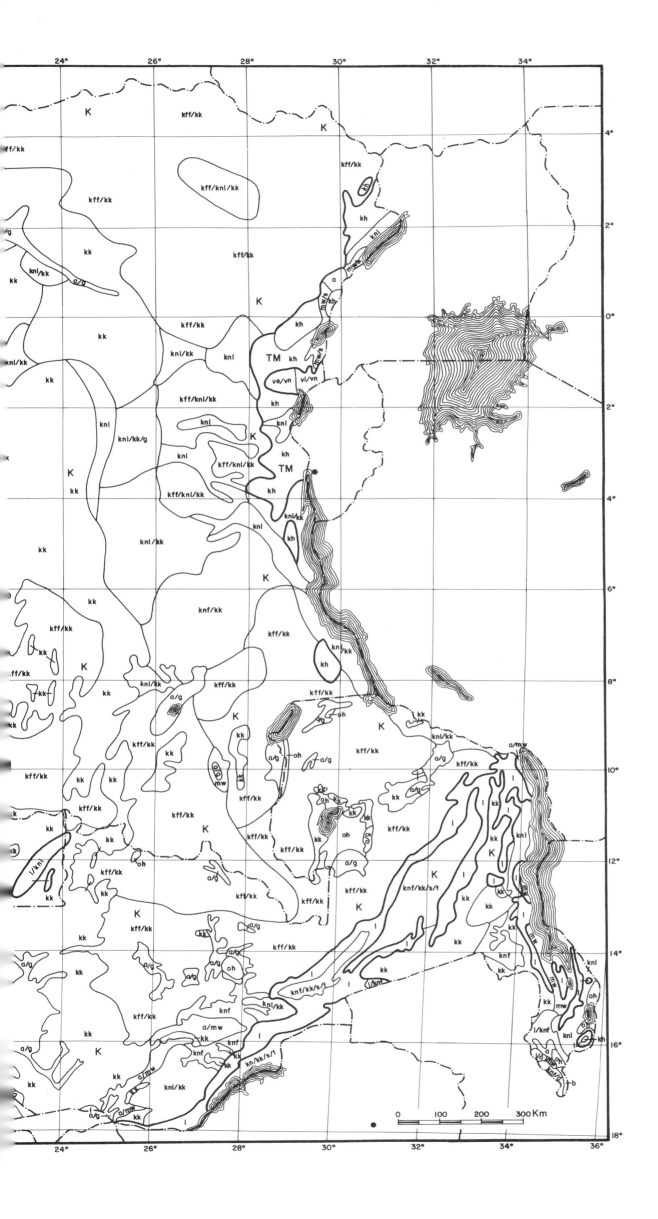

barre); *mw* (wet clays); *ve* (aeolian ando); *vn* (eutrophic ando).

(*3*) In the tropical mountains–kaolinitic region (*TM–K*) the principal soils are: *kk* (ferralitic); *kl* (latosolic kaolisols); *knf* (tropical ferruginous); *knl* (terres de barre); *l* (lithosols); *ll* (rock outcrops); *mw* (wet clays); *t* (solonetz).

(*4*) In the tropical mountains–ando region (*TM–A*) the principal soils are: *kh* (rubrozems); *knf* (tropical ferruginous); *lh* (lithosolic rankers).

The extension of ando soils is perhaps greater than shown in the maps. Due to the presence of ando and non-kaolinitic soils and to the fact that kaolisols are seldom dystrophic the mineral fertility of Tanzania soils is not bad.

Central Africa

Congo (Kinshasa) (Fig.35). Here two broad soil regions are in existence: (*1*) kaolinitic (*K*); and (*2*) tropical mountains (*TM*).

(*1*) In the kaolinitic (*K*) region the principal soils are: *a* (alluvial); *g* (gleisolic); *kff* (laterites); *kk* (ferralitic); *kl* (latosolic kaolisols); *knf* (tropical ferruginous); *knl* (terres de barre); *mw* (wet clays). This region contains many ferralitic (*kk*) soils of low fertility; but there are also some terres de barre (*knl*), called "ferrisols" by the Belgians.

(*2*) In the tropical mountains (*TM*) the principal soils are: *a* (alluvial); *kl* (rubrozems); *knl* (terres de barre); *mw* (wet clays); *s* (saline); *ve* (aeolian ando); *vl* (lithosolic ando); *vn* (eutrophic ando).

In the mountains the good soils (ando, terres de barre, dark clays, alluvial, rubrozems, etc.) abound; these regions might be classified as tropical mountains–ando.

Congo (Brazzaville) (Fig.35). This area has only one broad soil region: kaolinitic (*K*). Its principal soils are: *a* (alluvial); *g* (gleisolic); *kk* (ferralitic); *kl* (latosolic kaolisols); *knl* (terres de barre). The author lacks sufficient information concerning the base status (eutrophic–dystrophic) of the soils of this country.

Gabon (Fig.35). Here there is one broad soil region, which is kaolinitic (*K*), although there are a few tropical mountains (*TM*).

(*1*) In the kaolinitic region (*K*) the principal soils are: *a* (alluvial); *g* (gleisolic); *gs* (saline gleisols—mangrove); *kff* (laterite); *kk* (ferralitic); *knl* (terres de barre); *oh* (lenists); *s* (saline).

(*2*) In the tropical mountains (*TM*) there are many laterites (*kff*) and naturally lithosols.

Spanish Guinea (Fig.35). This country falls in a kaolinitic (*K*) region. The principal soils are: *gs* (saline gleisols—mangrove); *kk* (ferralitic).

Angola (Fig.35). Angola can be divided into three broad soil regions: (*1*) desert (*D*); (*2*) kaolinitic (*K*) and (*3*) tropical mountains (*TM*).

(*1*) In the desert region (*D*) the principal soils are: *a* (alluvial); *d* (arid brown); *ed* (desert sands); *g* (gleisolic); *kn* (eutrophic kaolisols); *l* (lithosols); *ll* (rock outcrops); *m* (dark clays); *mr* (rendzina clays); *r* (rendzinas); *s* (saline); *t* (solonetz).

(*2*) In the kaolinitic region (*K*) the principal soils are: *a* (alluvial); *d* (arid brown); *g* (gleisolic); *kff* (laterite); *kk* (ferralitic); *kn* (eutrophic kaolisols); *knf* (tropical ferruginous); *knl* (terres de barre); *kl* (latosolic kaolisols); *l* (lithosols); *m* (dark clays); *r* (rendzinas); *s* (saline); *t* (solonetz).

(*3*) In the tropical mountains (*TM*) the principal soils are: *kl* (latosolic kaolisols); *knl* (terres de barre); *l* (lithosols); *ll* (rock outcrops).

Zambia (Fig.35). Zambia lies in two broad soil regions: (*1*) kaolinitic (*K*) and (*2*) the tropical mountains (*TM*).

(*1*) In the kaolinitic region (*K*) the principal soils are: *a* (alluvial); *g* (gleisolic); *kff* (laterite); *kk* (ferralitic); *knf* (tropical ferruginous); *knl* (terres de barre); *mw* (wet clays); *oh* (lenists).

(*2*) In the tropical mountains (*TM*) the principal soils are: *kk* (ferralitic); *knf* (tropical ferruginous); *l* (lithosols); *s* (saline); and *t* (solonetz).

Malawi (Fig.35). Two broad soil regions are in existence here: (*1*) kaolinitic (*K*); and (*2*) tropical mountains (*TM*).

(*1*) In the kaolinitic region (*K*) the principal soils are: *a* (alluvial); *b* (recent brown); *d* (arid brown); *kk* (ferralitic); *knf* (tropical ferruginous); *l* (lithosols); *oh* (lenists).

(*2*) In the tropical mountains (*TM*) the principal soils are: *kh* (rubrozems) and *l* (lithosols).

Mozambique and Madagascar

Mozambique (Fig.36). Mozambique consists of four broad soil regions: (*1*) arid cinnamonic (*AC*); (*2*) kaolinitic (*K*); (*3*) tropical mountains (*TM*); and (*4*) dry mountains (*DM*).

(*1*) In the arid cinnamonic region (*AC*) the principal soils are: *a* (alluvial); *d* (arid brown); *g* (gleisolic); and *m* (dark clays).

(*2*) In the kaolinitic (*K*) region the principal soils are: *a* (alluvial); *d* (arid brown); *e* (aeolian sands); *g* (gleisolic); *gs* (saline gleisols–mangrove); *kff* (laterites); *kk* (ferralitic); *kl* (latosolic kaolisols); *kn* (eutrophic kaolisols); *knf* (tropical ferruginous); *knl* (terres de barre); *mr* (rendzina clays); *mw* (wet clays); *o* (peaty); *oh* (lenists).

(*3*) In the dry mountains (*DM*) the principal soils

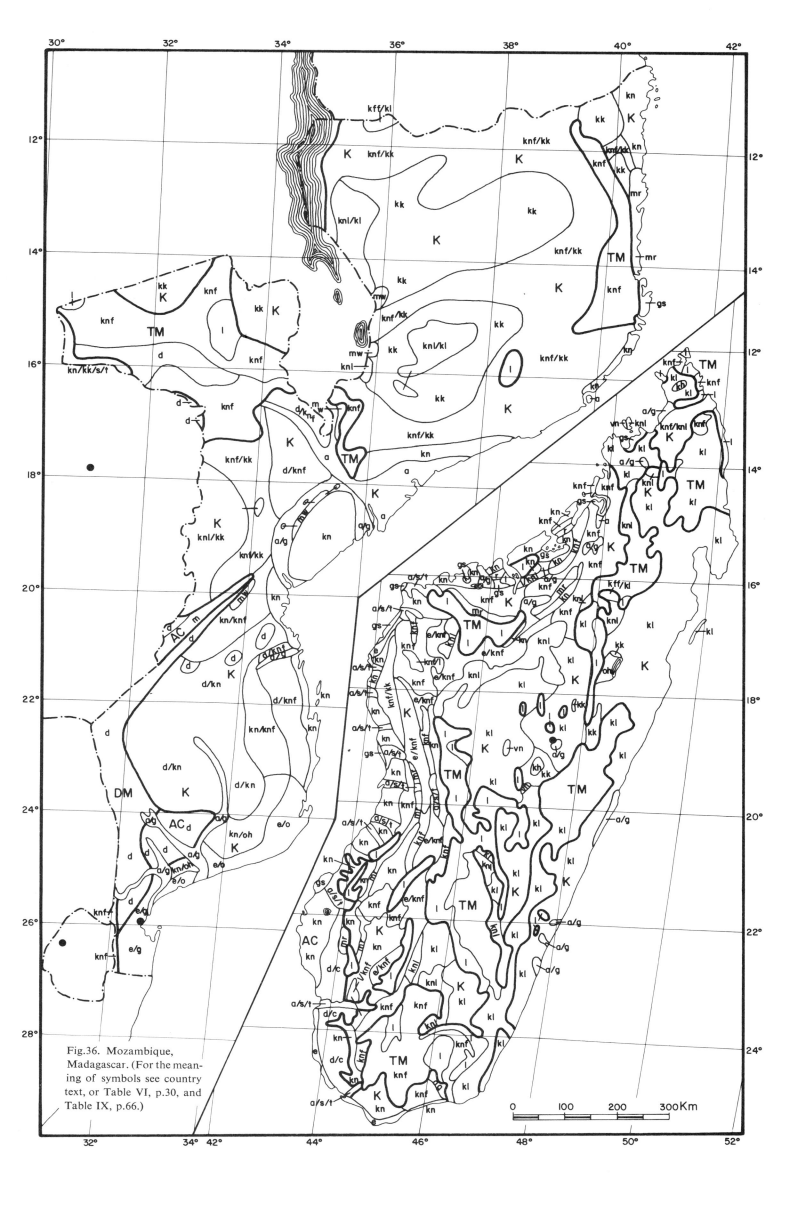

Fig.36. Mozambique,
Madagascar. (For the meaning of symbols see country
text, or Table VI, p.30, and
Table IX, p.66.)

0   100   200   300 Km

are: *a* (alluvial); *d* (arid brown); and *g* (gleisolic); lithosols abound.

   (*4*) In the tropical mountains (*TM*) the principal soils are: *d* (arid brown); *knf* (tropical ferruginous); and *l* (lithosols).

   Madagascar (Fig.36). Here three broad soil regions are in evidence: (*1*) arid cinnamonic (*AC*); (*2*) kaolinitic (*K*); and (*3*) tropical mountains (*TM*).

   (*1*) In the arid cinnamonic region (*AC*) the principal soils are: *a* (alluvial); *c* (cinnamonic); *d* (arid brown); *e* (aeolian sands); *kn* (eutrophic kaolisols); *mr* (rendzina clays); *s* (saline); *t* (solonetz).

   (*2*) In the kaolinitic region (*K*) the principal soils are: *a* (alluvial); *ag* (gleisolic alluvial); *e* (aeolian sands); *g* (gleisolic); *gs* (saline gleisols (mangrove); *kff* (laterites); *kh* (rubrozems); *kk* (ferralitic); *kl* (latosolic kaolisols); *kn* (eutrophic kaolisols); *knf* (tropical ferruginous); *knl* (terres de barre); *l* (lithosols); *mr* (rendzina clays); *oh* (lenists); *r* (rendzinas); *s* (saline); *t* (solonetz); *vn* (eutrophic ando).

   (*3*) In the tropical mountains (*TM*) the principal soils are: *a* (alluvial); *g* (gleisolic); *kl* (latosolic kaolisols); *knf* (tropical ferruginous); *knl* (terres de barre); *l* (lithosols).

Southern Africa

   South Africa (Fig.37). South Africa includes seven broad soil regions: (*1*) desert (*D*); (*2*) arid cinnamonic (*AC*); (*3*) mediterranean–cinnamonic (*MC*); (*4*) kaolinitic (*K*); (*5*) dry mountains (*DM*); (*6*) tropical mountains (*TM*); and (*7*) mediterranean mountains–kaolinitic (*M'–K*).

   (*1*) In the desert (*D*) region, the principal soils are: *d* (arid brown and other raw soils of the desert); *e* (aeolian sands); *kff* (laterites); *kn* (eutrophic kaolisols); *knf* (tropical ferruginous); *l* (lithosols); *mw* (wet clays); *s* (saline); *t* (solonetz).

   (*2*) In the arid cinnamonic (*AC*) region the principal soils are: *d* (arid brown).

   (*3*) In the mediterranean–cinnamonic region (*MC*) the principal soils are: *e* (aeolian sands); *m* (dark clays); *r* (rendzinas); *s* (saline).

   (*4*) In the kaolinitic region (*K*) the principal soils are: *d* (arid brown); *e* (aeolian sands); *g* (gleisolic); *kh* (rubrozems); *knf* (tropical ferruginous); *l* (lithosols); *ll* (rock outcrops); *m* (dark clays); *r* (rendzinas); *t* (solonetz); *u* (planosols).

   (*5*) In the dry mountains (*DM*) arid brown (*d*) and lithosols prevail.

   (*6*) In the tropical mountains (*TM*) the principal soils are: *knf* (tropical ferruginous); *m* (dark clays); and *u* (planosols).

   (*7*) In the mediterranean mountains–kaolinitic region (*M'–K*) the principal soils are: *cb* (brown cinna-

monic); *d* (arid brown); *knf* (tropical ferruginous); *l* (lithosols); *ll* (rock outcrops).

   Basutoland (Fig.37). There is only one broad soil region in existence: tropical mountains (*TM*). The principal soils are: *m* (dark clays); *u* (planosols); and naturally *l* (lithosols).

   Swaziland (Fig.37). There is only one broad soil region in existence, and that is kaolinitic (*K*). The principal soils are: *d* (arid brown); *kh* (rubrozems, perhaps not so acid); *knf* (tropical ferruginous).

   Bechuanaland (Fig.37). This country is composed of four broad soil regions: (*1*) desert (*D*); (*2*) arid cinnamonic (*AC*); (*3*) kaolinitic (*K*); and (*4*) tropical mountains (*TM*).

   (*1*) In the desert region (*D*) the principal soils are: *a* (alluvial); *d* (arid brown); *g* (gleisolic); *l* (lithosols); *m* (dark clays); *s* (saline); and *t* (solonetz).

   (*2*) In the arid cinnamonic (*AC*) region, arid brown (*d*) soils abound.

   (*3*) In the kaolinitic region (*K*) the principal soils are: *a* (alluvial); *d* (arid brown); *g* (gleisolic); *kk* (ferralitic); *kn* (eutrophic kaolisols); *knf* (tropical ferruginous); *mr* (rendzina clays); *s* (saline); *t* (solonetz).

   (*4*) In the tropical mountains (*TM*) the principal soils are arid brown (*d*) and lithosols (*l*).

   Rhodesia (Fig.37). This can be divided into four broad soil regions: (*1*) arid cinnamonic (*AC*); (*2*) kaolinitic (*K*); (*3*) dry mountains (*DM*); and (*4*) tropical mountains (*TM*).

   (*1*) In the arid cinnamonic (*AC*) region the principal soils are: *d* (arid brown); *knf* (tropical ferruginous); *m* (dark clays); *t* (solonetz).

   (*2*) In the kaolinitic (*K*) region the principal soils are: *d* (arid brown); *g* (gleisolic); *kk* (ferralitic); *kl* (latosolic kaolisols); *knf* (tropical ferruginous); *knl* (terres de barre); *l* (lithosols); *ll* (rock outcrops); *m* (dark clays); *mr* (rendzina clays); *r* (rendzinas); *s* (saline); *t* (solonetz).

   (*3*) In the dry mountains (*DM*) arid brown (*d*), and naturally lithosols, prevail.

   (*4*) In the tropical mountains (*TM*) *l* (lithosols) and *ll* (rock outcrops) abound.

   South West Africa (Fig.37). This area includes two broad soil regions: (*1*) desert (*D*); and (*2*) kaolinitic (*K*); the kaolinitic occupies a small area in the northeast.

   (*1*) In the desert region (*D*) the principal soils are: *a* (alluvial); *d* (arid brown); *e* (aeolian sands); *g* (gleisolic); *kff* (laterites); *l* (lithosols); *ll* (rock outcrops); *s* (saline); *t* (solonetz).

   (*2*) In the kaolinitic region (*K*) the principal soils are: *a* (alluvial); *d* (arid brown); *g* (gleisolic); *kk* (ferralitic); *kn* (eutrophic kaolisols).

Fig.37. Southern Africa.
(For the meaning of symbols
see country text, or Table
VI, p.30, and Table IX,
p.66.)

153

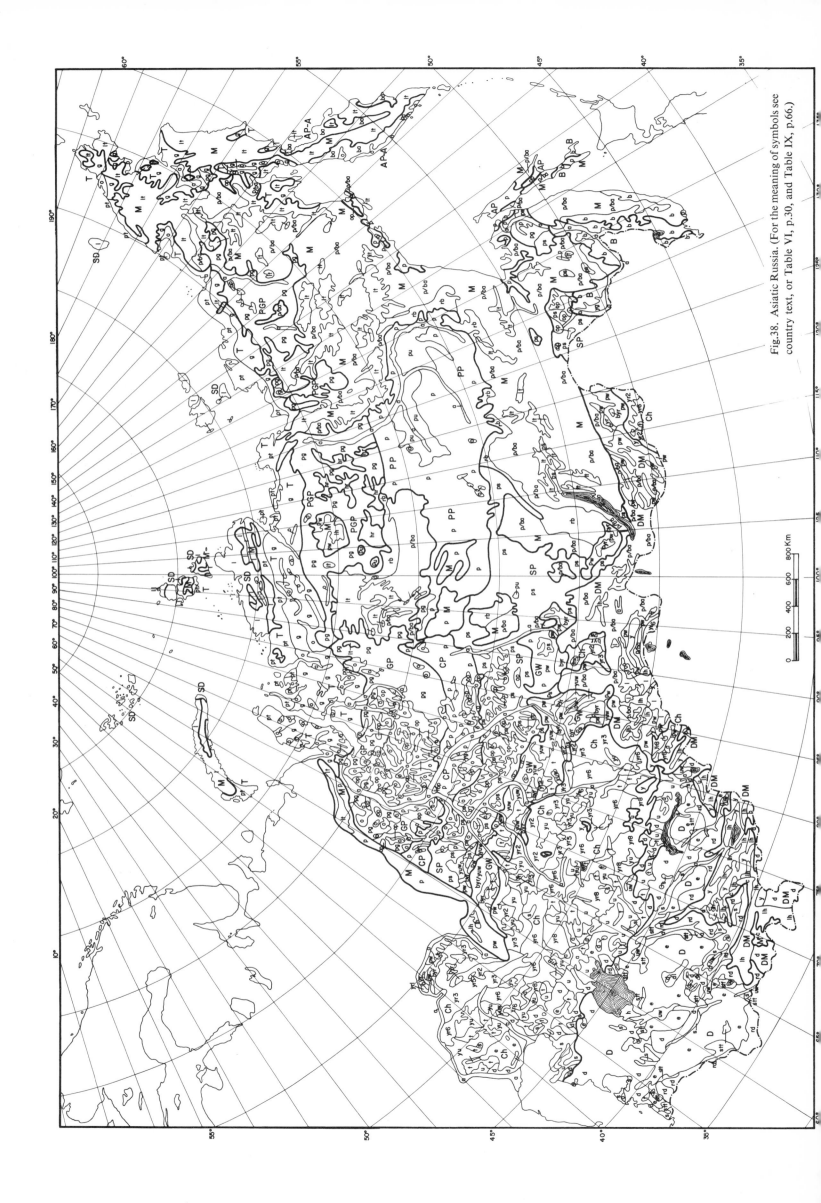

Fig.38. Asiatic Russia. (For the meaning of symbols see country text, or Table VI, p.30, and Table IX, p.66.)

ASIA

Asiatic Russia (Fig.38)

Asiatic Russia includes a great variety of broad soil regions: (*1*) sub-glacial desert (*SD*) and (*2*) tundra (*T*) in the north; (*3*) gley podsol (*GP*), (*4*) permafrost gley podsol (*PGP*), (*5*) continental podsol (*CP*), (*6*) permafrost podsol (*PP*), (*7*) sod podsol (*SP*), and (*8*) atlantic podsol–ando (*AP–A*) in the center; (*9*) grey-wooded (*GW*), (*10*) chernozemic (*Ch*), and (*11*) desert (*D*) in the south; (*12*) brunisolic (*B*), and (*13*) atlantic podsol (*AP*) in the southeast; moreover there are (*14*) high latitude humid mountains (*M*), (*15*) dry mountains (*DM*), and (*16*) mediterranean mountains (*M'*).

(*1*) In the sub-glacial desert (*SD*) the principal soils are lithosols (*l*).

(*2*) In the tundra (*T*) the principal soils are: *a* (alluvial); *g* (gleisolic); *o* (peaty); *pg* (gley podsols); and *pt* (tundra nanopodsols including arctic brown and tundra gleisols).

(*3*) In the gley podsol (*GP*) region, the principal soils are: *a* (alluvial); *o* (peaty); *op* (peaty podsols); *p* (podsols); *pg* (gley podsols).

(*4*) In the permafrost gley podsol region (*PGP*) the principal soils are: *a* (alluvial); *o* (peaty); *op* (peaty podsols); and *pg* (gley podsols).

(*5*) In the continental podsol region (*CP*) the principal soils are: *a* (alluvial); *o* (peaty); *op* (peaty podsols); and *p* (podsols).

(*6*) In the permafrost podsol region (*PP*) the principal soils are: *a* (alluvial); *o* (peaty); *op* (peaty podsols); *p* (podsols); *pg* (gley podsols); *ps* (sod podsols); *pu* (planosolic (solodized) podsols).

(*7*) In the sod podsol region (*SP*) the principal soils are: *a* (alluvial); *byr* (degraded chernozems); *o* (peaty); *op* (peaty podsols); *p* (podsols); *ps* (sod podsols); *pu* (planosolic (solodized) podsols); *pw* (grey-wooded); *yuw* (wet planosolic para-chernozems).

(*8*) In the atlantic podsol–ando region (*AP–A*) of Kamchatka the principal soils are: *a* (alluvial); *op* (peaty podsols); *p* (podsols); *pg* (gley podsols); *ps* (sod podsols).

(*9*) In the grey-wooded region (*GW*) the principal soils are: *a* (alluvial); *byr* (degraded chernozems); *o* (peaty); *p* (podsols); *ps* (sod podsols); *pw* (grey-wooded); *t* (solonetz); *yr2* (common chernozems); *yuw* (meadow planosolic para-chernozem).

(*10*) In the chernozemic region (*Ch*) the principal soils are: *a* (alluvial); *byr* (degraded chernozems); *d* (arid brown); *e* (aeolian sands); *pw* (grey-wooded); *s* (saline); *t* (solonetz); *u* (planosols); *yr1* (typical chernozems); *yr2* (common chernozems); *yr3* (southern chernozems); *yr5* (carbonaceous chernozems); *yr8* (light chestnut); *yu* (planosolic para-chernozems); *yuw* (meadow planosolic para-chernozems).

(*11*) In the desert region (*D*) the principal soils are: *a* (alluvial); *d* (arid brown); *dl* (desert lithosols); *e* (aeolian sands); *h* (organic); *o* (peaty); *rd* (serozems); *s* (saline); *stt* (takirs); *t* (solonetz); *u* (planosols); *uw* (meadow planosols).

(*12*) In the brunisolic (*B*) region the principal soils are: *a* (alluvial); *b* (braunerde); *yg* (humic gley).

(*13*) In the atlantic podsol region (*AP*) the principal soils are: *a* (alluvial); *op* (peaty podsols); *p* (podsols); *pg* (glei podsols); *ps* (sod podsols).

(*14*) In the high latitude humid mountains (*M*) the principal soils are: *b* (braunerde); *ba* (acid brown); *hr* (calcareous organic); *l* (lithosols); *lh* (lithosolic rankers); *lt* (tundra lithosols); *p* (podsols); *ps* (sod podsols); *pw* (grey-wooded); *rb* (brunified rendzinas). Lithosols abound.

(*15*) In the dry mountains (*DM*) the principal soils are: *a* (alluvial); *b* (recent brown); *byr* (degraded chernozems); *c* (cinnamonic); *d* (arid brown); *l* (lithosols); *lh* (lithosolic rankers); *lt* (tundra lithosols); *p* (podsols); *pb* (grey-brown podsolic); *pw* (grey-wooded); *u* (planosols); *y* (chernozemic); *yrb* (dark chestnut). Lithosols abound.

(*16*) In the mediterranean mountains (*M'*) there are many cinnamonic soils (*c*); and of course lithosols.

Because of the dry climate, or permafrost the soils of Asiatic Russia are in general little leached, and their mineral fertility varies from non-bad podsols to very good chernozems.

China (Fig.39)

China includes a great variety of broad soil regions: (*1*) brunisolic (*B*) and (*2*) chernozemic (*Ch*) in Manchuria; (*3*) brunisolic–cinnamonic (*B–C*) in northern China; (*4*) kaolinitic–alluvial (*K–Al*) in southern China; (*5*) chernozemic–desert (*Ch–D*), (*6*) arid cinnamonic (*A–C*) and (*7*) desert (*D*) in the arid west. Moreover China has many mountains: (*8*) high latitude humid (*M*) in Manchuria; (*9*) tropical (*TM*) in southern China; and (*10*) dry mountains (*DM*) in the west.

(*1*) In the brunisolic region (*B*) the principal soils are: *aw* (wet alluvial); *b* (braunerde); *bk* (jeltozems); *cb* (brown cinnamonic); *ck* (cinnamonic–kaolisol intergrades); *g* (gleisolic); *pb* (grey-brown podsolic); *s* (saline); *u* (planosols); *yu* (planosolic para-chernozems). Thus the brunisolic region of China differs substantially from those of Europe and the United States; there is more rubification and certain kaolinization, probably due to the monsoonal climate (summer rainfall).

(*2*) In the chernozemic region (*Ch*) the principal soils are: *a* (alluvial); *aw* (wet alluvial); *b* (recent brown); *g* (gleisolic); *h* (organic); *s* (saline); *u* (planosols); *uw* (wet planosols); *y* (chernozemic); *y3* (shallow chernozem?); *y8* (light chestnut); *yb* (brunisolic para-chernozems— prairie); *ye* (aeolian para-chernozems); *yu* (planosolic para-chernozems); *ywr* (rendzina meadow chernozems). Thus the chernozemic regions of Manchuria differ appreciably from those of Europe, due probably to the monsonic climate (summer rainfall) and the influence of rivers.

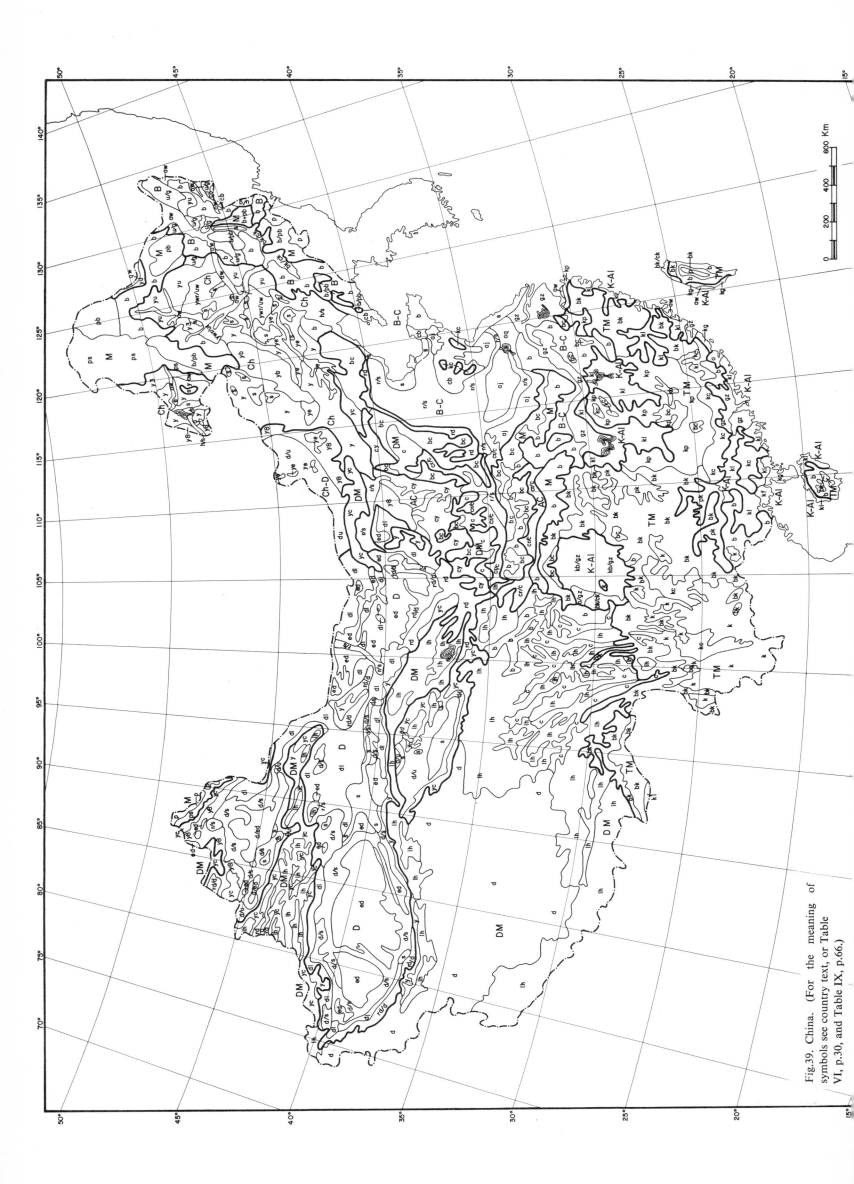

Fig.39. China. (For the meaning of symbols see country text, or Table VI, p.30, and Table IX, p.66.)

(3) In the brunisolic–cinnamonic region (*B–C*) of northern China the most important soils are: *aj* (alluvial with calcareous crust); *aw* (wet alluvial); *b* (recent brown); *c* (cinnamonic); *cb* (brown cinnamonic); *cr* (terra rossa); *gz* (paddy gleisolic); *r* (rendzina); *s* (saline). This region is highly influenced by the rivers; most of the materials are alluvial; some soils (paddy gleisolic for instance) have been affected by artificial flooding to grow rice.

(4) In the kaolinitic–alluvial (*K–Al*) region of southern China, the most important soils are: *aw* (wet alluvial); *b* (recent brown); *bk* (jeltozems); *ck* (cinnamonic–kaolisol intergrades); *gz* (paddy gleisolic); *kf* (concretionary kaolisols); *kg* (gleisolic–kaolisols); *kl* (latosolic kaolisols); *pk* (kaolinitic podsolic); *o* (peaty). The most important soils, from an agricultural point of view, are those flooded or irrigated to grow rice; and these soils are seldom kaolinitic. Moreover, it is presumed that the kaolisols of southern China are seldom dystrophic. Therefore, the panorama of soils is rather satisfactory in southern China.

(5) In the chernozemic–desert (*Ch–D*) region of Mongolia, the principal soils are: *d* (arid brown); *du* (red desert lessivé); *y8* (light chestnut?); *ye* (aeolian para-chernozems).

(6) In the arid cinnamonic (*AC*) region the principal soils are: *c* (cinnamonic); *cy* (reddish brown); *du* (planosolic red desert); *r* (rendzinas); *s* (saline); *u* (planosols); *ye* (aeolian para-chernozems); *y8* (light chestnut?).

(7) In the desert (*D*) region the principal soils are: *d* (arid brown and other raw soils of the desert); *dl* (desert lithosols); *ed* (desert sands); *r* (rendzinas); *rd* (serozems); *s* (saline); *u* (planosols); *y* (chernozemic); *y8* (light chestnut?); *yc* (reddish para-chernozems).

(8) In the high latitude humid mountains (*M*) of the north the principal soils are: *b* (recent brown); *bc* (cinnamonic brown); *kc* (kaolisol–cinnamonic intergrades); *p* (podsols); *pb* (grey-brown podsolic); *ps* (sod podsols). Naturally lithosols abound.

(9) In the tropical mountains of the south (*TM*) the principal soils are: *b* (recent brown); *bk* (jeltozems); *k* (kaolisols); *kc* (kaolisol–cinnamonic intergrades); *kl* (latosolic kaolisols); *lh* (lithosolic rankers); *p* (podsols); *pk* (kaolinitic podsolic). Kaolinization is not so advanced.

(10) In the dry mountains (*DM*) of the west the most important soils are: *a* (alluvial); *b* (recent brown); *bc* (brown cinnamonic); *c* (cinnamonic); *d* (arid brown); *lh* (lithosolic rankers); *rd* (serozems); *u* (planosols); *y* (chernozemic); *yc* (reddish para-chernozems).

As it is shown on the map the panorama of soils of China is rather satisfactory. Soils with 1:1 clays and/or dystrophic are exceptional. Naturally much of the land is too steep; but this fact also facilitates flooding irrigation and combined with a high summer rainfall permits to grow extensively rice.

Japan

Japan is a mountainous volcanic country and its soils cannot be shown in a small-scale map. It may be included in only one broad region: ando–brunisolic–mountainous (*A–B–M*); volcanic materials abound; the general tendency is towards the formation of brunisolic soils; relief is very mountainous; some areas in the north have podsols, and other areas in the south seem to have kaolinitic soils; but these areas do not change the whole panorama. The principal soils are ando, more or less gleisolic or brunisolic; lithosolic rankers; and lithosols.

The abundance of gleisolic soils is due to two causes: (*1*) leaching rainfall is high and in a mountainous country this causes waterlogging of the lowlands, and (*2*) the principal crop is rice; it is grown in the greater part of the agricultural land at least once each year; thus agricultural soils are usually gleisolic. Since ando soils abound in Japan, the terms ando and volcanic are only used for young ando. However, with the criteria proposed in this book the extension of ando soils in Japan is far greater.

Due to the abundance of ando soils in Japan, and still more to the mountainous relief, soils are usually young; in addition, agricultural soils receive waters which bring mineral substances from other areas; for all these reasons mineral fertility is usually high in Japan. However, ando soils immobilize phosphorus; moreover they contain free alumina which under certain conditions becomes toxic. But these problems have been already solved by Japanese scientists. The great density of the Japanese population, enormous in comparison to tillable area, is partly due to the volcanic origin of its soils; and natural productivity has been increased many times by the introduction of modern techniques. Japan has, certainly, the most intensive agriculture in the world.

Indochina (Fig.40)

Burma. Burma is composed of two soil regions: (*1*) kaolinitic–alluvial (*K–Al*); and (*2*) tropical mountains (*TM*).

(*1*) In the kaolinitic–alluvial (*K–Al*) region the principal soils are: *a* (alluvial); *c* (cinnamonic); *g* (gleisolic); *k* (kaolisols); *kf* (concretionary kaolisols); *kl* (latosolic kaolisols); *kp* (kaolinitic podsolic); *m* (dark clays); *ms* (cat clays); *s* (saline).

(*2*) In the tropical mountains (*TM*) many soils are kaolisols (*k*); naturally lithosols abound.

Thus, because of topography, and because the mountains from which the rivers descend are geologically young the panorama of soils in Burma is good; many soils are not kaolinitic; and many kaolinitic soils are eutrophic. Moreover from an agricultural point of view what imports is flooded or irrigated soils in which rice is grown; and such soils are seldom kaolinitic.

Thailand. Thailand consists of two soil regions: (*1*) kaolinitic–alluvial (*K–Al*); and (*2*) tropical mountains (*TM*).

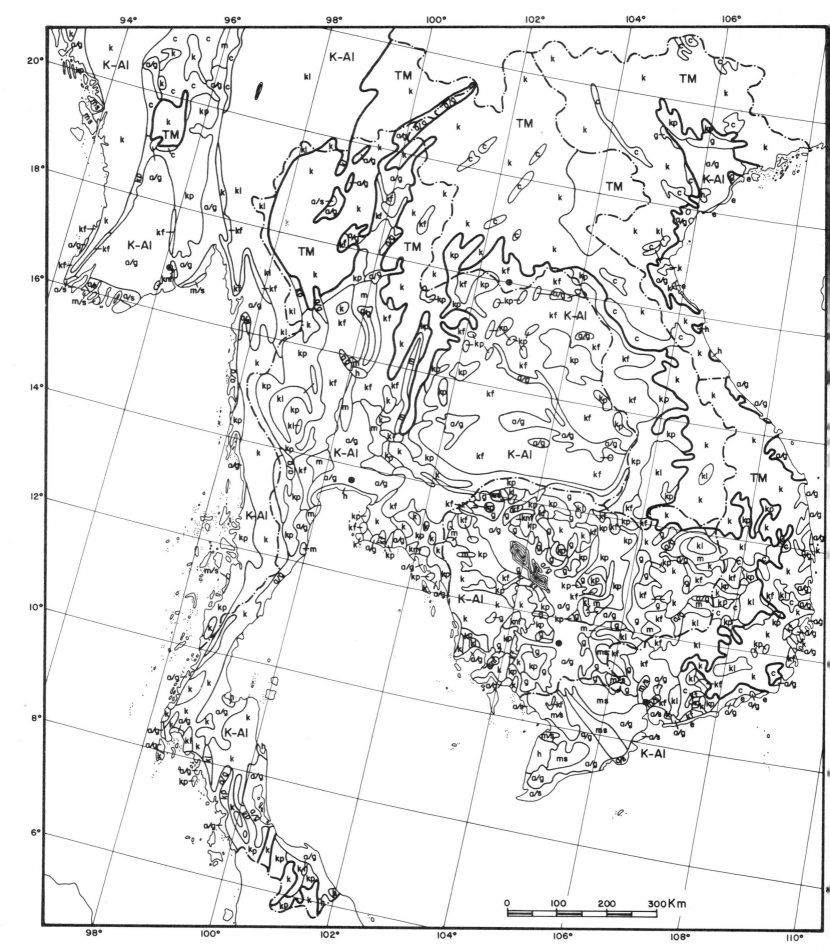

Fig.40. Indochina. (For the meaning of symbols see country text, or Table VI, p.30, and Table IX, p.66.

(*1*) In the kaolinitic–alluvial (*K–Al*) region the principal soils are: *a* (alluvial); *g* (gleisolic); *h* (organic); *k* (kaolisols); *kf* (concretionary kaolisols); *kl* (latosolic kaolisols); *kp* (kaolinitic podsolic); *m* (dark clays); *s* (saline).

(*2*) In the tropical mountains (*TM*) the principal soils are: *k* (kaolisols); *kf* (concretionary kaolisols); *kp* (kaolinitic podsolic); naturally lithosols abound.

Thus, in spite of the tropical climate the greater part of the soils of Thailand is not kaolinitic; and those that are seem to be eutrophic. Moreover, from an agricultural point of view, what is important are the soils that are flooded or irrigated and which are used to grow rice; and these soils are seldom kaolinitic. This is due to the fact that many of the rivers come from geologically young mountains.

Cambodia. In Cambodia two broad soil regions can be distinguished: (*1*) a kaolinitic–alluvial(*K–Al*); and (*2*) tropical mountains (*TM*).

(*1*) In the kaolinitic–alluvial (*K–Al*) region the most important soils are: *a* (alluvial); *g* (gleisolic); *k* (kaolisols); *kf* (concretionary kaolisols); *kl* (latosolic kaolisols); *knf* (tropical ferruginous); *kp* (kaolinitic podsolic); *m* (dark clays); *ms* (cat clays); *s* (saline).

(*2*) In the tropical mountains (*TM*) kaolisols (*k*) and lithosols abound. From an agricultural point of view the most important soils are those flooded or irrigated, in which rice is grown; and these soils are seldom kaolinitic. Although Cambodia is a tropical country, its soils are good, due to the fact that the rivers come from geologically young mountains.

Laos. Laos includes two kinds of broad soil regions: (*1*) kaolinitic–alluvial (*K–Al*); and (*2*) tropical mountains (*TM*).

(*1*) In the kaolinitic–alluvial (*K–Al*) region the principal soils are: *a* (alluvial); *g* (gleisolic); *kf* (concretionary kaolisols); *kl* (latosolic kaolisols); *kp* (kaolinitic podsolic).

(*2*) In the tropical mountains (*TM*) the principal soils are: *c* (cinnamonic); *k* (kaolisols); *kl* (latosolic kaolisols); naturally lithosols abound.

It is assumed that many of the kaolinitic soils are eutrophic. Moreover what really is important, from an agricultural point of view, is flooded or irrigated land, and these soils are seldom kaolinitic. So that, although Laos has a tropical climate, its soils are rather good.

North Vietnam. North Vietnam includes two broad soil regions: (*1*) kaolinitic–alluvial (*K–Al*); and (*2*) tropical mountains (*TM*).

(*1*) In the kaolinitic–alluvial (*K–Al*) region the most important soils are: *a* (alluvial); *e* (aeolian sands); *g* (gleisolic); *h* (organic); *k* (kaolisols); *kp* (kaolinitic podsolic).

(*2*) In the tropical mountains (*TM*) the principal soils are: *c* (cinnamonic); *g* (gleisolic); *k* (kaolisols).

From an agricultural point of view, the most important soils are those of the plains; and these soils seldom are kaolinitic. Moreover it is presumed that the kaolinitic soils of North Vietnam are seldom dystrophic. So that, in spite of its tropical climate, the soils of North Vietnam are rather good. This is due to the fact that the rivers come from geologically young mountains.

South Vietnam. South Vietnam includes two kinds of broad soil regions: (*1*) kaolinitic–alluvial (*K–Al*); and (*2*) tropical mountains (*TM*).

(*1*) The principal soils of the kaolinitic–alluvial (*K–Al*) region are: *a* (alluvial); *c* (cinnamonic); *e* (aeolian sands); *g* (gleisolic); *h* (organic); *k* (kaolisols); *kd* (dystrophic kaolisols); *kf* (concretionary kaolisols); *kl* (latosolic kaolisols); *m* (dark clays); *ms* (cat clays); *s* (saline).

(*2*) The principal soils of tropical mountains (*TM*) are: *a* (alluvial); *c* (cinnamonic); *g* (gleisolic); *k* (kaolisols); *kf* (concretionary kaolisols); *kl* (latosolic kaolisols); *kp* (kaolinitic podsolic).

From an agricultural point of view what really is important are the soils that can be flooded or irrigated to grow rice; and these soils are seldom kaolinitic. Moreover, it is presumed that the kaolisols of South Vietnam are usually "eutrophic", and plantation crops (rubber, etc.) grow well on such soils.

Indonesia (Fig.41)

Indonesia includes two kinds of broad soil regions: (*1*) kaolinitic–ando (*K–A*); and (*2*) tropical mountains–ando (*TM–A*).

(*1*) In the kaolinitic–ando (*K–A*) region, the principal soils are: *a* (alluvial); *c* (cinnamonic); *g* (gleisolic); *h* (organic); *k* (kaolisols); *kf* (concretionary kaolisols); *kl* (latosolic kaolisols); *m* (dark clays); *ph* (humus podsols); *r* (rendzinas); *v* (ando).

(*2*) In the tropical mountains–ando region (*TM–A*) the principal soils are: *a* (alluvial); *c* (cinnamonic); *g* (gleisolic); *h* (organic); *k* (kaolisols); *m* (dark clays); *r* (rendzinas); *v* (ando).

The great frequency of ando soils in many islands of Indonesia makes these islands fertile and permits the sustenance of a very dense population; one of the densest in the world. A re-classification of these soils on the basis of their cation exchange capacity and moisture tension, in relation to clay content, will probably show that the extension of ando soils is greater than shown in the maps. Borneo seems to be less volcanic than the other islands: it may be that it forms a kaolinitic region.

Sarawak. Sarawak includes two broad soil regions: (*1*) kaolinitic–ando (*K–A*); and (*2*) tropical mountains–ando (*TM–A*).

(*1*) In the kaolinitic–ando region (*K–A*) the principal soils are: *a* (alluvial); *g* (gleisolic); *h* (organic); and *ph* (humus podsols).

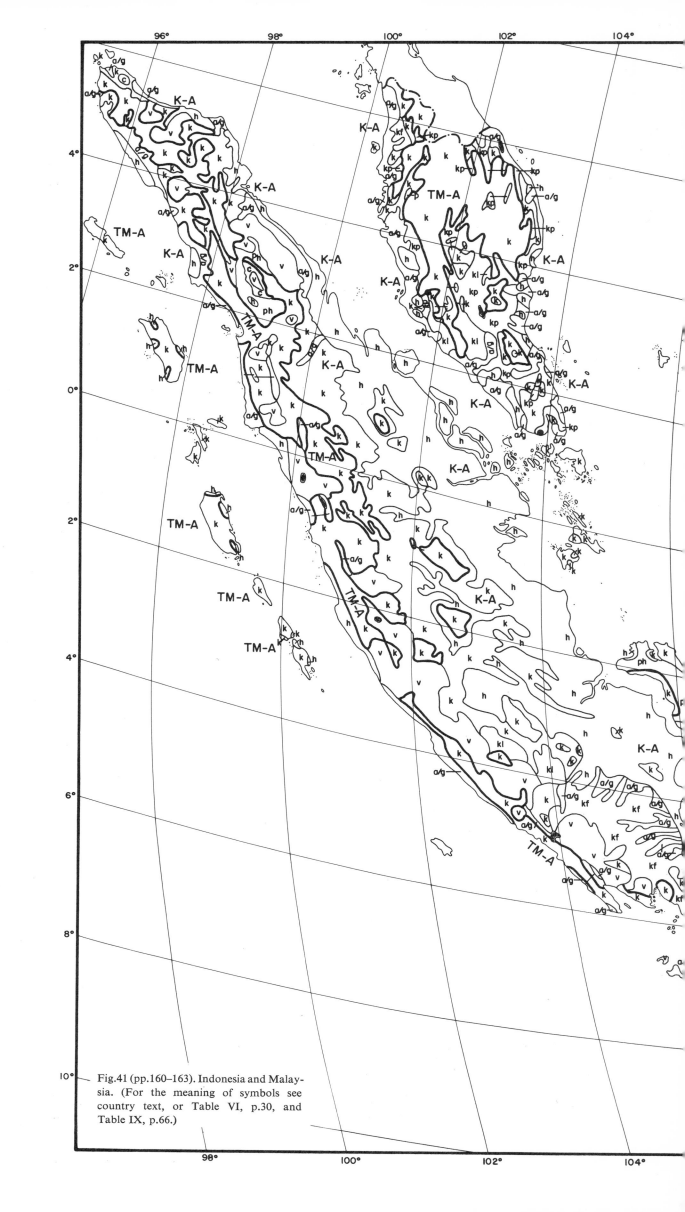

Fig.41 (pp.160–163). Indonesia and Malaysia. (For the meaning of symbols see country text, or Table VI, p.30, and Table IX, p.66.)

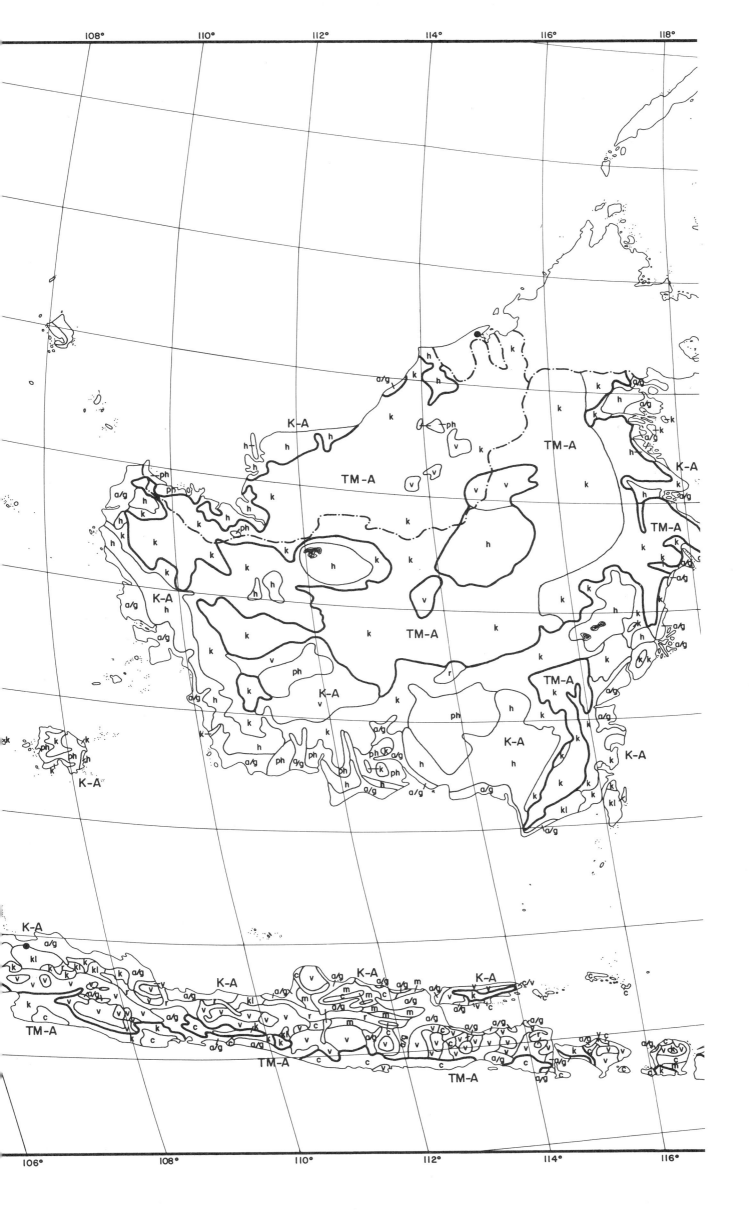

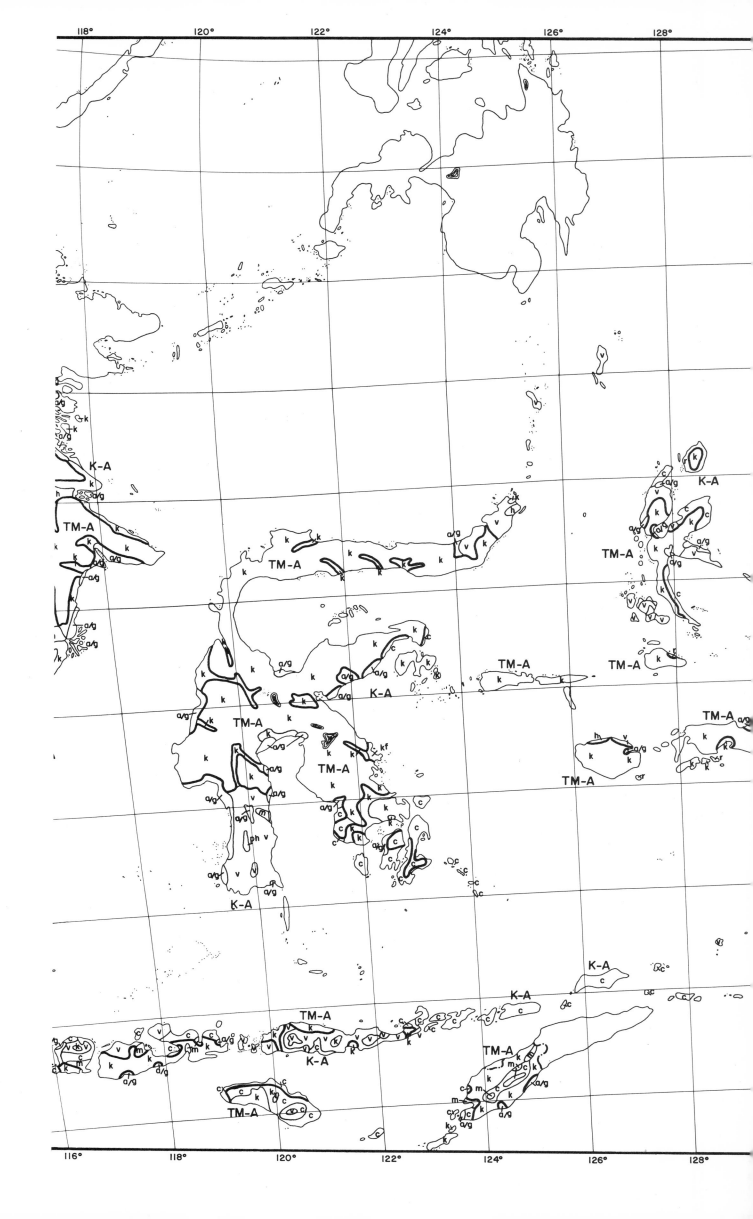

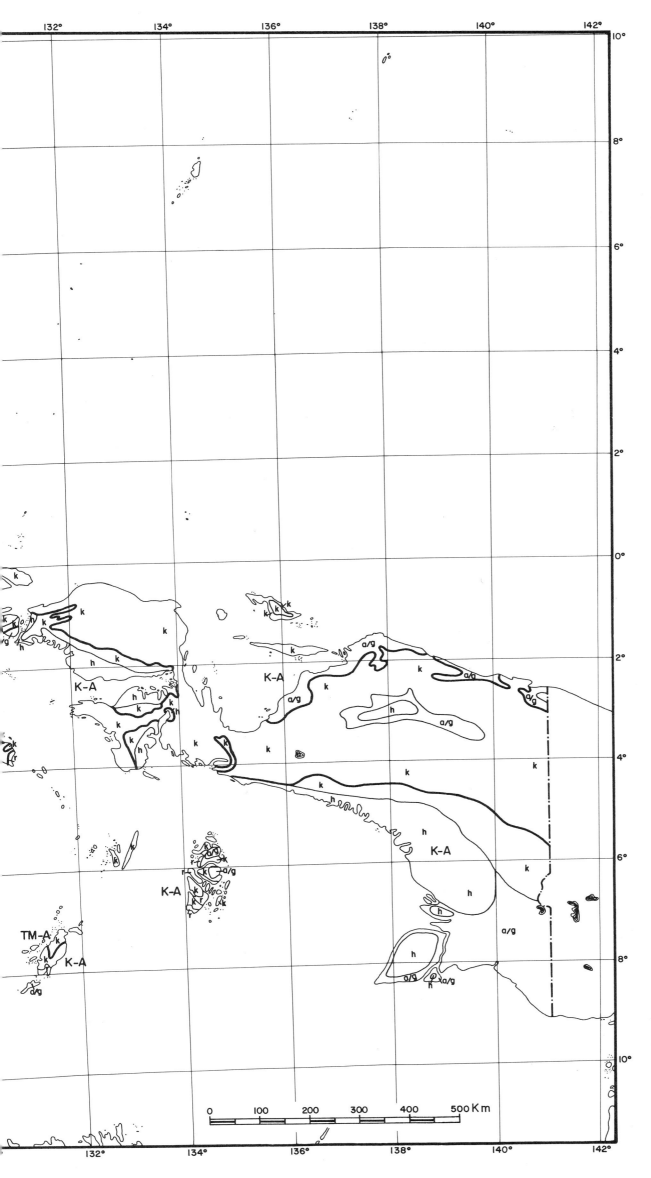

Fig.41 (continued).

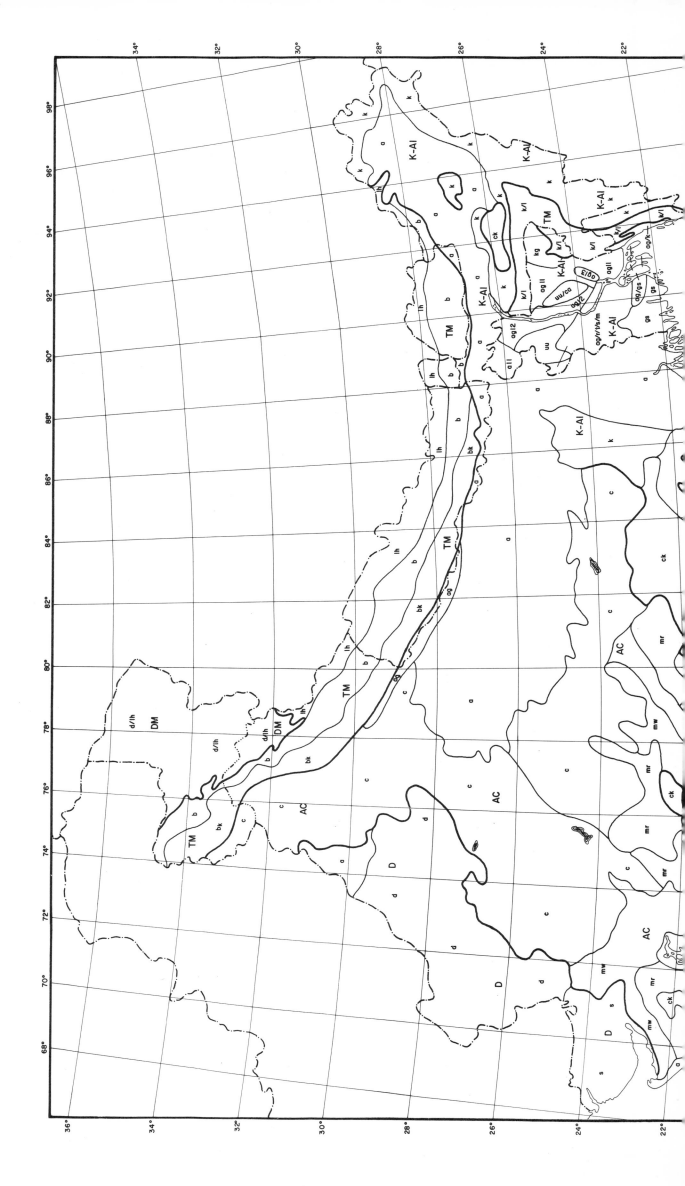

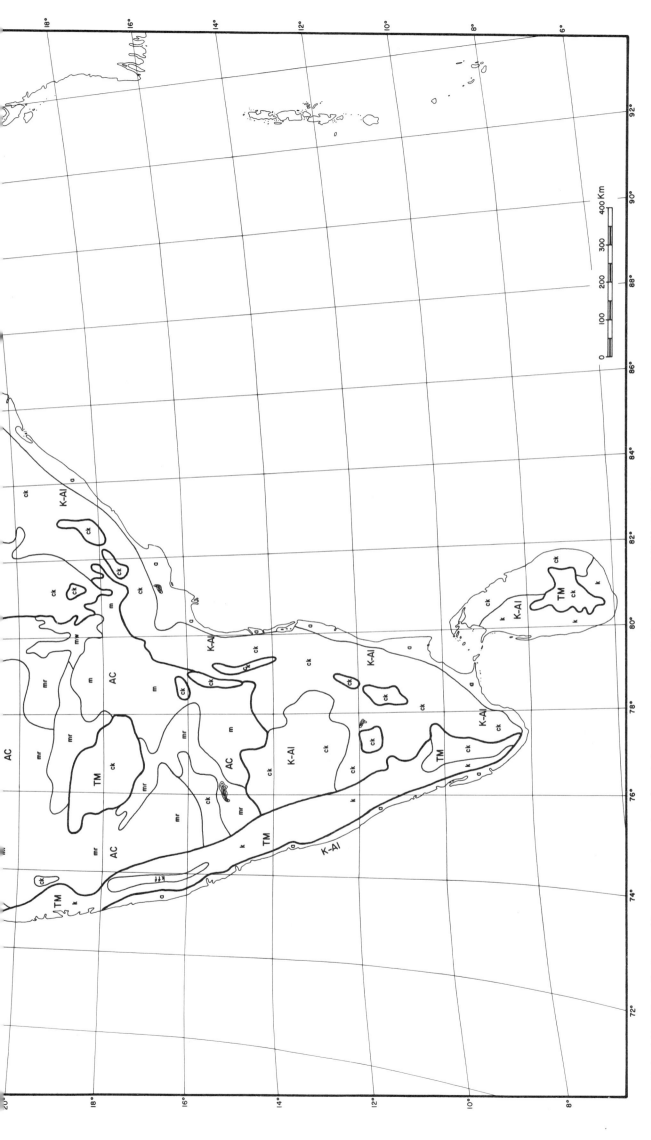

Fig. 42 (pp. 164, 165). India, East Pakistan. (For the meaning of symbols see country text, or Table VI, p. 30, and Table IX, p. 66.)

(2) In the tropical mountains–ando region *(TM–A)* the principal soils are: *k* (kaolisols); *ph* (humus podsols); *v* (ando).

Malaysia and Singapore. These areas include two broad soil regions: *(1)* kaolinitic–ando *(K–A)*; and *(2)* tropical mountains–ando *(TM–A)*.

(1) In the kaolinitic–ando region *(K–A)* the principal soils are: *a* (alluvial); *g* (gleisolic); *h* (organic); *k* (kaolisols); *kf* (concretionary kaolisols); *kp* (kaolinitic podsolic); *p* (podsols).

(2) In the tropical mountains–ando region *(TM–A)* the principal soils are: *k* (kaolisols); *kp* (kaolinitic podsolic).

Malaysia seems less volcanic than Indonesia; it may be a kaolinitic region.

India, East Pakistan and Ceylon

India (Fig.42). India includes the following broad soil regions: *(1)* desert *(D)* and *(2)* arid cinnamonic *(AC)* in the west; *(3)* kaolinitic–alluvial *(K–Al)* in the east and on the Malabar coast; *(4)* tropical mountains *(TM)*; and *(5)* dry mountains *(DM)* in Kashmir.

(1) In the desert *(D)* region the principal soils are: *a* (alluvial); *d* (arid brown); *s* (saline).

(2) In the arid cinnamonic *(AC)* region the principal soils are: *a* (alluvial); *c* (cinnamonic); *ck* (cinnamonic-kaolisol intergrades); *m* (dark clays); *mr* (rendzina clays); *mw* (wet clays). The dark clays of India *(m, mr, mw)* form one of the most extensive areas of dark clays of the world. Much cotton is grown on them.

(3) In the kaolinitic–alluvial *(K–Al)* regions the principal soils are: *a* (alluvial); *ag* (gleisolic alluvial); *ck* (cinnamonic–kaolisol intergrades); *k* (kaolisols); *l* (lithosols). There are also some laterites that are not shown in the map.

(4) In the tropical mountains *(TM)* the principal soils are: *b* (recent brown); *bk* (jeltozems); *ck* (cinnamonic–kaolisol intergrades); *k* (kaolisols); *kff* (laterites).

(5) In the dry mountains *(DM)* of Kashmir the principal soils are: *d* (arid brown); *lh* (lithosolic rankers).

Although India has a tropical climate the extension of kaolinitic soils is small. In the densely populated north, soils are alluvial; and the western part of Deccan has dark clays. Soils are not bad, and sustain a dense population.

Nepal and Bhutan (Fig.42). These countries include two broad soil regions: *(1)* kaolinitic-alluvial *(K–Al)*; and *(2)* tropical mountains *(TM)*.

(1) In the kaolinitic–alluvial *(K–Al)* region the most important soils are: *a* (alluvial); and *ag* (gleisolic alluvial).

(2) In the tropical mountains *(TM)* the most important soils are: *b* (recent brown); *bk* (jeltozems); and *lh* (lithosolic rankers).

East Pakistan (Fig.42). Practically all East Pakistan is included in a kaolinitic–alluvial *(K–Al)* region. But some hills of the east belong to tropical mountains *(TM)*.

(1) In the kaolinitic–alluvial *(K–Al)* region the principal soils are: *a* (alluvial); *ag* (gleisolic alluvial); *ca* (acid cinnamonic); *gs* (saline gleisols—mangrove); *k* (kaolisols); *kg* (gleisolic kaolisols); *m* (dark clays); *r* (rendzinas); *s* (saline); *t* (solonetz); *uu* (clay-pan planosols).

The agricultural possibilities of the alluvial and alluvial gleisolic soils depend greatly on the nature of the alluvium and on depth of flooding. BRAMMER (1965) classified the soils on this basis.

In the alluvial soils of the old Brahmaputra flood plain *(al1)* flooding is usually less than 6 ft. In the fresh-water deltaic flood plain *(ag11)* flooding is usually less than 6 ft.; soils are more gleisolic than in *al1*. In *ag12* flooding reaches more than 6 ft. In the middle Meghma flood plain *(ag13)* sudden flooding early in the monsoon season is the major agricultural limitation.

(2) In the tropical mountains *(TM)* the principal soils are: *k* (kaolisols) and *l* (lithosols).

As shown by the map the great majority of the soils of East Pakistan is alluvial and more or less gleisolic. They are good, but the main limitation is the danger of flooding or lack of water. They are intensively cropped; it is perhaps the most densely populated agricultural region of the world.

Ceylon (Fig.42). Ceylon includes two broad soil regions: *(1)* kaolinitic *(K)*; and *(2)* tropical mountains *(TM)*.

(1) In the kaolinitic *(K)* region the principal soils are: kaolisols *(k)* and *ck* (cinnamonic–kaolisol intergrades).

(2) In tropical mountains *(TM)* the principal soils are: *ck* (cinnamonic–kaolisol intergrades); naturally lithosols abound.

The author has insufficient information concerning the cation exchange capacity of these soils, to see what is the grade of their kaolinization; we presume eutrophic soils prevail, but we do not have precise information.

Southwestern Asia (Fig.43)

Syria. In Syria three broad soil regions may be recognized: *(1)* desert *(D)*; *(2)* mediterranean mountains *(M')*; and *(3)* mediterranean cinnamonic *(MC)* in the non-desert plains.

(1) In the desert *(D)* region the most important soils are: *a* (alluvial); *d* (arid brown); *l* (lithosols); *m* (dark clays); *rd* (serozems); *s* (saline); *t* (solonetz); *u* (planosols).

(2) In the mediterranean mountains *(M')* cinnamonic soils *(c)* abound; but there are also terra rossa, recent brown, rendzinas, cinnamonic brown, etc.

(3) In the non-desert plains *(MC)* the principal soils are: *a* (alluvial); *s* (saline); *t* (solonetz); *u* (planosols); but there are also rendzinas, terra rossa, recent brown, etc.

All these soils originated from geologically young

materials and suffered little leaching. That is why their mineral fertility is high. The principal limiting factors of agricultural production, from an edaphic point of view, are stoniness, shallowness, steep slopes in the mountains; waterlogging–salinity–alkalinity in certain parts of the plains. Naturally fertilizers increase yields.

Lebanon. Two broad soil regions can be recognized: (*1*) mediterranean mountains (*M′*); and (*2*) mediterranean cinnamonic (*MC*) in the plains.

(*1*) In the mediterranean mountains (*M′*) the principal soils are: *c* (cinnamonic) and *d* (arid brown); but there are also many other soils not mentioned in the map: rendzina, terra rossa, recent brown, cinnamonic brown (brun méditerranéen), grey–brown podsolic, etc.

(*2*) In the plains (*MC*), it seems that dark clays (*m*) abound, but there are also many alluvial soils, recent brown, etc.

All these soils arose from geologically young materials and suffered little leaching, thus their mineral fertility is usually high. The chief problems are stoniness, shallowness, steep slopes in the mountains; waterlogging, salinity in the plains. Naturally fertilizers increase yields.

Jordan. Three broad soil regions can be recognized: (*1*) desert (*D*); (*2*) mediterranean mountains (*M′*); and (*3*) mediterranean cinnamonic (*MC*).

(*1*) In the desert (*D*) the most important soils are: *a* (alluvial); *d* (arid brown and other raw soils of the desert); *ed* (desert sands); *l* (lithosols); *m* (dark clays); *rd* (serozems); *s* (saline); *t* (solonetz).

(*2*) In the mediterranean mountains (*M′*) the principal soils are: *c* (cinnamonic); *l* (lithosols); *m* (dark clays); *s* (saline); *t* (solonetz). All these soils are interspersed with lithosols and rock outcrops.

(*3*) In the non-desert plains (*MC*) the main soils are: *a* (alluvial); *b* (recent brown); *s* (saline); *t* (alkaline); *u* (planosols). All show signs of rubification.

All Jordan soils arise from geologically young materials and have suffered little leaching. That is why their mineral fertility is usually satisfactory. The principal limiting factors of agricultural production, from an edaphic point of view, are stoniness, shallowness, steepy slopes in the mountains; salinity–alkalinity–waterlogging in some plains. Naturally the use of fertilizers increases yields.

Israel. Israel is so small and mountainous that its soils cannot be shown in a small-scale map. It includes three broad soil regions: (*1*) *M′* (mediterranean mountains); (*2*) *MC* (mediterranean cinnamonic); and (*3*) *D* (desert); between *MC* and *D* there is naturally a transition region. The principal soils are: cinnamonic (*c*); slightly leached cinnamonic; non-calcic brown (*pc*); arid brown (*d*); serozem (*rd*); desert lithosols or hammada (*dl*); planosolic reddish desert (*ud*); dark clays (*m*); terra rossa (*cr*); rendzina (*r*); desert sands (*de*); alluvial (*a*); planosols (*u*); gleisolic (*g*); solonchak (*s*). Their distribution follows the patterns of the corresponding regions.

Since all these soils are not intensively leached, mineral fertility is usually satisfactory in Israel. Naturally many soils have steep slopes, are shallow, suffer from drought or waterlogging.

The Sinai Peninsula (U.A.R.). This area is virtually a more or less mountainous desert; but the U.A.R. extends a little into Palestine where agriculture is possible without irrigation. The principal soils are: *d* (arid brown and other raw soils of the desert); *ed* (desert sands); *l* (lithosols); *s* (saline); *t* (solonetz); *u* (planosols).

Iraq. Iraq consists of two broad soil regions: (*1*) *D* (desert) and (*2*) *DM′* (dry mediterranean mountains). The latter occupies only a small area in the north.

(*1*) In the desert region (*D*) the principal soils that have been recognized are: *a* (alluvial); *b* (recent brown); *d* (arid brown and other raw soils of the desert); *ed* (desert sands); *l* (lithosols); *rd* (serozems); *s* (saline); *t* (solonetz); *u* (planosols).

(*2*) In the dry mediterranean mountains (*DM′*) the principal soils are: *b* (recent brown); *cy* (reddish brown); *pb* (grey–brown podsolic); the latter at the higher parts with relatively humid climate; naturally all these soils are interspersed with lithosols and rock outcrops.

Excepting some soils of the desert which are far from the rivers and are agriculturally of little importance, all these soils originated from geologically new materials, the mountains of Kurdistan, Armenia and Anatolia. They have suffered little leaching. That is why the main limiting factors of the agricultural production, from an edaphic point of view, are stoniness, shallowness, steep slopes in the mountains; and salinity–alkalinity, too much or too low permeability in the irrigated plains. Naturally the use of fertilizers increases yields.

Concerning salinity it is to be noted that some of these plains are irrigated without drainage since four or more millennia; an enormous amount of salts has been accumulated; and leaching is necessary; fortunately the composition of the water of the rivers and soil texture make drainage not so difficult.

Iran. Iran consists of four broad soil regions: (*1*) brunisolic (*B*) on the western coast of Caspian Sea; (*2*) mediterranean mountains-cinnamonic (*M′C*) on the eastern coast of the same sea; (*3*) the extensive desert (*D*) of central and southern Iran; (*4*) the dry mediterranean mountains (*DM′*) of the west and north of the country; and (*5*) the less dry mountains (*M′–DM′*) that border the Caspian Sea.

(*1*) In the brunisolic region (*B*) the principal soils are: *a* (alluvial); *oh* (lenists, rich in organic matter but not so peaty); *s* (saline); *t* (solonetz).

(*2*) In the mediterranean mountains and dry mediterranean mountains (*M′–DM′*) that border the Caspian Sea, the main soils are recent brown (*b*) and lithosols (*l*).

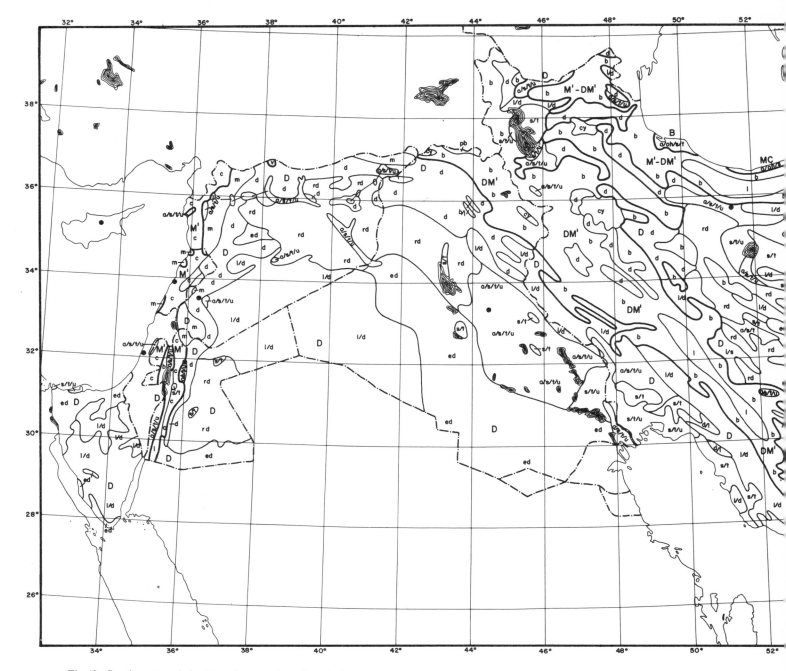

Fig.43. Southwestern Asia. (For the meaning of symbols see country text, or Table VI, p.30, and Table IX, p.66.)

(3) In the desert (D) region the most important soils are: a (alluvial); b (recent brown); cy (reddish brown); d (arid brown and other raw soils of the desert); l (lithosols); s (saline); t (solonetz); u (planosols).

(4) In the dry mediterranean mountains (DM') that abound in western Iran, the most important soils are: a (alluvial); b (recent brown); cy (reddish brown); d (arid brown and other raw soils of the desert); l (lithosols); s (saline); t (solonetz); u (planosols).

All these soils arose from geologically young materials and have suffered very little leaching. That is why their mineral fertility is usually high. The limiting factors of agricultural production, from an edaphic point of view, are stoniness, shallowness, steep slopes in the mountains; and salinity–alkalinity–waterlogging in the irrigated plains. Naturally the use of fertilizers increases yields.

Afghanistan. Here two broad soil regions can be recognized: (1) D (desert); and (2) DM' (dry mediterranean mountains).

(1) In the desert region (D) the main soils that have been identified are: a (alluvial); d (arid brown and other raw soils of the desert); ed (desert sands); l (lithosols); rd (serozems); s (saline); t (solonetz); and u (planosols).

(2) In the dry mediterranean mountains (DM') the main soils that have been identified are: a (alluvial); b (recent brown); c (cinnamonic); l (lithosols); lh (lithosolic rankers); r (rendzinas); s (saline); t (solonetz); u (planosols). Naturally all these soils are interspersed with lithosols and rock outcrops.

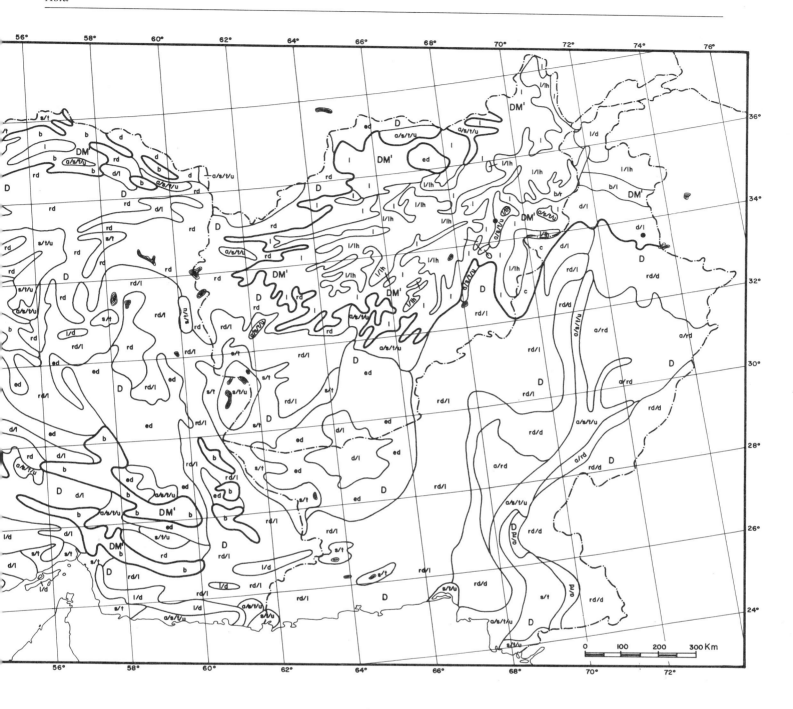

All these soils arose from geologically young materials, and have suffered little leaching. That is why their mineral fertility is usually high. The main limiting factors of the agricultural production from an edaphic point of view are shallowness, stoniness, steep slopes, salinity and alkalinity. In summary, the problems of Afghanistan are those of deserts and dry mountains. Naturally fertilization increases yields.

West Pakistan. Here two broad soil regions can be recognized: (*1*) the desert (*D*); and (*2*) the dry mediterranean mountains (*DM'*); the slopes of Kashmir are humid, but they occupy a very small area.

(*1*) In the desert (*D*) the most important soils are: *a* (alluvial); *d* (arid brown); *l* (lithosols); *rd* (serozems);

s (saline); *t* (solonetz); *u* (planosols).

(*2*) In the dry mediterranean mountains (*DM'*) the main soils are: *b* (recent brown); *c* (cinnamonic); *d* (arid brown and other raw soils of the desert); *l* (lithosols); *lh* (lithosolic rankers); *r* (rendzinas).

All these soils arose from geologically young materials and have suffered little leaching. So that their mineral fertility is usually high. The main limiting factors of agricultural production, from an edaphic point of view, are stoniness, shallowness, steep slopes in the mountains; waterlogging–salinity–alkalinity in some parts of the plains. Naturally fertilizers increase yields.

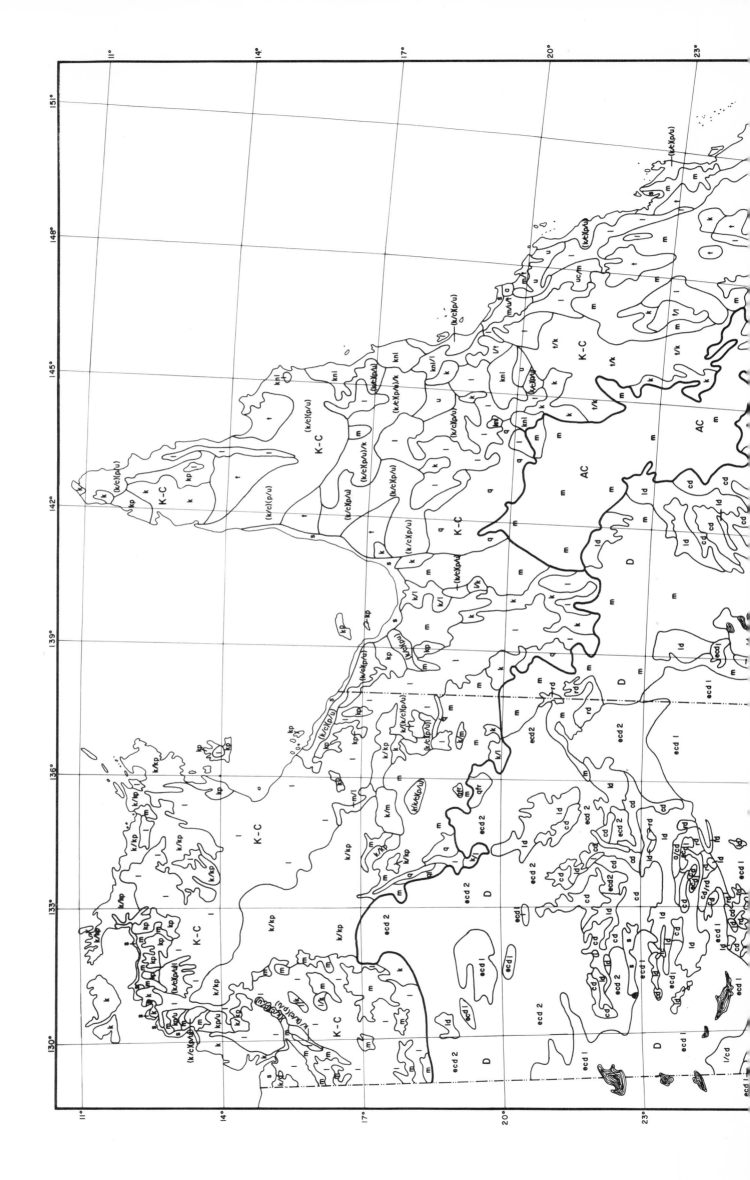

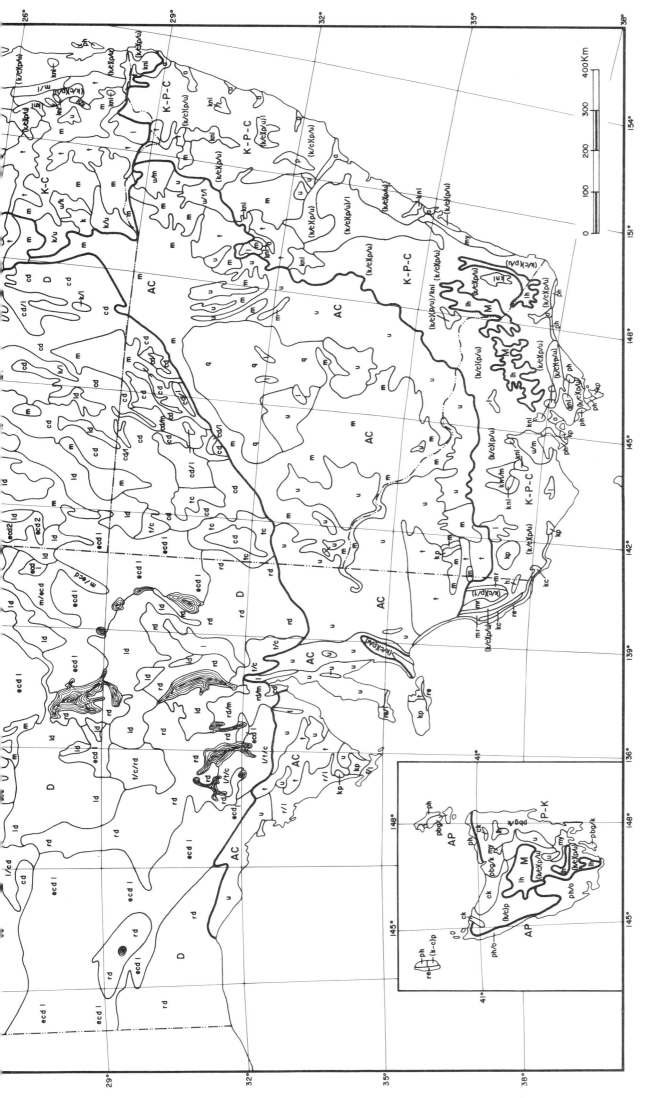

Fig. 44. Australia (Western Australia not included). Inset: Tasmania. (For the meaning of symbols see country text, or Table VI, p.30, and Table IX, p.66.)

AUSTRALIA AND NEW ZEALAND

Australia (Fig.44, 45)

The greater part of Australia is a desert and forms a desert region (*D*). Some parts have a dry climate and form and arid cinnamonic region (*AC*). However, Australia is a very old continent and some of the soils of the dry regions are palaeo-lateritic, palaeo-kaolisols or kaolisols.

Kaolisols are still more frequent in less dry areas, and we may recognize a kaolinitic region (*K*), a kaolinitic–cinnamonic region (*K–C*), a kaolinitic–podsolic region (*K–P*), a kaolinitic–podsolic–cinnamonic region (*K–P–C*) and a podsolic–kaolinitic region (*P–K*). Australia has very few mountains; but in the southeast and on Tasmania a high latitude humid mountainous region (*M*) may be recognized. There is also a region of atlantic podsol (*AP*) on Tasmania.

(*1*) In the desert region (*D*) the more important soils are: *a* (alluvial); *c* (cinnamonic); *cd* (red desert); *ecd* (red desert sands) in hills or plains; *k* (kaolisols); *l* (lithosols); *ld* (desert lithosols); *m* (dark clays); *qfr* (calcareous palaeo-lateritic); *rd* (serozems); *s* (saline); *t* (solonetz); *tc* (cinnamonic solonetz); *u* (planosols); *uc* (cinnamonic planosols).

(*2*) In the arid cinnamonic (*AC*) region the more important soils are: *cd* (red desert); *kp* (kaolinitic podsolic); *l* (lithosols); *m* (dark clays); *q* (palaeo-kaolisols); *r* (rendzinas); *re* (aeolian rendzinas); *t* (solonetz); *u* (planosols).

(*3*) In the kaolinitic–cinnamonic region (*K–C*) the main soils are: *a* (alluvial); *k* (kaolisols); *knl* (terres de barre); (*k/c*)(*p/u*) (kaolinitic–cinnamonic) (podsolic–planosolic); *kp* (kaolinitic podsolic); *l* (lithosols); *ld* (desert lithosols); *lk* (kaolinitic lithosols); *m* (dark clays); *p* (podsols); *ph* (humus podsols); *q* (palaeo-kaolisols); *qfr* (calcareous palaeo-lateritic); *ql* (lithosolic palaeo-kaolisols); *rd* (serozems); *s* (saline); *t* (solonetz); *u* (planosols); *uc* (cinnamonic planosols).

(*4*) In the kaolinitic–podsolic–cinnamonic region (*K–P–C*) the more important soils are: *a* (alluvial); *h* (organic); *kc* (kaolisol–cinnamonic intergrades); (*k/c*) (*p/u*) (kaolinitic–cinnamonic) (podsolic–planosolic); *knl* (terres de barre); *kp* (kaolinitic podsolic); *l* (lithosols); *mr* (rendzina clays); *my* (chernozemic clays); *ph* (humus podsols); *re* (aeolian rendzinas); *t* (solonetz); *u* (planosols).

(*5*) In the kaolinitic–podsolic region (*K–P*) the principal soils are: *e* (aeolian (sea dunes); (*k/c*)(*p/u*) (kaolinitic–cinnamonic) (podsolic–planosolic); *knl* (terres de barre); *kp* (kaolinitic podsolic); *o* (peaty); *pb* (grey–brown podsolic); *pc* (non-calcic brown); *ph* (humus podsols).

(*6*) In the podsolic–kaolinitic (*P–C*) region the more important soils are: *ck* (cinnamonic–kaolisol intergrades); *k* (kaolisols); (*k/c*)*p* (kaolinitic–cinnamonic–podsolic); (*k/c*) (*p/u*) (kaolinitic–cinnamonic) (podsolic–planosolic); *lh* (lithosolic rankers); *my* (chernozemic clays); *pbg* (gleyed grey–brown podsolic); *u* (planosols).

(*7*) In the atlantic podsol region (*AP*) of Tasmania, the principal soils are: (*k/c*)*p* (kaolinitic–cinnamonic) podsolic; *o* (peaty); *pbg* (gleyed grey–brown podsolic); *ph* (humus podsols); *re* (aeolian rendzinas).

(*8*) In the high latitude humid mountains (*M*), the more important soils are: *lh* (lithosolic—alpine—rankers).

To sum up, many Australian soils are rich in 1:1 clays and have low cation exchange capacities in spite of the dry climate; many soils may be classified as kaolinitic-cinnamonic intergrades (*k/c*); their adjusted *CEC* (cation exchange capacity)/clay ratio is just on the limit between kaolisols and non-kaolisols. Moreover soils with eluvial horizon are frequent; however, it is often difficult to decide whether clay eluviation has been produced by organic matter (podsolization) or by natrium (solonization). That is why many soils are (*k/c*)(*p/u*) (kaolinitic–cinnamonic) (podsolic–planosolic).

Since many soils of Australia are kaolinitic or nearly kaolinitic their fertility is low; mineral deficiencies are frequent; phosphorus deficiency is almost universal. That is why the problem of mineral fertilization is more acute in Australia than in other parts of the world. Australians have highly contributed to the study of mineral deficiencies.

New Zealand (Fig.46)

New Zealand includes six broad soil regions: (*1*) atlantic podsol (*AP*), (*2*) high latitude humid mountains (*M*), and (*3*) brunisolic (*B*) on the South Island; (*4*) podsol–ando (*P–A*), (*5*) high latitude humid mountains–ando (*M–A*) and (*6*) brunisolic–ando (*B–A*) on the North Island.

(*1*) In the atlantic podsol (*AP*) regions, the principal soils are: *a* (alluvial); *aw* (wet alluvial); *ba* (brun acide); *pb* (grey–brown podsolic); *pg* (gley podsols); and *vl* (lithosolic ando).

(*2*) In the high latitude humid mountains (*M*) the principal soils are: *a* (alluvial); *b* (braunerde); *ba* (brun acide); *l* (lithosols); *lh* (lithosolic rankers); *pb* (grey–brown podsolic); *pg* (gley podsols); and *vl* (lithosolic ando).

(*3*) In the brunisolic regions (*B*) the principal soils are: *a* (alluvial); *aw* (wet alluvial); *b* (braunerde); *ba* (brun acide); *g* (gleisolic); *o* (peaty); *pb* (grey–brown podsolic); *u* (planosols).

(*4*) In the podsol–ando regions (*P–A*), the principal soils are: *aw* (wet alluvial); *b* (braunerde); *ba* (brun acide); *e* (aeolian sands); *g* (gleisolic); *o* (peaty); *pb* (grey-brown podsolic); *pho* (peaty humus podsols); *u* (planosols); *v* (ando); *vl* (ando—red–brown loam from basic volcanic ash); *v2* (ando—red–brown from volcanic ash); *v4* (ando–recent from ash).

(*5*) In the high latitude humid mountains–ando regions (*M–A*) the principal soils are: *a* (alluvial); *ba* (brun acide); *l* (lithosols); *pg* (gley podsols); *vl* (lithosolic ando); *v2* (ando—red–brown from volcanic ash); *v3* (ando—yellow–brown from pumice); *v4* (ando—recent from ash).

(*6*) In the brunisolic–ando regions (*B–A*): *a* (allu-

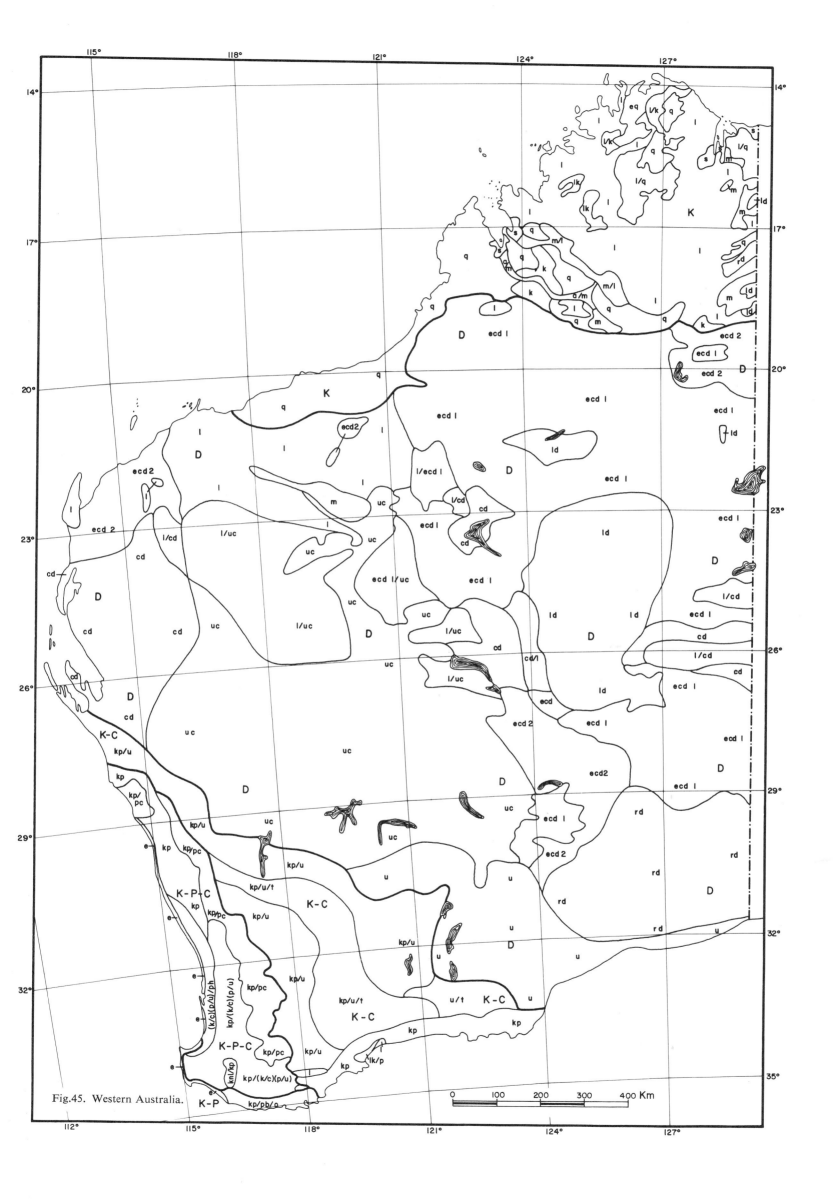

Fig.45. Western Australia.

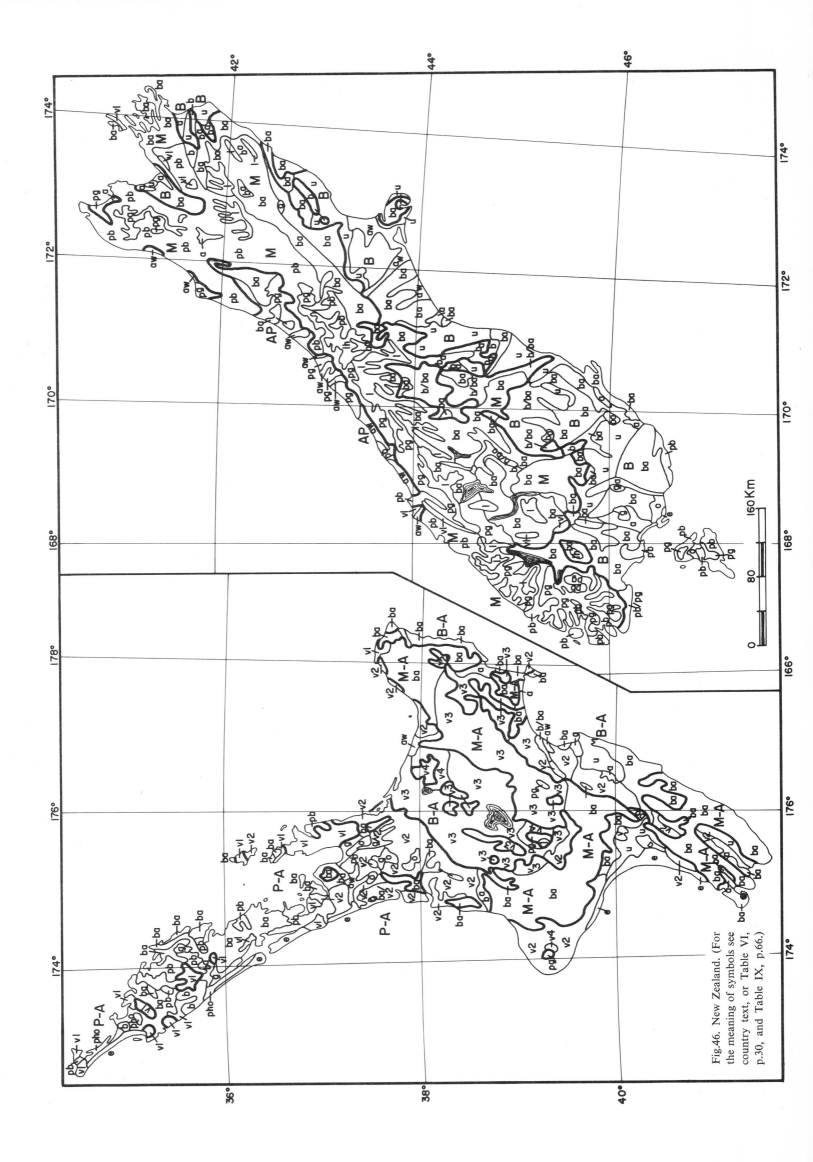

Fig.46. New Zealand. (For the meaning of symbols see country text, or Table VI, p.30, and Table IX, p.66.)

vial); *aw* (wet alluvial); *b* (braunerde); *ba* (brun acide); *u* (planosols); *v2* (ando—red–brown from volcanic ash); *v3* (ando—recent from ash).

The soils of New Zealand, more especially those of the North Island, are to a great extent ando; volcanic materials form a more or less large part of their parent material. That is why their mineral fertility is higher than might be expected from the climate under which they have been formed; but their ando characters also pose some problems.

Chapter 8 | Agricultural Potentialities

It is not the author's intention to give here a detailed discussion of soil relations of plants. Only a general outline will be given, which is necessary in order to understand the differences in agricultural possibilities between different types of soils. One should always keep in mind that soil is only one part of the environment in which plants grow; and environment acts as a whole. That is why one cannot separate soil action from the other factors. Many failures in understanding soil relations of plants can be attributable to the fact that not due attention has been given to the interaction of soil with other factors, more especially climate.

Soil acts on plants in three different ways:

(*1*) It absorbs the toxins that are produced in the rhizosphere, and helps their destruction. When plants are grown in solutions, the medium becomes rapidly saturated with toxins and we are obliged to change it very frequently in order to obtain a normal growth; but by adding to the solution substances that have an absorbing capacity (artificial resins, etc.), we obtain good growth, even if the solution is not changed frequently. In soil it is soil colloids that serve as absorbing substances.

(*2*) It serves as a reservoir of water; it absorbs rain or irrigation, and delivers it gradually to plants.

(*3*) It provides the nutrients that are necessary for plant growth; it is only C, H, and O that come directly or indirectly from the air.

Soil as neutralizer of toxins

Experiments carried out all over the world have shown that between certain limits the yield of one plant (one individual) is directly proportional to available space (area × depth). This is not due to limitations in water or nutrients, because by increasing nutrients, the space required per plant increases instead of decreases; in fertile soils the rate of sowing is less than in poor soils; in a small container you cannot obtain a normal plant growth, even by maintaining the best concentration of water and nutrients; as the experiments of the author (PAPADAKIS, 1941) have shown space acts as an independent factor. The only explanation of these facts is that toxins (injurious substances) are produced in the rhizosphere, that is why the amount of roots that can live in a certain space is limited and roots search always for new soil. The existence of these toxins has been demonstrated

by PICKERING (1917), who advanced the theory; and it has been confirmed by more modern methods by PAPADAKIS (1941, 1949, 1962c). In these experiments the writer has also shown that when fine earth is mixed with gravel, the yield is proportional to the fine earth; the experiment was factorial with various levels of the nutrients and water, and the influence of clay (fine earth) was always observed, with high or low levels of nutrients or water and with all possible combinations of levels. Thus, colloids act as an independent factor; they absorb and facilitate the destruction of the toxins produced in the rhizosphere, either by the plants themselves, or by micro-organisms. It is possible that these toxins are antibiotics produced by micro-organisms living on root excretions (PAPADAKIS, 1963b).

PICKERING (1917) maintains that such toxins are easily oxidized. If this is so, and there is abundant evidence supporting Pickering's assumption, soil should be, at the same time, rich in colloids and permeable. Increase in clay reduces permeability so that above a certain limit yields do not increase by increasing clay content. It can be said that the principal characteristic of soil, that which makes it an adequate medium for plant growth, is its absorbing capacity. Soil is of value for its absorbing capacity, but such capacity should be accompanied by permeability. Ando (volcanic) soils have a high cation exchange capacity and are at the same time very permeable. This is probably the explanation why, since immemorial time men cluster around volcanoes, in spite of the great catastrophes they cause. Population densities around volcanoes are often many times greater than in other soils. It may be that the high productivity of ando soils is due, to a considerable extent, to their high absorbing capacity.

On the other hand kaolisols have a low cation exchange capacity, and it may be that their low productivity is partly due to this cause. Unfortunately ecologists and soil scientists have paid little attention to rhizosphere toxins, and very little is known about the capacity of various colloids and soils to absorb and help destroy these toxins.

Soil as a reservoir of water

Rain is rapidly absorbed by soil. How rapid this absorption is is shown by irrigation practice. Very often irrigated fields receive at one time a quantity of water that corresponds to 100 mm (4 inches), i.e., to the rainfall of a rainy month; and all this water is absorbed in one hour or so. Irrigation practice shows also that when water runs as a sheet, on the surface of the soil, it cannot go very far, it becomes absorbed. But water absorption depends on water capacity, which is the difference between field capacity (the amount soil holds against gravity), and wilting point. Under natural conditions soil seldom dries below the wilting point, because except for a few centimeters of the surface, soil is dried by plant roots and such absorption stops when the wilting point is reached; moreover, there is little difference between the water content of a soil dried at usual temperatures (e.g., 30°C) and wilting point. Penetration of water may be naturally stopped by an impermeable layer, bed-rock for instance. But when a horizon is rich in swelling clays, it swells and water penetration is virtually stopped. In very clayey soils much of the water that penetrates deeply descends through cracks before they are closed by clay swelling.

Capillary rise of water is relatively slow. When dry weather prevails the amount of water that arrives by capillarity from lower layers is less than that lost by evaporation, and soil surface dries out. This fact virtually stops capillary rise. The water that remains in the soil is lost by root absorption. However, if weather is moist and soil surface is not dried out, capillary ascension continues. However, in this case soil is usually covered by vegetation and losses by soil evaporation are low. Thus during dry periods, soil evaporation is responsible for the loss of the water that remains in soil surface (the upper 10–20 cm) after each rain. The water that has penetrated more rapidly is lost by plant transpiration; but no water can penetrate more profoundly until soil surface has reached its water-holding capacity.

Thus, the behaviour of soil as water reservoir depends on weather conditions. If rains are scarce (desert conditions) a sandy soil is preferable, because the amount retained in the surface and lost by evaporation is small; the percentage entering more deeply and used in plant transpiration is greater. In a clayey soil small rains moisten only the surface and this water is lost by evaporation. In the deserts we often encounter a water table below active dunes, because the water which penetrated below the surface cannot be lost either by soil evaporation, or by plant transpiration, since these dunes are barren of vegetation. On the other hand, if there is an alternation of a humid and a dry season, the quantity of water which the soil can store during the humid season is important. A sandy, shallow soil can store little water. A clayey soil of the same depth may store more. But when both soils are deep a medium texture soil is better, because in a clayey soil water cannot penetrate deeply; a deep ando soil is the best because water penetrates easily and water capacity is also high.

Soil as provider of nutrients: "potential", "actual" and "mineral" fertility

All the nitrogen that the soil contains is in organic form. It has been proved experimentally that it is impossible to increase soil total nitrogen content without a corresponding increase of its carbon content. Except in the case of legumes and mineral fertilizers, all the nitrogen plants absorb arises from the decomposition of organic matter (a little is provided by rain).

Because of its origin, soil organic matter contains all bio-elements. So that what happens with nitrogen happens also with the other nutrients, which for reasons of convenience we shall call "mineral". But in this case a part of the nutrient can be absorbed directly from the mineral part of the soil. The origin of mineral nutrients is in the mineral fraction; the first plants that grew in the soil absorbed mineral nutrients from the mineral fraction and little by little the organic reserve has been built up.

Since organic matter is the source of all the nitrogen and a great part of the other nutrients, soil fertility depends greatly on its organic matter content. That does not mean that soils can be classified as more or less fertile according to their organic matter content; but when a given soil is considered its fertility increases or decreases with its organic matter.

Organic matter content is greatly affected by the cropping system. It is highest when the land is used as pasture, being mowed or grazed, but seldom plowed. It is minimum when soil is maintained continuously barren of vegetation by frequent tillage operations. Annual crops that rapidly cover the soil and are not intertilled (wheat for instance) tend to maintain the soil at a high content of organic matter, although lower than that corresponding to pasture. Crops that do not rapidly occupy the soil and are intertilled (potato, maize, cotton for instance), tend to maintain organic matter at a low level, although not so low as barren fallow. Usually crops alternate and the result of the rotation depends on the relative frequency of sown pastures (e.g., alfalfa), rapid occupation crops (e.g., wheat), slow occupation crops (e.g., maize) and barren fallow.

However, the amount of nitrogen which at a certain moment is available in a soil does not only depend on organic matter content, it depends also on the rapidity with which this organic matter decays. In this respect a barren fallow accelerates organic matter decay and results in the accumulation of great quantities of nitrates, various times greater than what plants would absorb from the soil were it not maintained barren of vegetation. On the contrary when a soil bears a dense vegetation, organic matter decay is slowed. The result is that when wheat is sown after oats ("rapid occupation" crop), without intercalation of a barren fallow, the yield is low; when it is sown after maize ("slow occupation" crop) the yield is better; and when sown after a barren fallow the yield is best.

For all these reasons the writer (PAPADAKIS, 1938, 1952a, 1954, 1960a) introduced the concept of "potential" and "actual" fertility. "Potential" fertility depends on organic matter content; it is maximum by using the soil as mown or grazed pasture; it is lowest when a soil is maintained continuously barren of vegetation (this has been done experimentally). On the other hand, "actual" fertility depends on immediately available nutrients (nitrates, etc.); it is minimum in a pasture of grasses; it is maximum after a barren fallow that lasted several warm months. "Potential" fertility can be determined by determining total N; "actual" fertility by determining nitrates. The influence of crop system on these contents and crop yields has been proved by experiments carried out all over the world (PAPADAKIS, 1938, 1954). For instance, pastures bring the soil to its highest potential fertility (highest organic matter content); continuous barren fallow brings the soil to its lowest level of potential fertility (lower organic matter content). But increase or decrease do not continue indefinitely; after a certain number of years an equilibrium is attained; "potential" fertility cannot be further increased; or soil cannot be further impoverished (PAPADAKIS, 1938, 1952a, 1954).

Naturally, if parent material is poor in a certain "mineral" nutrient, and/or during soil formation this nutrient has been leached away or immobilized, the organic matter of this soil will be *relatively* poor in this nutrient, and plant growth will suffer from such poverty. The deficiency may not be so apparent when the soil is not cropped, because the organic reserve that is built even in a poor soil is sufficient to maintain a normal growth. It becomes apparent after certain years of cropping, when organic matter content has been reduced by the frequent repetitions of intertilled crops or barren fallow. Therefore, besides "potential" fertility and "actual" fertility, which depend on the cropping system, another aspect of fertility exists, "mineral" fertility, which is also affected by cropping but depends greatly on parent material and soil formation processes. Mineral fertility should be considered in relation to each element. A soil may be rich in zinc and poor in molybdenum; sometimes richness in one element may be the cause of the deficiency of another.

Mineral fertility depends greatly on parent material and soil formation, but unfortunately many pedologists are not interested in soil fertility; it is not their job. On the other hand, fertility specialists are not interested in pedology; it is not their job. That is why progress has been very slow in this field. The content in each nutrient of parent materials should be known; and what happens to each of them during the several types of soil formation; but our knowledge is still very limited. However, a few pedologists were interested in fertility problems; and a few fertility specialists were understanding pedology; so that some progress has been achieved.

AGRICULTURAL POTENTIALITIES OF THE VARIOUS SOIL GROUPS: A. SOILS WITH DEGREE OF LEACHING 1 AND 2 (DOMINATED BY 2:1 CLAYS AND NOT STRONGLY ELUVIATED)

Rendzinas

Many limestones are of organic origin; or have been the seat of life during their formation; life contains all bio-elements. Thus limestones are usually more or less well provided with all micro-elements. Rendzinas have suffered very little leaching, so little that even the surface horizon is calcareous. Such little leaching is unlikely to severely impoverish soil. Since parent material is not poor, and soil forming processes do not severely impoverish it, mineral fertility is usually fair in rendzinas. Some rendzinas (para-rendzinas) are formed from silicates under a dry climate; the lime released by weathering is not leached away by weathering and accumulates. In this case also mineral fertility is usually fair. The material is often igneous and consequently well provided with minerals; and leaching has been very slight. What often happens is that high Ca content makes some elements unavailable to plants. It was known long ago that some plants (lupins, some American grapes, chestnut, tea, etc.) suffer iron deficiency when grown in calcareous soils. What happens with iron, happens also with other elements similar to iron in their chemical behaviour. On the other hand, high lime content makes rendzina very good for many plants (legumes except lupin, etc.). The vegetation of calcareous soils is often different from that of other soils, because of the abundance of calcicolous plants, and the low frequence of calcifuge ones. Calcicolous plants are those that live in calcareous soils, and calcifuges those that avoid them. This is often attributed to pH, but pH usually acts by modifying the availability of plant nutrients. A high content in Ca favours the accumulation of organic matter and this is an advantage. On the other hand, it slows down weathering, and that may be a disadvantage. BRAMMER (1965) observed in East Pakistan that Ganges alluvia, rich in lime, are less fertile than those of the Brahmaputra, which are not.

In humid climates, rendzinas are often shallow, because leaching is intense and old soils usually have their surface horizon decalcified. It is erosion that keeps soil young and permits the existence of rendzinas. Being shallow, rendzinas are exposed to drought; but rendzinas can be due to other factors than erosion; when their water capacity is good, rendzinas are very good soils; in podsolic regions they are the best soils. In kaolinitic regions rendzinas are rare; but when they exist and they have no other defects (salinity, waterlogging, etc.) they are the best soils. Rendzinas may be at the same time saline, gleisolic, etc. These soils will be discussed in the corresponding paragraphs.

Dark clays

Dark clays have often been formed from basalts or other ferromagnesian igneous rocks; sometimes they arise from volcanic materials; or from accumulation of clays in a lowland. Leaching has always been very moderate in the case of these soils, which have S higher than 30 mequiv./100 g of soil. So that dark clays seldom show mineral deficiencies.

High clay content creates some inconveniencies. A

light rain is retained on the surface and is entirely lost if it is followed by dry weather. Heavy rains cannot penetrate with sufficient rapidity and may cause runoff or water-logging. Tillage operations require more power. In regions with shifting agriculture, where all tillage operations are done by hand, dark clays are not used, because they require too much labour. When an extensive region is covered by dark clays, water does not penetrate and there is no water table; drinking water for livestock is often difficult to find; and this is a problem. However, modern technology permits the solution of these problems (tractors for tillage, bringing water from neighbouring regions, etc.).

In irrigated agriculture very heavy clays are a difficulty, because water absorption is very slow. When a clay is alkaline, reclamation is difficult or costly. Some crops (peanut, sesame, etc.) cannot be grown in clayey soils. Many clays are saline, alkaline, waterlogged or flooded. The better dark clays are rendzina clays and chernozemic clays; and the best are rendzina–chernozemic clays. Organic matter well provided with calcium gives dark clays a good structure and corrects their physical defects.

Ando

Ando are young soils from volcanic materials. Volcanic materials are igneous and usually contain all bio-elements. Moreover these materials weather rapidly, and the soil is young; otherwise amorphous clays crystallize and soil ceases to be ando. Thus ando soils are usually well provided with bio-elements. However, free iron and alumina fix phosphorus and make it relatively unavailable to plants. That is why ando soils very often require phosphorous fertilizers, and the quantities required are high. It may be that foliar fertilization will reduce such quantities. Another method to reduce the amount of phosphorus needed is to use fused phosphate instead of soluble phosphate, mix the phosphate with compost so as to increase the percentage of it that is transformed into organic phosphorus, localized applications, pellets, or pellets covered with preforated plastic material.

When leaching rainfall is very high, as in Japan and northern New Zealand, ando soils may show signs of deficiency of certain elements, such as potassium, cobalt, sulphur, copper and boron (for Brassica root crops). But that depends greatly on the composition of parent material. Dry climate ando seldom show such deficiencies. (See DURING, 1964.)

In many countries, e.g., Japan, ando soils are encountered under a podzolizing environment; slightly polymerized and acid organic matter mobilizes alumina, and alumina is toxic to plants. But cropping stops the production of slightly polymerized substances, so that this defect is observed in newly reclaimed soils; but it disappears with cropping; manuring accelerates this process; irrigation with water containing bases and neutralizing

acidity is also effective (KOBO, 1964); phosphorous fertilizers reduce exchangeable alumina.

Some ando soils have unweathered pumice scoria at low depth, and this fact limits their depth; breaking this layer and mixing it with surface soil has excellent results (KOBO, 1964); not only roots can penetrate deeply, but weathering is accelerated.

In the case of paddy (irrigated rice) the high permeability of ando soils is a defect, because it increases water consumption; a high percolation of water reduces also soil temperature, because the water does not remain a sufficient time at the surface to be warmed by the sun's rays. But except for the case of irrigated rice, for upland rice and all other crops the high permeability of ando soils, combined with high water capacity is a great advantage. On the island of Santorini (Greece), with 370 mm (15 inches) of rain, yields of 2,000 kg of barley per hectare are obtained and tomatoes are grown; in spite of the rainless summer, grapes also grow very well, due to the deep ando soils; population density is high.

Everywhere in the world, volcanic soils are considered as highly productive and population is attracted to volcanoes, in spite of frequent catastrophes. In the tropics they are the best soils. This is chiefly due to their high absorbing capacity, combined with high permeability, which permits them to absorb and oxidize the toxins of the rhizosphere and sustain a very dense vegetation. The best ando soils are naturally the "chernozemic ando".

Raw soils

Raw soils are usually young and that is why mineral deficiencies are infrequent. But much depends on parent materials. From a mineral fertility point of view lithosols are usually good soils, but absorbing capacity per unit area is very low. The same defect is encountered in dunes; but dunes are often deep, and absorbing capacity per unit area is often satisfactory; moreover they are permeable. Dunes formed from volcanic materials, from sea shells, etc., are usually fertile. On the other hand, podsolized dunes (landes, coconut sands, etc.) are very poor. Naturally dunes are often exposed to wind erosion; and their unlevel surface creates some problems. In the case of irrigation they may be too permeable. Alluvial soils often have fair mineral fertility; they seldom are leached; on the contrary, they receive nutrients from other soils. The origin of the material is very important. When they arise from the erosion of volcanic materials, or of young soils formed on consolidated rocks, they usually have a high mineral fertility, but when they have already been leached or de-silicated in their place of origin (kaolinitic materials) they are poor.

Naturally texture both of surface and lower horizons varies greatly from one alluvial soil to another and that has a great influence on their productivity, irrigation capability, etc. Moreover, many alluvial soils are waterlogged, flooded, saline, alkaline, etc. These soils will be dealt with later.

From a mineral fertility point of view recent brown are usually good. In the tropics they are the best soils, but their depth may be shallow, or they may be too stony. Arid brown and para-serozem are also usually good soils. Given irrigation they give a good yield. Naturally that depends also on their texture, parent material, etc.

Rankers

Rankers are usually young soils and their mineral fertility is satisfactory; they are also well provided with organic matter. Their defect is their shallowness; their value depends on their depth; naturally texture is also important; until a certain point clay content compensates depth. But in acid rankers (podsol rankers or intergrades to them) the nutrients released by weathering are leached away; and mineral fertility is low. As long as the soil is used as forest or pasture the organic reserve of nutrients masks this poverty. But when the soil is cropped with frequent intertilled crops and barren fallows, which lower "potential" fertility (organic matter) serious deficiencies appear. Due to their topography these soils are also exposed to erosion. Chernozemic rankers are naturally the best and podsol rankers the worst.

Brunisolic soils

Braunerde are usually fairly well provided with bases. They usually have a good absorbing capacity per unit area; due to the presence of iron, their structure is not bad, even when their clay content is high. They may be grouped with the best soils of podsolic regions (including brunisolic). The situation is different concerning brun acide, which are usually poorer in nutrients and of lower absorbing capacity. But productivity of these soils is highly increased by liming and by using fertilizers.

Cinnamonic soils

The mineral fertility of "neutral" cinnamonic soils is usually satisfactory. When dry they are often too hard for their texture; but when neither too dry nor too wet, they are easy to work. Their structure is usually good. Their absorbing capacity per unit of clay is usually high. So that they may be considered as fairly good soils. They are often poor in organic matter but this is a question of climate. "Acid" cinnamonic soils are usually intergrades to kaolisols, and their fertility is also intermediate.

Chernozemic soils

Chernozemic soils usually arise from calcareous or volcanic materials, loess, glacial drift, etc. All these materials are geologically young. Moreover chernozemic soils

had not suffered severe leaching. Therefore their mineral fertility is usually good. Their absorbing capacity is also usually high. So that chernozemic soils may be considered as good soils, and chernozemic ando even better. Some chernozemic soils (meadow chernozems, humic glei) are transitorily or permanently waterlogged; some may be saline, alkaline or gleisolic; we shall discuss these soils in other paragraphs.

AGRICULTURAL POTENTIALITIES OF THE VARIOUS SOIL GROUPS: B. SOILS WITH DEGREE OF LEACHING 2–6 (DOMINATED BY 1:1 CLAYS AND/OR PODSOLIZED)

Kaolisols

Kaolisols have low absorbing capacities. This is perhaps a cause of low productivity, more especially when organic matter content is reduced by cropping. We do not know what is the capacity of kaolinitic clays to absorb and help destroy rhizosphere toxins. Moreover kaolisols are the result of allitic weathering, which leaches away the greater part of bases and the greater part of silica. Such intense weathering leaches away many bio-elements. That is why mineral deficiencies are common in kaolisols. High content of free iron immobilizes phosphorus. As a consequence, soils are poor in available phosphorus, and high quantities are needed. In order to avoid such immobilization it is advisable to localize fertilizers, not mix them with all soil, but to give them in pellets, possibly covered with perforated plastic, etc. The true solution will probably be foliar application. As long as the soil is rich in organic matter, mineral deficiencies do not have grave consequences, because organic matter is a reserve of all nutrients. But when organic matter content is reduced by cropping (intertilled crops and barren fallow) deficiencies appear. A distinction should be made between "eutrophic" and "dystrophic" soils. The latter are much poorer than the former. Certain crops, cocoa and yams for instance, are grown only in "eutrophic" soils. Others, rubber, oil palm, coconut grow fairly well in dystrophic soils. The best kaolisols are perhaps terres de barre, because they are "eutrophic" and have a "latosolic" profile; in tropical ferruginous soils the existence of a concretionary horizon is a defect of more or less importance. Ferralitic soils, by having very low cation exchange capacity and being dystrophic, are perhaps the worst soils. Exchangeable aluminium is often a problem in kaolinitic soils, but the presence of bases and iron reduces it. That is why "eutrophic" kaolisols, terra roxa and ferrisols are among the best kaolisolic soils. Kaolisols are usually deep, well structured, permeable; these are great advantages.

Podsolic soils

In podsolic soils (grey–brown podsolic, etc.) the only effect of podsolization is clay eluviation. The eluvial horizon is often "acid"; but the illuvial one is often "neutral". Mineral fertility is lower than in brunisolic soils, but higher than in podsols. The "eluvial" horizon is greatly improved by cropping, more especially when lime and mineral fertilizers are added, and since it is this horizon that is poor, podsolic soils are transformed into good soils. Prairie lessivé have a chernozemic horizon; that is why prairie lessivé are the better podsolic soils. Grey-wooded and non-calcic brown have usually a neutral "eluvial" horizon; they have been much less impoverished than grey-brown podsolic with an acid eluvial horizon.

Kaolinitic podsolic, more especially red–yellow podsolic, combine the defects of kaolisols and those of podsolic soils; that is why their mineral fertility is usually still lower than that of grey–brown podsolic. However, they are improved by liming and fertilizers. Neutral kaolinitic podsolic are better than red–yellow podsolic.

Ground-water laterites. In ground-water laterites the presence of an impermeable horizon, which begins at a distance of less than 50 cm from the surface, is a serious handicap. Soil is often waterlogged. Moreover the absorbing capacity of the soil above the laterite is often low; and the soil is sometimes dystrophic. For these reasons laterites are seldom good soils, although they can be used to grow rice; by adding nitrogen rather good yields can be obtained.

Podsols

Many podsols are the result of "podsolic weathering"; others were so poor in clay that not only clay but also iron has been eluviated from the "eluvial" horizon; others may be considered as degraded podsolic; after the eluviation of clay has impoverished the soil in clay, iron also has been eluviated. Leaching with organic acids is very severe, that is why the mineral fertility of podsols is usually low. Moreover these soils are usually sandy and their mineral absorbing capacity low. When leaching is moderate the nutrients leached from the eluvial horizon have a chance to be retained in the eluvial horizon. However, when leaching is intense such retention is very little. That is why atlantic podsols (those of regions with high leaching rainfall) are usually poorer than continental podsols (those of the regions with low leaching rainfall). But leaching depends also on other factors, more especially drainage. Podsols arising from the degradation by leaching of podsolic soils are naturally very poor. Degradation is often the result of planting coniferous forest (DUCHAUFOUR, 1959).

Humus podsols have usually suffered a very intense leaching with organic acids; although drainage is usually bad, some drainage takes place—otherwise the land would be transformed into a lake, since rainfall greatly exceeds evapotranspiration. This is why humus podsols are usually poor in mineral fetility.

It is to be noted that cropping impedes podsoliza-

tion, replaces it by humification and improves podsols. The improvement is accelerated by liming, using mineral fertilizers and manures. That is why sod podsols are rather good soils, and many podsols of Germany, Belgium, The Netherlands, etc. are now fertile soils.

AGRICULTURAL POTENTIALITIES OF THE VARIOUS SOIL GROUPS: C. HALOMORPHIC, ORGANIC AND GLEISOLIC SOILS

Saline soils

Saline soils are usually alluvial; materials are young; moreover instead of being leached, they receive waters which have leached other soils. That is why their mineral fertility is usually rather good; and when de-salinized they are transformed into good soils. But when the saline soil is also alkaline, leaching destroys its structure; gypsum, sulfur or organic matter permits the avoidance of this inconvenience.

Many saline soils are used to grow rice; salts are diluted and do not harm the crop; such cropping helps also to de-salinize the soil when some water is lost by drainage. Some crops can better resist a low alkalinity; for instance, barley, some varieties of wheat, sorghum, beet, cauliflower, asparagus, grapes, and olives.

Solonetz

The great defect of solonetz is the impermeability and alkalinity of the illuvial horizon. It is a very bad medium for root growth. Moreover structure is deficient, permeability low, and soil suffers alternatively from waterlogging and drought. Gypsum, sulfur and organic matter permit to improve them.

Planosols

The main defect of planosols is impermeability of the illuvial horizon; it seems also that the eluvial horizon is impoverished. However, there are great differences in these respects between planosols. Planosolic para-chernozems are rather good soils although not so good as non-planosolic para-chernozems. Planosolic humic–glei suffer from waterlogging, and the same is true for many other planosols. Drainage, when necessary, pasture crops, and fertilizers improve greatly these soils.

Organic soils

"Eutrophic" organic soils are rather rich in mineral nutrients and organic matter; their problem is waterlogging. After drainage, they become good soils. "Dystrophic" organic soils are poor in mineral nutrients; very often the soil is a podsol with a thick peaty or organic horizon. In this case, besides waterlogging, we have the problem of mineral defficiencies and very often low absorbing capacity per unit area after the destruction of organic matter.

Gleisolic soils

Some gleisolic soils have a high base status; they even may be calcareous or saline. In this case mineral fertility is seldom a problem; waterlogging is the usual defect; salinity in certain cases. But acid gleisolic soils have been often leached by acid organic substances and have lost a great part of their mineral nutrients; they are often podzolized. In this case besides the problem of waterlogging, that of mineral deficiencies is also serious.

Some gleisolic soils, reclaimed from the sea (mangrove soils, etc.), contain considerable quantities of sulfur. When soil is drained, this sulfur is oxidized and transformed into sulfuric acid; soil is leached with sulfuric acid; and a "cat clay" is formed.

AGRICULTURAL POTENTIALITIES ACCORDING TO SOIL REGION

Within each soil region the choice of soils is limited; so that a soil that is rather bad in a region with good soils may be good in another, where other soils are worse than it. Moreover ecologic conditions change from region to region and the problem of soil productivity takes another form. That is why it is necessary to discuss the problem of agricultural potentialities not only soil by soil, as was done in the foregoing sections, but region by region.

Podsolic regions (P)

In podsolic regions the best soils are those that are less podsolized: rendzinas (when not too lithosolic), braun erde, alluvial and other raw soils. Organic and gleisolic soils are usually dystrophic and more or less podsolized, but when this is not so they become good after drainage.

In atlantic podsol regions (*AP*) the choice of soils is often so limited that grey–brown podsolic and brun acide may be considered as good. But cropping combined with liming, fertilizers and manure has transformed many soils in very productive.

In continental podsol regions (*CP*) the best soils are grey-wooded and naturally brunisolic and alluvial.

In sod podsol regions (*SP*) grey-wooded, sod podsol, degraded chernozems, calcareous rendzinas and alluvial are among the best soils.

In glei-podsol regions (*GP*) the less a soil is gleisolic the better; alluvial soils, when well drained, fulfil this requirement.

In permafrost podsol regions (*PP*) alluvial and sod podsols are among the best.

In atlantic podsol–ando regions (*AP–A*) ando soils and sod podsols (probably they are sod podsol because they are ando) are among the best.

In podsol–ando regions (*P–A*) ando, alluvial, and acid brown, usually of volcanic origin, are good soils; they support over-grassing.

In grey–brown podsolic regions (*GB*) prairie soils, braun erde, rendzinas, planosolic para-chernozem are preferable.

In brunisolic regions (*B*) the panorama of soils is usually good; rendzinas (when not lithosolic), braun erde, cinnamonic brown, para-chernozem lessivé, degraded chernozem, calcareous braun erde (old rendzinas), terra rossa (when not lithosolic), and naturally alluvial are the best soils.

In brunisolic–alluvial regions (*B–Al*) the panorama of soils is still better, because of the abundance of alluvial soils, which when well drained are very good.

Still better are the soils of the brunisolic ando regions (*B–A*), where ando and braun erde are among the best soils.

Mediterranean cinnamonic regions (MC)

In the mediterranean cinnamonic regions (*MC*) all soils are good when not lithosolic and well drained; naturally saline soils solonetz and rather pure sands are not included. Some alluvial soils need drainage; and some dark clays are alkaline or suffer waterlogging. Deep rendzinas and terra rossa are very good soils.

Mediterranean cinnamonic–ando (MC–A) regions. These regions have still better soils than the other parts of the Mediterranean. When leaching rainfall is high, acid soils (classified as brun acide) abound; but they are easily transformed into good soils by cropping, liming and phosphorous fertilizers.

Cinnamonic regions (C). Cinnamonic regions also have good soils, except, of course, when drainage is bad or they are lithosolic; but many of these soils intergrade to kaolisols.

Brunisolic cinnamonic (B–C) regions. These regions also have good soils, except when the soil is saline, alkaline, waterlogged, or lithosolic.

Arid cinnamonic regions (AC)

Arid cinnamonic regions have also very good soils, except when the soil is saline, solonetz, waterlogged or lithosolic. Serozems, arid brown and reddish brown are usually good soils; dark clays when heavily clayey or waterlogged present certain difficulties; but except this case they are productive soils.

Arid cinnamonic–brunisolic regions (AC–B). These regions also have good soils, except when they have obvious defects (salinity, waterlogging, etc.).

Chernozemic regions (Ch)

These are probably the best soil regions of the world. All soils are good except saline, alkaline, solonetz and some planosols, waterlogged or lithosolic soils.

Cinnamonic chernozemic regions (C–Ch). Soils are good but organic content is less than in the chernozemic region; planosols are more abundant (non-calcic brown, etc.).

Sub-antarctic chernozemic. Grassland rankers are rather good but their absorbing capacity is rather low (too sandy).

Chernozemic–desert regions (Ch–D). Usually soils are very good, but some of them are too sandy, or have an illuvial horizon at shallow depth (reddish desert lessivé).

Mediterranean mountain–chernozemic regions (M'–Ch). Soils are usually very good, but lithosols abound.

Desert regions (D)

Desert lithosols are usally too stony, and many desert sands are too sandy; planosolic reddish desert are often too shallow; desert pavement creates also some difficulties. Saline soils, solonetz and waterlogged soils need reclamation to be used. But soils that do not present these obvious defects are good, and with irrigation give high yields.

Kaolinitic regions (K)

The best soils of kaolinitic regions are those which are not kaolinitic (ando, dark clays, recent brown, alluvial, etc.). It is in these soils that sugar cane and most of the cotton and rice are grown. The most densely populated regions of the tropics have ando, alluvial or other not-kaolinitic soils. Among kaolinitic soils the best are "terres de barre" ("eutrophic" with "latosolic" profile); they sustain dense populations. Terra roxa are also very good and much coffee is grown on them. Other "eutrophic" kaolisols (tropical ferruginous) are also good, although less than terres de barre; it is on them that cocoa and yams are grown. Dystrophic soils (acid ferruginous, ferralitic, etc.) are bad; but some plantation crops (coconut, rubber, oil palm) and cassava grow rather well on them; with long fallows rice is grown. Ground-water laterites suffer from waterlogging; moreover they often

have low absorbing capacity. However, with long fallows and flooding rice is grown. It is to be noticed that in soils in which phosphorus is immobilized by iron, waterlogging reduces iron and releases phosphorus, creating a high "actual" fertility concerning this element, which is often limiting in kaolisols. As stated before mineral deficiencies are frequent in kaolinitic soils.

Subtropical kaolinitic regions (SK). In this region (southeastern United States) de-silication is not severe; soils are near the limit between kaolinitic and non-kaolinitic soils; podsolization has led to very low base status. The best soils are naturally those that are not kaolisols and are not acid, without presenting other defects as, for instance, waterlogging. Many soils have been improved by liming, adding fertilizers and growing pasture crops.

Young kaolinitic regions (YK). In these regions many soils have high cation exchange capacities, some are rendzinas or old rendzinas and the panorama of soils is far better than in kaolinitic regions (K). They usually sustain dense populations.

Kaolinitic–ando regions (K–A). In many kaolinitic regions volcanic materials abound and many soils are ando; the proportion of ando soils is perhaps higher than reported, because the similarity of profile between latosolic kaolisols and ando led to a confusion between them; many pedologists rely too much on morphology; many kaolisols reported as rich in organic matter and mapped in our maps as rubrozems (*kh*) are probably ando. That is why many soils reported as kaolisols are good. Rubrozems, dark clays, terres de barre, and, naturally, ando are very good soils; there are also organic soils rich in amorphous clays, that are very good after reclamation.

Kaolinitic–alluvial regions (K–Al). These regions have many alluvial soils that are good; moreover many soils are on the limit between kaolisols and cinnamonic.

Kaolinitic–brunisolic regions (K–B). On the northwestern coast of Caucasia many soils are reported as krasnozems and jeltozems. These soils are on the limit between kaolisols on the one hand, cinnamonic and brunisolic on the other; that is why they are not bad; but they are usually acid. For tea and citrus they are very good.

Kaolinitic–cinnamonic regions (K–C). In the transition between kaolinitic and cinnamonic regions many soils are well provided with bases and moderately desilicated; dark clays are frequent. These soils are better than those of kaolinitic regions.

Kaolinitic–chernozemic (K–Ch) regions. Because of lower temperatures and waterlogging, soils of these regions are richer in organic matter than those of kaolinitic regions, and mineral fertility is better; but many soils suffer from waterlogging. The best soils are prairie, and humic glei after drainage; terra roxa also are good soils.

Mountains (M)

Mountainous soils are rejuvenated by erosion and their mineral fertility is usually good; but many of them are too shallow and too stony, and their topography makes cropping difficult. When the mountain is cold allitic weathering is not common, but podsolic weathering may take place, and results in very poor soils.

High latitude humid mountains (M). The best soils of these mountains, when not lithosolic, are rendzinas, braun erde, terra rossa, cinnamonic brown, neutral rankers and eutrophic organic soils; the less fertile are podsol rankers, dystrophic peat. The humus nano-podsols of mountainous tundra have limited agricultural potentialities.

High latitude humid mountains-ando (M-A). These regions have better soils; ando abound and many reported as brun acide are also ando; but acid ando soils present some problems (see p.180).

Mediterranean mountains (M'). In the mediterranean mountains podsolization is difficult, because of the dry summer; it only takes place in certain areas with a cold humid summer. Soils with good mineral fertility, rendzina, terra rossa, braunerde, braun cinnamonic, cinnamonic, neutral rankers, reddish brown abound; when sufficiently deep and not stony, they are good soils.

Some mountains in southwestern Asia are drier (*dry mediterranean mountains (DM')*). Serozems and arid brown are frequent; they are very good soils, when sufficiently deep and not stony, but naturally their yields without irrigation are low, for climatic reasons.

Mediterranean mountains-ando regions (M'-A). These regions have still better soils; ando and dark clays abound; some soils reported as brun acide are also ando; all these soils are good but those acid present some problems (see p.180).

Dry mountains (DM). Mineral fertility is usually high in dry mountain soils; they may be shallow, they may be stony, they may be poor in organic matter, but mineral deficiencies are not common; some soils are chernozemic, others serozems or arid brown. Naturally their yields without irrigation are low for climatic reasons. Some of these mountains (*DM-A*) have ando soils; but little atten-

tion has been hitherto paid to dry climate ando soils. Naturally ando soils are even better.

Tropical mountains (TM). Many tropical mountains are geologically young and kaolinitic soils are rare, but others are old plateaux, and kaolinitic soils, often laterite rocks are frequent; moreover in low mountains allitic weathering is perfectly possible. The best soils of tropical mountains are naturally those that are neither kaolinitic nor podsolic (recent brown, prairie, cinnamonic, alluvial, neutral rankers, eutrophic, organic, etc.). Rubrozems are good soils for tea; and after liming and fertilizing for all crops. In some low tropical mountains, kaolinitic soils are so frequent, that they are classified as *tropical mountains-kaolinitic (TM-K)*.

Tropical mountains-ando (TM-A). Volcanic tropical mountains have very good soils and population heaps up on them. Many soils reported as rubrozems or terres de barre (ferrisols and humic ferrisols) are ando. Naturally acid ando have their problems (see p.180).

Palaeo-kaolinitic regions

In these regions many soils have suffered allitic weathering, although they are now encountered under a climate that does not favour it. Moreover these regions are geologically very old. Mineral deficiencies are very common. Australia is perhaps the continent where phosphorus is more needed than in any other part of the world; and this is perhaps true for other minerals too. These regions have been studied in Australia, but they are also common in Africa, although not shown in the maps; kaolinitic soils and laterite rocks are common in the African deserts.

Kaolinitic-podsolic regions (K-P and P-K). In these regions the best soils are those that are neither kaolisols nor podsols or podsolic. In kaolinitic and podsolic soils mineral deficiencies are common, but they have been overcome by fertilization.

Kaolinitic-podsolic-cinnamonic regions (K-P-C). In these regions the best soils are those that are neither kaolinitic nor podsolic; chernozemic dark clays are among them. In kaolinitic and podsolic soils mineral deficiencies are common but they have been overcome by fertilization.

Mediterranean mountains-kaolinitic (M'-K) regions. In these mountains laterite is frequent and is naturally of low productivity; much better are arid brown and cinnamonic brown soils.

Chapter 9 | Soil Survey

SOIL CLASSIFICATION IN THE LOWER CATEGORIES

Until recently the number of great groups recognized in soil classification was small, a few tens; and variation within each group was great. Moreover, the definitions of each of these groups were not precise; frequently pedologists of the same school classified the same profile differently. Under such conditions classification into great groups was of little use for detailed soil surveys. Many soil survey services virtually ignored these groups and used series. Series were established in a more or less empirical way, with little reference to soil formation and without relationships to classification in higher categories. Often a series included soils belonging to different major groups and it was a matter of indifference to the surveyor where his series belonged.

In many cases rigid rules were drawn up for the establishment of series, and this was inconvenient, because, in a natural classification, criteria should vary from group to group even when these groups are related. Moreover, classification was based on morphology below the plow layer and thus the upper part of the profile was virtually excluded. The upper horizon is very important in soil classification; as we have said, W. L. Kubiena and the Canadians, for example, give considerable importance to the type of humus of A_0 and A_1; the distinction between eutrophic and dystrophic tropical soils (the "ochrosols" and "oxysols" of Charter) is very important and is based on characteristics of the upper horizon. Moreover, we cannot understand and classify a profile, when omitting that part of it, in which all soil forming processes originate. Too little attention has been paid to analytical, mineralogical and micropedological data. Until recently soil series included soils that varied considerably in texture of the upper horizon, and were subdivided in *types* on the basis of this characteristic. But at present it has been recognized that a series with so wide a variation in texture is an artificial assemblage; now series are homogeneous in texture of the upper horizon and do not need subdivision into types.

In spite of all these shortcomings soil series so established are more or less natural, and they permitted detailed soil survey when pedology was little advanced and the great groups were few and loosely defined. This is because when a man of good judgment classifies things with which he is well acquainted he classifies them correctly. But being empirical, the method was laborious and rather inefficient.

Soil types—now series—are subdivised into *phases*. Soils which have suffered temporary or artificially induced changes, not accompanied by significant changes in the morphology of the soil body, are considered as phases of the corresponding series. For instance, drainage phases are recognized where drainage has been improved artificially; water-logged phases, when the soil has become poorly drained owing to waters diverted from adjacent sites; burned phases, when peat has been burned. In a similar way we may recognize salinized, solonized, de-salinized and de-solonized phases; low potential fertility phases, when frequent repetition of barren fallow or inter-tilled crops has lowered organic matter content; virgin phases; limed phases; eroded phases when man-induced erosion has eliminated a part of the upper horizon; recently silted phases; and so on. Soils differing in slope, stoniness, depth, salinity, etc., are often considered as phases. But such differences are neither temporary nor artificially induced; in the author's opinion soils differing in such respects are "classes".

As has been pointed out by TAYLOR and POHLEN (1962), phases may be recognized in any category of soil classification; phase is not a category.

Soils can be classified into *classes* according to a single character considered separately—for instance texture, or thickness of the humic horizon, or depth, etc. In this way we can have texture classes, depth classes, salinity classes, etc. Classification into classes is artificial and the subdivisions that are obtained are not natural units. However, when applied within a genetic group of limited variation (for instance planosolic para-chernozem or tropical ferruginous slightly lessivé cuirassé) such classification is very useful. By saying that a soil is "plano-solic para-chernozem, from preponderantly volcanic materials, sandy loam on clay, 30 cm humic, 40 cm eluvial", our soil is well defined and can be distinguished from any other soil; other specifications may be added, if necessary. It is to be noted that in all sciences natural classification stops at a certain level; below that it becomes artificial, based on quantitative characters, habitat, etc. By using classes we avoid arbitrariness, obscure definitions, and laborious and tedious correlations. LEEPER (1956) is correct, in this respect; artificial classification is useful, but only below a certain level within a minor genetic group. It cannot replace natural classification, but it completes it. No definite rules can be given concerning the criteria that should be used for subdivision into classes. They should vary according to group; in certain groups thickness of the "eluvial" horizon is important, in other groups there is no such horizon; in certain groups depth to laterite, pan, bed-rock, etc., is important, in others all soils are deep; and so on. The criteria vary also from one region to another; if all the para-chernozems of the area surveyed are sands, texture is of little use for their subdivision, it serves only to their correlation with soils of other areas.

Textural classes are widely used; but in the case of podsolic, solonetzic, planosolic and slightly lessivé soils, texture of the clay accumulation horizon is also considered by using terms as "sand on sandy loam" or "loamy sand on clay", etc. Classes according to thickness of the "humic", "eluvial" or "slightly lessivé" horizon depth of "pan", "laterite" or "bed-rock", etc., are also widely used. But instead of dividing by means of arbitrary intervals it is better to specify thickness or depth in centimeters or inches; a 1 cm interval is often too small, but 1 inch is usually convenient.

Soils formed from different materials, when the difference is important from a pedo-genetic point of view, are separated into different classes. We should always keep in mind that soil is parent material more or less altered, so that soils formed from different materials are usually different; moreover, such differences are often important from the point of view of mineral deficiencies.

Features external to the soil (morphological position, climatic region, altitude, etc.) are usually excluded from soil classification and this is correct; soils should be classified on the basis of their own features. But in the lower categories, it is useful to consider, as separate classes, soils which are encountered in substantially different "habitats", unless appropriate research has shown that they are identical. In this respect the Russians are right in classifying soils according to their habitat, but "habitat" should be used only in the lower categories. We should follow the example of botanists; for a botanist a *Trifolium repens* is *Trifolium repens* whatever the "habitat" may be; but they consider them as separate ecotypes, unless their identity has been duly proved.

By classifying into minor genetic groups, and then subdividing the soils of each minor group into classes according to various single characters, we are enabled to define each individual soil perfectly and to point out all differences and similarities existing between soils. However, the number of class combinations is enormous, and at first sight this may appear to be inconvenient. However, it is not. The creation of a great number of series, arbitrarily and obscurely defined, causes confusion and should be avoided. But class combinations obtained by subdividing according to texture, thickness of the humic horizon, depth, etc., are so clearly defined that they do not require definition. If we use the textural classes proposed in Chapter 2, and centimeters or inches instead of arbitrary fixed intervals, their definition is self-evident. The greater number of class combinations does not create confusion; on the contrary, the similarites and differences existing between soils are pointed out by their classes, and their correlation is self-evident.

The use of classes permits simplification of soil mapping. Instead of defining series more or less arbitrarily, and obscurely, and then attempting to delimit the areas occupied by each—a laborious work which is seldom carried out accurately—the reverse procedure is used. On the basis of crop-ecologic, morphological, geological and other considerations, and their relation to soil distribution, as shown by profiles that have been studied, we divide the entire surveyed area into a large number of small areas, that are homogeneous or cannot be further subdivided for one reason or another: the legend gives soil distribution for each of these "individual" areas. For instance, for area 1 soils are referred to as planosolic para-chernozem clay loam on clay 25–35 cm eluvial, the thicker horizon being encountered where drainage is better; for area 2 planosolic para-chernozem sandy clay loam on clay 25–40 cm humic 25–60 cm eluvial; for area 3 planosolic para-chernozem loamy sand on clay 25–30 cm humic 25–35 cm eluvial occupies 80% of the total area, planosol sand on clay 0–25 cm humic 10–30 cm eluvial is encountered in depressions, and so on. To illustrate the method the following example is given. Suppose we have to prepare the soil map of a farm. The map shows the fields or enclosures, given that from a practical point of view it is not necessary to use smaller "individual" areas; and for each of them the legend gives the soil or soils encountered as in the preceding example. More details concerning mapping are given in the section on soil survey method.

THE CONCEPT OF LAND TYPE AND ITS USE IN SOIL SURVEY

A "land type" (PAPADAKIS, 1954) includes all land where crop-ecologic conditions (climatic, edaphic, etc.) are so uniform that the same crops can be grown—pastures and forests also being considered as crops. The crops can be treated in the same way and the same yields can be obtained. Land types differ in the crops grown; for instance, if on certain lands because of the wetness, oat is preferred to wheat, while on others wheat is preferred, these two kinds of land form two separate land types; if on certain lands, because they are sandy, rye is preferred to wheat, while on others, for opposite reasons, wheat is preferred, these two kinds of land also form two separate land types. If certain lands need to be sown early, because later they are wet, while others, for opposite reasons, can be sown later, these two kinds of land again form two separate land types. If a pasture is used for putting working horses to grass, while another pasture can only be used for sheep, these two pastures form two separate types of land; if one land area, for reasons that are not due to management, gives substantially higher yields than another, or a better quality of wine, tobacco, etc., these two areas belong to separate land types; and so on.

Land types constitute small crop-ecologic regions.

According to the crops grown, their behaviour and yield, a country, or the world, may be divided into crop-ecologic regions, and these may be further and further subdivided. The main differences between broad ecologic regions are climatic; in the lowest units they are edaphic. Land type is the term used for the lower units of such a classification.

Except in certain areas still empty of population, land types are well known to farmers. In each village farmers have identified the various land types that exist, and in many cases they have given them special names. Naturally, the same land type may have different names in distant villages, and the same name may be used with different meanings. Farmers know also the area occupied by each land type in their village. Land types differ in soil conditions. It is practically impossible for two soils to differ substantially from a pedological point of view and behave similarly concerning crops; pedological differences result in ecological differences. It may be that two soils are very similar and that in spite of this they belong to different land types, because they are encountered under different climates. However, such cases are rare in detailed soil surveys, and they are obvious. For all these reasons land types are very homogeneous from a pedological point of view. Exaggerating a little, we may say that farmers have already identified and delimited the kinds of soil that exist in their village. The task of the soil surveyor consists in finding the pedologic differences between these land types, in describing and classifying them pedologically and in systematizing the empirical classification made intuitively by the farmers. In some cases the soil surveyor may depart from the land types as recognized by farmers; farmers' opinions are not accepted indiscriminately. Opinions vary from farmer to farmer, they are checked and collated with aerial photographs, statistical information, morphologic data, data on vegetation and soil profiles. They serve as starting points; but farmers' information considerably simplifies the problem and may reduce the time and cost of a survey. It is to be noted that this method uses the crop-ecological behaviour of soils as a basis for the survey. As a consequence, this aspect is studied exhaustively and the information concerning agricultural potentialities is detailed and sound.

The method is applicable even in countries of primitive agriculture. Primitive farmers know their land equally well, or better, than those of developed countries; the more "primitive" a man is, the nearer he is to nature and the better he understands it; in this respect the Indian farmers of Peru and Bolivia and the African farmers of West Africa are unequalled.

Unfortunately, this method is little used. Crop-ecologic behaviour is studied after the map has virtually been finished. The reason is that land types are crop-ecologic units, and many pedologists lack the necessary crop-ecologic training. Moreover, many of them do not duly appreciate the experience of farmers. Finally large-scale mapping has often been done as a routine work, with field personnel only concerned with a few soil features fixed by headquarters.

SOIL SURVEY METHOD

The fundamental problem in a soil survey is to understand formation and distribution of the soils in the area of the survey. This is arrived at by studying a certain number of profiles at crucial points, and relating the data so obtained to land types, and variation of land forms, parent material, vegetation, etc. The crop ecologist of the team—it may be a pedologist with sufficient knowledge of crop ecology—determines and delimits the land types. This is easy, because the same land types are repeated even in distant villages of the same broad crop-ecologic region. The necessary information is obtained by interviewing county agents and farmers. Questionnaires, even direct questions, should be avoided. The surveyor talks with farmers about the agriculture of the region, and while talking the necessary information is obtained. Each farmer is asked information about the whole village, not especially his own land. In this way the information obtained is more truthful; and the information given by one farmer can be checked by that provided by others. Among farmers we encounter people who have a very good picture of the nature of the lands of the village; naturally the information is not given in technical terms, and the ecologist should be able to understand the farmer's language. Aerial photographs show land use and serve to complete and check the picture. At the same time the geo-pedologist—a pedologist with knowledge in land forms and geology, or vice-versa—identifies and delimits morphological regions, drainage regions, parent material regions, etc., on the basis of the aerial photographs and other information. And the vegetation specialist—who should have sufficient knowledge of pedology—identifies and delimits vegetation units. The crop-ecologic, pedological, morphological and vegetation studies begin at the same time and are carried out in close connection. The crop ecologist should follow the work of the pedologist, geo-pedologist, vegetation specialist and try to understand the relation of his land types to soils, land forms, etc. The geo-pedologist and vegetation specialist should act in a similar way. The pedologist should follow the work of all the others and with this background formulate a hypothesis concerning soil formation and distribution, and check these hypotheses by studying profiles at crucial points. He should ask his collaborators definite questions, and plan their collaboration. Rigid plans are to be avoided; the plan is modified continuously as some questions are elucidated and others arise.

As a result of this work the survey area can be divided into a great number of small, homogeneous "individual" areas, which is not necessary or convenient to subdivide. Two similar, but not contiguous areas are considered as separate, although they may belong to the same land type, morphological division, etc. The areas are numbered and for each of them the legend gives soil distribution, agricultural capability, etc. For instance, area 1 is classified as planosolic para-chernozem sand loam on clay 25–35 cm humic 25–50 cm eluvial; area 2 as 80%

planosolic para-chernozem sandy loam on clay 25–35 cm humic 25–40 cm eluvial with planosol sand on clay 0–25 cm humic 10–25 cm eluvial in depressions, and so on.

In order to point out the similarities and differences between areas, and to shorten the legend, similar areas are grouped and the group is designed by a letter. For instance, all areas with planosolic para-chernozem are shown by a letter, those with 50 % or more para-chernozem mixed with planosol by another letter and so on. These groups can be grouped into higher category groups; for this a combination of two letters can be used, the first letter showing the higher group, the second the lower one.

It is often useful to classify the "individual" areas according to various criteria and prepare different maps: of soils, land types, drainage, landforms, etc. The individual areas are the same, and they are numbered in the same way in all maps, but they are grouped differently according to the map. That is why the basic map is one; on it each "individual" area is shown by its number. Letters showing groups of similar areas and the lines that delimit such groups differ from map to map, and are printed in red, or another colour. A small area is never split, because it is supposed that it is impossible to map the differences existing within it; otherwise the area would have been split into two in the basic map.

The advantages of this method are obvious:

(*1*) Variation within a soil series is great, it is still greater within a soil association, which is often used as mapping unit. However, many of the "individual" areas belonging to a soil series or association are more homogeneous; existing information permits to describe their soil more precisely. Unfortunately such information is lost, because with the system of series and associations, all the "individual" areas belonging to a series or association are shown on the map as identical; the legend does not provide separate information for each of them. Our system gives separate information for each "individual" small area; that permits to define better the soils of each one and show the differences that exist between them.

(*2*) To identify soil series, define and delimit them is an arbitrary, difficult, laborious and tedious work; series are often arbitrary, their definition obscure, and errors are unavoidable. On the contrary, to divide the area surveyed into small areas as homogeneous as possible and give a good picture of the soils of each one, using minor genetic groups and classes is easier.

(*3*) The laborious work of correlating series is avoided. We do not assert that the soils encountered in two "individual" areas are identical; we merely say that both are planosolic para-chernozems, the first sandy loam on clay 30–40 cm humic 35–50 cm eluvial, the second sandy clay loam on clay 25–35 cm humic 25–40 cm eluvial, both from predominantly volcanic material; the agricultural possibilities of each one are given; moreover, there is a classification into land types; that is all. Such information permits comparison of soils with soils encountered in any other part of the world, but their identity with other soils is not given as certain.

(*4*) This method is more rapid. As has been said, farmers have already classified soils into land types and delimited them. By taking advantage of their experience time and money are saved.

(*5*) The information concerning land capability is much sounder, because the crop-ecologic behaviour of the land has been taken as a basis for the preparation of the map. With the conventional method mapping units are often too heterogeneous from a crop-ecologic point of view, and information is obtained by a mechanical elaboration of questionnaires, an unreliable method.

(*6*) All information is sounder, because it comes from different sources (various farmers, morphology, geology, vegetation, soil profiles) and each type of information is critically collated and checked at the very moment it is gathered, with all the others. Usually the data corroborate one another, but when a discrepancy appears, the soil surveyor can try to explain it or find the error.

(*7*) The method has great flexibility. A small-scale map may be prepared, in which the "individual" areas corresponding to broad crop-ecologic and/or morphological regions, are extensive, and where the soils assigned to each of them present considerable variation; then a larger-scale map, in which the individual areas are smaller and the soils assigned to each offer less variation; and so on. Such a method of closer and closer approximation is very convenient; it is sounder, more rapid and at any phase of the survey there is more or less detailed information for all of the territory; a specially selected staff is used for the first approximation (small-scale maps); once these maps have been prepared junior staff can be used for the last approximations under the direction of the former.

The inconvenience of this method is that it requires good judgment and solid, although not necessarily detailed, knowledge of crop ecology and pedology of all the surveyors; but judgment and knowledge are strengthened by practising the method.

Addendum

Pedro has recently carried out experiments on clays (PEDRO, 1966) which have shown that intense leaching (500,000 mm annually) with warm water (65°C) can alter kaolinite, removing silica and leaving in situ bohemite. The same happens with 2 : 1 clays. Such an alteration is possibly slow under natural conditions, but it cannot be excluded. Another interesting result of Pedro's experiments is the relatively high pH (6.8–8.0) of drainage waters, even when weathering takes place under a leaching of 500,000 mm annually. This supports the author's theory that under dry conditions weathering of volcanic ashes may produce a concentration of Na sufficient to cause clay eluviation; it is to be noted that in the experiment of Pedro with warm water some alumina has been eluviated, more especially in the case of Volvic lava; clay eluviation is easier than that of alumina.

As Pedro points out the results of weathering are determined by the relative solubility of the different products (bases, SiO_2, Al_2O_3, Fe_2O_3) under the conditions of the experiment. This fact supports the concept of the degree of leaching (PAPADAKIS, 1962c), which he uses as a basis of soil classification at the highest level.

In the text it has not sufficiently been emphasized that the rapidity with which crystalline clays are synthesized seems to depend on the presence of bases. That may explain why ando soils, rich in amorphous clays, seem to be more frequent under humid climates.

CORRELATION WITH 7TH APPROXIMATION (ITS SUPPLEMENT OF MARCH, 1967)

In the text reference is very frequently made to the 7th Approximation, and correlations have frequently been given with it. Recently, however, the 7th Approximation has been modified substantially, and in March 1967, the U.S. Soil Conservation Service published a supplement to which reference is not prohibited. In writing this addendum this supplement is always referred to.

Diagnostic horizons

The terms *petrocalcic horizon* and *duripan* have been introduced from the 7th Approximation and have practically the same meaning in this book.

The *mollic epipedon* of the 7th Approximation is a neutral "dark humic" horizon of specified thickness; but mollic epipedon has 0.58% C or more; in this book a humic horizon has 1% or more.

The *umbric epipedon* is an acid "dark humic" horizon of specified thickness; but umbric epipedon has 0.58% C or more; in this book a "humic" horizon has 1% or more.

Histic horizon is an "organic" or "peaty" horizon of specified thickness.

Ochric epipedon is not the result of a definite process of soil formation and the author has no diagnostic horizon that corresponds to it.

Argillic horizon includes the author's "textural" and "slightly lessivé" horizons; the differences between "textural" and "slightly lessivé" is that in the former a higher clay content is accompanied by higher cation exchange capacity.

Natric is a horizon that is simultaneously "textural" and "natric" or "magnesic".

The term *spodic horizon* has been introduced from the 7th Approximation and has approximately the same meaning in this book.

The *cambic* horizon is not the result of a definite soil formation process and the author has no diagnostic horizon that corresponds to it; some "gley", "braunified" and "rubified" horizons are termed cambic in the 7th Approximation.

Theoretically the *oxic horizon* corresponds to the "kaolinitic" and "superkaolinitic" horizons in this book; but the criteria used for identification are different.

The *calcic* horizon corresponds more or less to the "calcareous accumulation" in this text.

Soil groups

Theoretically the *entisols* of the 7th Approximation correspond to the raw soils in this book; but entisols include some gleisols and some light texture kaolisols; on the contrary some dry climate raw soils (arid brown) are not included in the entisols. Many aquents are gleisols; fluvents correspond to alluvial; psamments to desert sands and recent dunes; orthents to recent brown. Further subdivision of entisols is based chiefly on soil temperature; the author's classification is based exclusively on soil properties.

Vertisols correspond to our dark clays; but dark clays have at least 30 mequiv. of bases. The classification of vertisols is based on climate and chroma; that of dark clays on humus and carbonate content (chernozemic clays, rendzina clays, etc.).

Inceptisols include ando and rankers; some gleisolic, cinnamonic, brunisolic and raw are also included. Andepts correspond to ando; durandepts to ando with pan; dystrandepts to acid ando; eutrandepts to chernozemic ando; hydrandepts to hydrol ando. Cryochrepts and cryumbrepts to arctic brown. Dystrochrepts to brun acide; eutrochrepts to braunerde; halaquepts to saline gley and gleyed saline; plinthaquepts to ground-water laterites; haplumberts include many rankers; humaquepts include many low humic gley.

Aridisols is rather an environmental than a systematic term. They include many solonchacks, many solonetz, many planosols, some raw (arid brown), some rendzinas (serozems). Camborthids correspond to arid brown; haplargids to planosolic red desert; calciorthids to serozems; natrargids are solonetz; salorthids are solonchacks; paleargids planosols.

Mollisols correspond to chernozemic in this book; but they also include many cinnamonic, rendzinas and raw soils, some gleisols, some solonetz and some planosols. Natralbolls, natraquolls, natriborols, natrustolls, natrixerolls, are usually solonetz. Many argialbolls, argiaquolls, argiborols, argiudolls, argiustolls and argixerolls are planosols. Many aquolls are humic gley. Many rendolls are rendzinas. The classification of mollisols is based, to a great extent, on climate. The author classifies chernozemic soils on the basis of profile characteristics.

Spodosols correspond to podsols. Humods to our humus podsols; but many humus podsols are included in aquods. Orthods correspond to iron and iron-humus podsols; fragiaquods and fragihumods to fragipan podsols. Here again the 7th Approximation uses temperature in soil classification.

Alfisols include both podsolic and planosolic soils; some gleisols, solonetz, brunisolic, cinnamonic, tropical ferruginous, lateritic podsolic and groundwater laterites are also included. Natraqualfs, natriboralfs, natrudalfs, natrustalfs, natrixeralfs are solonetz. Eutroboralfs grey wooded; many ustalfs and xeralfs are planosols; some rhodoxeralfs are cinnamonic; some hapludalfs, haplustalfs, and haploxeralfs are cinnamonic; some haplustalfs and tropudalfs are tropical ferruginous, or lateritic podsolic; some plinthustalfs and plinthoxeralfs groundwater laterites. The classification of alfisols is greatly based on climate.

Ultisols include the red–yellow podsolic krasnozems, acid ferruginous, and rubrozems of this book; but many other soils are included. Plinthaquults, plin-

thudults and plinthustults are ground-water laterites; humults rubrozems; hapludults red–yellow podsolic some illitic podsolic and illitic gleisols are included in ultisols. The classification of alfisols is to a great extent based on climate.

Theoretically *oxisols* correspond to kaolisols; but many kaolisols are classified in other orders of the 7th Approximation. Acrohumox, acrorthox and acrustox correspond to ferralitic; eutrorthox and eutrustox to eutrophic kaolisols; plinthaquox include ground-water laterites; acrohumox correspond to ferralitic rubrozems; gibbsiaquox, gibbsihumox and gibbsiorthox are kaolisols with concretionary or cuirassé profile, in which the gravel size fraction contains 30% or more gibbsite. Possibly gibbsitic kaolisols should be introduced in our classification.

Histosols correspond to organic soils.

POTENTIAL FERTILITY

Experiments in Senegal (FAUCK et al., 1967) on the influence of cropping on soil characteristics, have shown that organic matter decreases very rapidly; 30% has been lost during the first 2 years. But after 10 years an equilibrium is reached; no further losses are observed. Total loss in 10 years has been 40–60%. These experiments confirm that both the descent of potential fertility and its stabilization at a new level are more rapid under tropical conditions, with high soil surface temperatures.

AGRICULTURAL POTENTIALITIES AND LAND USE MAPS

In the World Soil Resources Office of FAO, Dr. Bramão devised a simple and very effective method of mapping agricultural potentialities. Red lines and numbers showing the climatic regions are printed on a soil map. The climatic classification used is that of the author (PAPADAKIS, 1966b), which is sufficiently detailed; moreover it is based on ecologic criteria so that each unit has definite agricultural potentialities and other ecological implications; finally it is decimal, each unit is shown by a number, which has always the same meaning, and may be easily printed on maps. The legend shows the agricultural possibilities of each climatic unit, and their variation according to soil type and conditions. Since both climatic and pedologic classifications are world wide, such maps facilitate also the transfer of experience from one part of the world to another.

Printing of climatic regions on land use maps would also be very useful

References

AGARWALL, R. R. and MUKERJI, S. K., 1951. Gangetic alluvium of India: pedochemical characters in the genetic soil types of Gorakhpur district in the United Provinces. *Soil Sci.*, 72: 21–32.

AHN, P. M., 1961. Soils of the lower Tang Basin, southwestern Ghana. *Ghana Min. Agr. Soil, Land Use Survey, Mem.*, 2: 266 pp.

AKHTYRSTEV, B. P., 1962. Provincial characteristics of soils of the sub-zone of broad-leaved forests of the central Russian upland. *Pochvovedenie*, 1: 26–40.

AMERYCK, J., 1960. La pédogenèse en Flandre sablonneuse. *Pédologie*, 10: 124–190.

ANONYMOUS, 1958. *Levantamiento de Reconocimiento dos Solos do Estado Rio de Janeiro e Distrito Federal*. Min. Agr., Rio de Janeiro, 634 pp.

ANONYMOUS, 1960a. A look at Canadian soils. *Agr. Inst. Rev.*, 15 (2): 10–60.

ANONYMOUS, 1960b. *Levantamiento de Reconocimiento dos Solos do Estado de São Paulo*. Min. Agr., Rio de Janeiro, 352 pp.

AUBERT, G., 1963. Soils with ferruginous or ferralitic crusts of tropical regions. *Soil Sci.*, 95: 235–242.

AVERY, B. W., STEPHEN, I., BROWN, G. and YAALON, D. H., 1959. The origin and development of brown earths on clay-with-flints and coombe deposits. *J. Soil Sci.*, 10: 177–195.

BALDWIN, M., KELLOGG, CH. E. and THORP, J., 1938. Soil classification. In: *Soils and Man—Yearbook Agr.*, *U.S. Dept. Agr.*, *1938*, pp. 979–1001.

BARSHAD, B., 1964. Chemistry of soil development. In: F. E. BEAR (Editor), *Chemistry of the Soil*. Reynolds, New York, N.Y., pp.1–70.

BARSHAD, I., 1946. A pedologic study of California prairie soils. *Soil Sci.*, 61: 423–442.

BARSHAD, I. and ROJAS-CRUZ, L. A., 1950. A pedological study of a podsol soil profile from the equatorial region of Colombia, South America. *Soil Sci.*, 70: 221–236.

BASINSKI, J. J., 1955. The Russian approach to soil classification and its recent development. *J. Soil Sci.*, 10: 14–26.

BAUMGART, I. L., 1953. Outline of New Zealand soils and their classification. *Australian Conf. Soil Sci.*, 1: 6.

BEAR, F. E., 1955. *Chemistry of the Soil*. Reynolds, New York, N.Y., 373 pp.

BECKETT, P. H. T., 1961. Some Sarawak soils, 2. Soils of the Bintulu coastal area. *J. Soil Sci.*, 12: 218–233.

BENGTSON, N. A. and VAN ROYEN, W., 1957. *Fundamentals of Economic Geography*. Prentice-Hall, Englewood Cliffs, N.J., 611 pp.

BLOOMFIELD, C., 1954. A study of podzolization, 3–5. The mobilization of iron and aluminium. *J. Soil Sci.*, 5: 39–56.

BLOOMFIELD, C., 1955. A study of podzolization, 6. The immobilization of iron and aluminium. *J. Soil Sci.*, 6: 284–292.

BLOOMFIELD, C., 1956. The deflocculation of kaolinite by igneous leaf extract. *Congr. Intern. Sci. Sol.*, *6e*, *Rappt.*, B: 27–32.

BOBKOV, V. P., 1962. Characteristics of some physico-chemical properties of dark compact meadow soils of the Volga–Akhtubin bottomland. *Pochvovedenie*, 7: 67–72.

BONFILS, C. G., CALCAGNO, J. E., ETCHEVEHERE, H., IPUCHA AGUIRRE, J., MIACZYNSKI, C. R. O. and TALLARICO, L. A., 1959. Suelos y erosión en la región pampeana semiárida. *Rev. Invest. Agr.*, 13: 321–404.

BONNET, J. A., 1939. The nature of lateritization as revealed by chemical, physical and mineralogical studies of a lateritic soil profile from Puerto Rico. *Soil Sci.*, 48: 25–40.

BOTELHO DA COSTA, J. V., AZEVEDO, A. L., CARDOSO FRANCO, E. P. and PINTO RICARDO, R., 1959. Grey brown and reddish brown semiarid soils of southern Angola. *Proc. 3rd Interafrican Soils Conf.*, *Dalaba*, *1959*, pp.245–252.

BRAMAO, D. L. and SIMONSON, R. W., 1956. Rubrozem—a proposed soil group. *Congr. Intern. Sci. Sol.*, *6e*, *Rappt.*, E: 25–30.

BRAMMER, H., 1965. The soils of East Pakistan in relation to agricultural development. *Symp. Soil Sci. Soc. Pakistan. Dacca*, *1965*.

BROWN, J. and TEDROW, J. C. F., 1964. Soils of the northern Brooks range, Alaska, 4. Well-drained soils of the glaciated valleys. *Soil Sci.*, 97: 187–195.

CLARK, J. S., BRYDON, J. E. and FARSTAD, L., 1963. Chemical and clay mineralogical properties of the concretionary brown soils of British Columbia, Canada. *Soil Sci.*, 95: 344–352.

CLINE, M. C., 1949. Profile studies of normal soils of New York: I. Soil profiles sequences involving brown forest, gray-brown podsolic and brown podsolic soils. *Soil Sci.*, 68: 259–272.

CLINE, M. C., 1953. Major kinds of profiles and their relationships in New York. *Soil Sci.*, 17: 123–127.

COSTA LIMA, J. W., 1953. *Anais 4a Reunião Brasileira de Ciencia do Solo*, pp.403–427.

CROMPTON, E., 1960. The significance of the weathering/leaching ratio in the differentiation of major soil groups, with particular reference to some very strongly leached brown earths of the hills of Britain. *Trans. Intern. Congr. Soil Sci.*, *7th*, *1960*, *Madison*, 4: 406–412.

DABIN, B., LENEUF, N. et RIOU, G., 1960. *Carte pédologique de la Côte-d'Ivoire*. Secrétariat d'Agriculture, Direction des Sols, Abidjan.

DAMMAN, A. W. H., 1962. Development of hydromorphic humus podsols and some notes on the classification of podsols in general. *J. Soil Sci.*, 13: 92–97.

DAN, J. and KOYUMDJISKY, H., 1962. Principles of a proposed classification for the soils of Israel. *Proc. Intern. Soil Conf.*, *New Zealand*, *1962*, pp.410–421.

DA SILVA TEIXEIRA, A. J., 1959. *Interafrican Soil Conf.*, *3rd*, *1959*, *Dalaba*, 1: 467–477.

DEL VILLARS, E. H., 1944. The tirs of Morocco. *Soil Sci.*, 57: 313–399.

DEWAN, M. L. and FAMOURI, J., 1964. *The Soils of Iran*. F.A.O., Rome, 320 pp.

D'HOORE, J. L., 1960. La carte des sols d'Afrique au sud du Sahara. *Pédologie*, 10: 191–204.

D'HOORE, J. L., 1964. *La Carte des Sols d'Afrique au 1/5.000.000*. Commission de Coopération Technique en Afrique, Lagos, 2 vol., 210 pp.

DOUGLAS, L. A. and TEDROW, J. C. F., 1960. Tundra soils of arctic Alaska. *Trans. Intern. Congr. Soil Sci.*, *7th*, *1960*, *Madison*, 4: 291–304.

DUCHAUFOUR, C., 1957. *Pédologie; Tableaux descriptifs et analytiques des Sols*. Ecole Natl. Eaux Forêts, Nancy, 87 pp.

DUCHAUFOUR, C., 1959. *La Dynamique du Sol forestier en Climat atlantique*. Presses Universitaires Laval, Quebec, Que., 77 pp.

DUCHAUFOUR, C., 1965. *Précis de Pédologie*. Masson, Paris, 482 pp.

DUCHAUFOUR, C. et JACQUIN, F., 1964. Résultats de recherches récentes sur l'évolution de la matière organique dans les sols. *Compt. Rend. Acad. Agr. France*, 50: 376–387.

DUDAL, R. and MOORMANN, F. R., 1962. *Characteristics of Major Soil Groups of South East Asia and Considerations on their Agricultural Potential*. F.A.O., Rome, 25605/E.

DURING, C., 1964. The amelioration of volcanic ash soils in New Zealand. *Rept. Meeting on the Classification and Correlation of Soils from Volcanic Ash, Tokyo, 1964*, pp.129–133.

ENGLAND, C. B. and PERKINS, H. F., 1959. Characteristics of three reddish brown lateritic soils of Georgia. *Soil Sci.*, 88: 294–302.

FAGUNDES, A. B., ARAUJO, W. A., RAMOS, F. and KEHRIG, A. G., 1951. *Anais Tercera Reunião Brasileira de Ciencia do Solo, Recife, Brasil, 1951*, pp.649–674.

FAUCK, R., 1962. *Etude des Sols, Région de Dongas, N. Dahomey*. Office Rech. Sci. Territ. Outre Mer, Paris.

FAUCK, R., MOUREAUX, C. et THOMANN, C., 1967. Bilans de l'évolution des sols de Séfa après quinze ans de culture continue. *Compt. Rend. Acad. Agr. France*, 53: 698.

FREI, E. and CLINE, M. G., 1949. Profile studies of normal soils of New York, 2. Micromorphological studies of the gray-brown podsolic–brown podsolic soil sequence. *Soil Sci.*, 68: 333–334.

FRIDLAND, V. M., 1961. Soils of hilly territories of northern Vietnam. *Pochvovedenie*, 12: 57–74.

GERASIMOV, I. P., 1962. The new American classification of soils. *Pochvovedenie*, 6: 34–46.

GILE JR., L. H., 1958. Fragipan and water-table relationships of some brown podsolic and low humic-glei soils. *Soil Sci. Soc. Am. Proc.*, 22: 560–565.

GLENTWORTH, R., 1962. The principal genetic soil groups of Scotland. *Proc. Intern. Soil Conf., New Zealand, 1962*, pp.480–486.

GOLLAN, J. e LACHAGA, D., 1939. *Algunos Suelos Tipos de Santa Fe–Inst. Exptl. Invest. Fom. Agr. Gan. (Santa Fe), Publ*, 16.

GOLLAN, J. e LACHAGA, D., 1944. Algunos suelos tipos de Santa Fe. *Inst. Exptl. Invest. Fom. Agr. Gan., Santa Fe, Publ. Tech.*, 55.

GOLLAN, J., CRUELLAS, J. e NICOLLIER, V., 1936. *Suelos de Misiones—Inst. Exptl. Invest. Fom. Agr. Gan. (Santa Fe), Publ.*, 3.

GOUVEIA, D. H. and GOUVEIA, J., 1959. Guijá grey soils. *Proc. 3rd Interafrican Soils Conf., Dalaba, 1959*, pp.479–482.

HALLSWORTH, E. G., 1963. An examination of some factors affecting the movement of clay in an artificial soil. *J. Soil Sci.*, 14: 360–371.

HAPSTEAD, M. and RUST, R. H., 1964. A pedological characterization of five profiles in gray-wooded soils area of Minnesota. *Soil Sci. Soc. Am. Proc.*, 28: 113–118.

HARDY, F. and RODRIGUEZ, G., 1939. Soil genesis from andesite in Grenada, British West Indies. *Soil Sci.*, 48: 361–364.

HARRADINE, FR., 1963. Morphology and genesis of noncalcic brown soils in California. *Soil Sci.*, 96: 277–287.

HOGAN, J. D. and BEATTY, M. T., 1963. Age and properties of Peorian loess and buried paleosols in southwestern Wisconsin. *Soil Sci. Soc. Am. Proc.*, 27: 345–350.

HUBERT, P., 1961. A reconnaissance soil map of northeastern Congo. *Pédologie*, 11: 74–88.

HUTCHESON JR., T. B., 1963. Chemical and mineralogical characterization and comparison of Hagerstown and Maury soil series. *Soil Sci. Soc. Am. Proc.*, 27: 74–78.

HUTCHESON JR., T. B., LEWIS, R. J. and SEAY, W. A., 1959. Chemical and clay mineralogical properties of certain Memphis catena soils of western Kentucky. *Soil Sci. Soc. Am. Proc.*, 23: 474–478.

IVANOVA, Y. N. et al., 1962. New information on the general geography and classification of the soils of the polar and boreal belt of Siberia. *Pochvovedenie*, 11: 7–23.

JACKMAN, R. H., DURING, C., ANDREWS, E. D. and LYNCH, P. B., 1962. An evaluation of the usefulness of the New Zealand soil classification. *Proc. Intern. Soil Conf., New Zealand, 1962*, pp.453–461.

JACKSON, M. L., 1956. *Advan. Soil Chem.*, 1956: 47.

JACKSON, M. L. and SHERMAN, G. C., 1953. Chemical weathering of minerals in soils. *Advan. Agronomy*, 5: 219–318.

JANZEN, W. K., 1962. Exchangeable cations and mechanical composition of solodic soils of Saskatchewan. *J. Soil Sci.*, 13: 116–123.

JANZEN, W. K. and MOSS, H. C., 1956. Exchangeable cations in solodized-solonetz and solonetz-like soils of Saskatchewan. *J. Soil Sci.*, 6: 203–212.

JARVIS, L. L., ROSCOE JR., E. and BIDWELL, O. W., 1959. A chemical and mineralogical characterization of selected brunizem, reddish prairie, grumusol, and planosol soils developed in pre-Pleistocene materials. *Soil Sci. Soc. Am. Proc.*, 23: 234–259.

JENNY, H., 1941. *Factors of Soil Formation*. Mc-Graw Hill, New York, N.Y., 281 pp.

JENNY, H., 1948. Great soil groups in the equatorial regions of Colombia, South America. *Soil Sci.*, 66: 5–24.

JENSEN, S. T., 1962. Soil classification and soil mapping in Denmark. *Proc. Intern. Soil Conf., New Zealand, 1962*, pp.349–354.

KAMOCHITA, Y., 1962. Soil classification as a basis for increasing the productivity of agricultural land. *Proc. Intern. Soil Conf., New Zealand, 1962*, pp.428–433.

KANO, I., 1956. A pedological investigation of Japanese volcanic-ash soils. *Congr. Intern. Sci. Sol., 6e, Rappt., E*: 105–109.

KANO, I., HONJO, Y. and ARIMURA, S., 1963. Genesis and characteristics of red yellow soils derived from gabbro in northern Kyushu. *Bull. Kyushu Agr. Exptl. Sta.*, 9: 15–26.

KARIM, A. and KHAN, D. H., 1955. Soils of the Nanokhi series, East Pakistan, 1. Morphology, textural separates, and exchangeables cations. *Soil Sci.*, 80: 139–146.

KARIM, A. and KHAN, D. H., 1956. Soils of the Nanokhi series, East Pakistan. 2. Chemical investigation and classification. *Soil Sci.*, 81: 389–398.

KARIM, A. and QUASEM, A., 1961. A study of the soils of the Barind tract, East Pakistan. *Soil Sci.*, 91: 406–412.

KELLOGG, CH. E., 1941. *The Soils that Support Us*. Macmillan, New York, N.Y., 370 pp.

KING, L. C., 1962. *The Morphology of the Earth; a Study and Synthesis of World Scenery*. Oliver and Boyd, Edinburgh, 699 pp.

KOBO, K., 1964. Amelioration of volcanic ash soils and their potentialities. *Rept. Meeting on the Classification and Correlation of Soils from Volcanic Ash, Tokyo, 1964*, pp.126–128.

KOVDA, V. A., 1961. Principles of the theory and practice of reclamation and utilization of saline soils in the arid zones. *Salinity Probl. Arid Zones, Proc. Teheran Symp.*, pp.201–213.

KUBIENA, W. L., 1953. *The Soils of Europe*. Murby, London, 317 pp.

KUBIENA, W. L., 1956. Red-earth formation and lateritization (their differentiation by micromorphological characteristics). *Congr. Intern. Sci. Sol., 6e, Rappt.*, E: 247–249.

LEAHEY, A., 1963. The Canadian system of soil classification and the seventh approximation. *Soil Sci. Soc. Am. Proc.*, 27: 224–255.

LEE, C.-K., 1943. Chemical characteristics of the great soil groups of China. *Soil Sci.*, 55: 343–350.

LEEPER, G. W., 1956. The classification of soils. *J. Soil Sci.*, 7: 59–64.

LENEUF, N., 1959. *L'Altération des Granites Calco-Alcalins et des Granodiorites en Côte-d'Ivoire Forestière et les Sols qui en sont Dérivés*. Office Rech. Sci. Territ. Outre Mer, Paris, 210 pp.

LENEUF, N. and AUBERT, G., 1960. Attempt to measure the rate of ferralitization. *Trans. Intern. Congr. Soil Sci., 7th, 1960, Madison*, 4: 225–228.

LICHMANOVA, A. I., 1962. Some properties of mechanical fractions of pale gray forest soil. *Pochvovedenie*, 6: 68–69.

LOBOVA, E. V. and KOVDA, V. A., 1960. The soil map of Asia (scale 1/6,000,000). *Trans. Intern. Congr. Soil Sci., 7th, 1960, Madison*, 4: 27–35.

MACKNEY, D., 1961. A podsol development sequence in oak-woods and heath in central Europe. *J. Soil Sci.*, 12: 23–40.

MANIL, G., 1955. General considerations on the problem of soil classification. *J. Soil Sci.*, 10: 5–13.

MARBUT, C. F., 1936. *Atlas of American Agriculture, 3. Soils of the United States*. U.S. Dept. Agr., Washington, D.C., 98 pp.

MARTIN, D., 1959. Yellow ferralitic soils developed from metamorphic rocks in southwest Cameroons. *Proc. 3rd Interafrican Soil Conf., 1959, Dalaba*, 1: 227–232.

McCLELLAND, J. E., 1951. The effect of time, temperature and particle size on the release of bases from some common soil-forming minerals of different crystal structure. *Soil Sci. Soc. Am. Proc.*, 15: 301–307.

McCLELLAND, J. E., MOGEN, C. A., JOHNSON, W. M., SHROER, F. W. and ALLEN, J. S., 1959. Chernozems and associated soils of eastern North Dakota. *Soil Sci. Soc. Am. Proc.*, 23: 51–56; 56–60.

McCONAGHY, S. and McALESEE, D. M., 1957. Studies on the basaltic soils of Northern Ireland. *J. Soil Sci.*, 8: 127–134.

MILLER, E. V. and COLEMAN, N. T., 1952. Colloidal properties of soils from western equatorial South America. *Soil Sci. Soc. Am. Proc.*, 16: 239–244.

MILLER, F. T. and MEHLICH, A., 1960. Charge characterization as a criterium for classification of some equatorial soils. *Trans. Intern. Congr. Soil Sci., 7th, 1960, Madison*, 4: 432–442.

MOHR, E. C. J. and VAN BAREN, F. A., 1954. *Tropical Soils*. Van Hoeve, The Hague, 498 pp.

MOLFINO, R. H., 1956. Ensayo edafologico sobre la Antartida Argentina. *Fac. Agron. Univ. La Plata*, 32: 1–41.

MOLTHAM, H. D. and GRAY, F., 1963. A characterization and genetic study of two modal reddish prairie soils. *Soil Sci. Soc. Am. Proc.*, 27: 69–74.

MOSS, H. C. and ARNAUD, J. ST., 1955. Grey-wooded (podzolic) soils of Saskatchewan, Canada. *J. Soil Sci.*, 6: 293–311.

MUIR, A., ANDERSON, B. and STEPHENS, I., 1957. Characteristics of some Tanganyika soils. *J. Soil Sci.*, 8: 1–18.

MUIR, J. W., 1962. The general principles of classification with reference to soils. *J. Soil Sci.*, 13: 22–30.

NEJGEBAUER, V., CIRIC, M. and ZIVKOVIC, M., 1961. *Soil Map of Yugoslavia*. Beograd.

NEW ZEALAND SOIL BUREAU. *Atlas*. Dept. Sci. Ind. Res., Wellington.

NYE, P. H., 1955. Some soil-forming processes in the humid tropics, 4. The action of the soil fauna. *J. Soil Sci.*, 6: 73–83.

NYUN, M. A. and McCALEB, S. B., 1955. The reddish-brown lateritic soils of the North Carolina region. Davidson and Hiwasser Series. *Soil Sci.*, 80: 27–41.

OAKES, H., 1957. *The Soils of Turkey*. F.A.O., Ankara, 180 pp.

OBENG, H. B., ENDREDY, A. S., ASAMOA, G. K. and SMITH, G.K., 1963. *Classification of the Soils of Ghana, 2. Ground Water Laterites, Red Earths and Yellow Earths (Latosols)*. Agr. Res. Inst., Kumasi.

PAHAULT, P. and SOUGNEZ, N., 1961. Some pedological and plant-sociological aspects of the northwestern region between Vesdre and Meuse. *Pédologie*, 11: 125–137.

PANABOKKE, C. R., 1959. A study of some soils in the dry zone of Ceylon. *Soil Sci.*, 87: 67–74.

PANIN, P. S. and ARISTARKHOV, A. N., 1962. Characteristics of the chemistry and permeability to water of soils salinized by soda in the Karabakh plain of Azerbaidzhan. *Pochvovedenie*, 5: 12–21.

PAPADAKIS, J., 1938. *Ecologie Agricole*. Bibliothèque Agr. Belge, Gembloux / Paris, 312 pp.

PAPADAKIS, J., 1941. An important effect of soil colloids. *Soil Sci.*, 51: 219–220.

PAPADAKIS, J., 1949. El especio (volumen de tierra fina) como factor de crecimiento de las plantas. *Lilloa*, 18: 215–224.

PAPADAKIS, J., 1952a. *Agricultural Geography of the World*. J. Papadakis, Buenos Aires, 118 pp.

PAPADAKIS, J., 1952b. *Mapa Ecologico de la República Argentina*. 2nd ed. Min. Agr. Gan., Buenos Aires, 1: 254 pp.; 2: Atlas, 24 maps, 6 pp.

PAPADAKIS, J., 1954. *Ecologia de los Cultivos, 1. Ecologia General; 2. Ecologia Especial*. Min. Agr. Gan., Buenos Aires, 223 pp.; 463 pp.

PAPADAKIS, J., 1960a. *Geografia Agricola Mundial*. Salvat Editores, Barcelona, 649 pp.

PAPADAKIS, J., 1960b. A note on the formation of clay horizons. *Soils Fertilizers*, 23: 7.

PAPADAKIS, J., 1960c. Considerations on the formation of bleached horizons and iron crusts. *Soils Fertilizers*, 23: 240–241.

PAPADAKIS, J., 1960d. Avances recientes en pedologia. *Inform. Invest. Agr., Supl.*, 1: 135–147.

PAPADAKIS, J., 1961. *Climatic Tables for the World*. J. Papadakis, Buenos Aires, 175 pp.

PAPADAKIS, J., 1962a. Some considerations on soil classification: the 7th approximation. *Soil Sci.*, 94: 115–119.

PAPADAKIS, J., 1962b. Grado de lavado de los suelos Argentinos con algunas consideraciones sobre suelos de otros paises de América Latina. *2ª Reunión Argentina de la Ciencia del Suelo, Mendoza*.

PAPADAKIS, J., 1962c. Degree of leaching as distinct from degree of weathering, and its implications in soil classification and fertility. *Proc. Intern. Soil Conf., New Zealand, 1962*, pp.45–49.

PAPADAKIS, J., 1963a. Soils of Argentine. *Soil Sci.*, 95: 356–366.

PAPADAKIS, J., 1963b. Plant population stress and antibiotics of the rhizosphere. *Soil Sci.*, 96: 257–260.

PAPADAKIS, J., 1964. *Soils of the World*. J. Papadakis, Buenos Aires, 141 pp.

PAPADAKIS, J., 1966a. *Crop-Ecologic Survey in West Africa (Liberia, Ivory Coast, Ghana, Togo, Dahomey, Nigeria)*. F.A.O., Rome, 1: 103 pp.; 2: 45 pp.

PAPADAKIS, J., 1966b. *Climates of the World and their Agricultural Potentialities*. J. Papadakis, Buenos Aires, 174 pp.

PAPADAKIS, J., CALCAGNO, J. E. e ETCHEVEHERE, P. H., 1960.

Regiones de Suelos de la República Argentina, Mapa Esquemàtico. Inst. Suelos Agrot., Buenos Aires, 41 pp.

PEDRO, G., 1964. *Contribution à l'Etude expérimentale de l'Altération géochimique des Roches cristallines.* Inst. Natl. Rech. Agron., Paris, 344 pp.

PEDRO, G., 1966. Sur l'altération expérimentale de la kaolinite et sa transformation en boehmite par lessivage à l'eau. *Compt. Rend.*, 262: 729–732.

PENDLETON, R. L., 1943. Analyses of some Siamese laterites. *Soil Sci. Soc. Am. Proc.*, 8: 403–407.

PENDLETON, R. L. and SHARASUVAVA, S., 1946. Analyses of some Siamese laterites. *Soil Sci.*, 12: 423–440.

PIAS, J. and GUICHARD, E., 1959. *Interafrican Soil Conf., 3rd, 1959, Dalaba*, 1: 227–232.

PICKERING, S., 1917. The effect of one plant on another. *Ann. Botan.*, 31: 181–187.

PIÑEIRO, A. R. and ZUKARDI, R., 1959. *Rev. Agr. Noroeste Argentino*, 3 (1/2): 259–285.

POLYNOV, B. B., 1937. *The Cycle of Weathering.* Thomas Murby, London, 220 pp.

RADEKE, R. E. and WESTIN, FR. C., 1963. Gray wooded soils of the black hills of South Dakota. *Soil Sci. Soc. Am. Proc.*, 27: 573–576.

RADWANSKI, S. A. and OLLIER, C. D., 1959. A study of an East African catena. *J. Soil Sci.*, 10: 149–168.

RAMANN, E., 1905. *Bodenkunde*, 2 Aufl. Springer, Berlin.

RAYCHOUDHURI, S. P., 1961. New systems of classification and nomenclature of soils. *J. Indian Soc. Soil Sci.*, 9: 1–8.

RAESIDE, J., 1961. Letters to the Editor. *Bull. Soil Sci. Intern. Soc.*, 19: 20–21.

REED, W. E., 1951. Reconnaissance soil survey in Liberia. *U.S. Dept. Agr., Bull.*, 66: 107.

RETZER, J. L., 1956. Alpine soils of the Rocky Mountains. *J. Soil Sci.*, 7: 22–32.

RIQUIER, J., 1960. Phytoliths of certain tropical soils and of podsols. *Trans. Intern. Congr. Soil Sci., 7th, 1960, Madison*, 4: 425–431.

RODE, A. A., 1955. *Soil Science.* Translation by A. Gourevitch, Jerusalem, 517 pp.

ROY, B. B. and BARDE, N. K., 1962. Some characteristics of the black soils of India. *Soil Sci.*, 93: 142–147.

ROZANOV, V. G. and ROZANOVA, I. M., 1961. Soils of the arid monsoon tropical zone in Burma. *Pochvovedenie*, 12: 75–84.

ROZANOV, V. G. and ROZANOVA, I. M., 1962. Soils of the dry monsoon tropical zone of Burma. *Pochvovedenie*, 3: 73–82.

RUBLIIN, E. V., 1962. Soils of the North American prairies. *Pochvovedenie*, 6: 45–57.

SACADURA GARCIA, J. A. and CARVALHO CARDOSO, J., 1960. The Soils of S. Tomé and Principe Islands. *Trans. Intern. Congr. Soil Sci., 7th, 1960, Madison*, 4: 88–96.

SALAM, M. A. A., 1962. Soils of the Amereya–Maryut area, Mediterranean littoral, United Arab Republic. *Proc. Intern. Soil Conf. New Zealand, 1962*, pp.355–361.

SHERMAN, G. D. and ALEXANDER, L. T., 1959. Characteristics and genesis of low humic latosols. *Soil Sci. Soc. Am. Proc.*, 23: 168–170.

SHERMAN, G. D. and UEHARA, G., 1956. The weathering of olivine basalt in Hawaii and its pedogenic significance. *Soil Sci. Soc. Am. Proc.*, 20: 337–340.

SHERMAN, G. D., FOSTER, Z. C. and FUJIMOTO, C. K., 1948. Some of the properties of the ferruginous humic latosols of the Hawaiian Islands. *Soil Sci. Soc. Am. Proc.*, 13: 471–476.

SILVA CARNEIRO, L. R., 1951. *Anãis Tercera Reunião Brasileira de Ciencia do Solo, Recife.* 675 pp.

SIMONSON, R. W., 1959. Outline of a generalized theory of soil genesis. *Soil Sci. Soc. Am. Proc.*, 23: 152–155.

SIUTA, J., 1962. Effects of reducing processes and acidification on solubility of mineral compounds of soil. *Pochvovedenie*, 5: 62–72.

SLAVNYI, YU. A., 1961. Characteristics of processes of soil formation in the central area between the Rivers Amur and Zeya. *Pochvovedenie*, 7: 73–84.

SMITH, G. D. and RIECKEN, F. F., 1950. Brunigra soils, a new name for prairie soils. *Soil Sci. Soc. Am. Proc.*, 15: 335.

SMITH, G. K., 1962. *Report on Soil Agricultural Survey of Sene-Obosum River Basins.* U.S. Agency Intern. Development Ghana.

STACE, H. C. T., 1961. *The Morphological and Chemical Characteristics of Representative Profiles of the Great Soil Groups of Australia.* C.S.I.R.O., Adelaide, 230 pp.

STEFANOVITS, P., 1962. Principles of soil classification used in the preparation of the genetic soil map of Hungary. *Proc. Intern. Soil Conf. New Zealand, 1962*, pp.399–403.

STEPHENS, C. G., 1956. *A Manual of Australian Soils.* C.S.I.R.O., Adelaide, 48 pp.

STEPHENS, C. G. and DONALD, C. M., 1958. Australian soils and their response to fertilizers. *Advan. Agronomy*, 10: 167–256.

STEVENSON, F. Y., MARKS, V. Y. and MARTIN, W., 1952. Electrophoretic and chromatographic investigations of clay-adsorbed organic colloids, 1. Preliminary investigations. *Soil Sci. Soc. Am. Proc.*, 16: 69–73.

STOBBE, P. C., 1962. Classification of Canadian soils. *Proc. Intern. Soil Conf. New Zealand, 1962*, pp.318–324

STORIE, R. E., 1946. Soil regions of California illustrated by twenty-four dominant soil types. *Soil Sci. Soc. Am. Proc.*, 11: 425–430.

STORIE, R. E., 1953. Preliminary study of Bolivian soils. *Soil Sci. Soc. Am. Proc.*, 17: 128–131.

STORIE, R. E. and HARRADINE, F., 1958. Soils of California. *Soil Sci.*, 85: 207–227.

STORRIER, R. R. and MUIR, A., 1962. Characteristics and genesis of a ferritic brown earth. *J. Soil Sci.*, 13: 259–270.

SWANSON, C. L. W., 1946. Reconnaissance soil survey in Japan. *Soil Sci. Soc. Am. Proc.*, 11: 493–507.

SYS, C., 1960. *Carte des Sols du Congo et du Ruanda Urundi.* Inst. Natl. Étude Agr. Congo, Bruxelles, 84 pp.

SYS, C., 1961. *La Cartographie des Sols au Congo.* Inst. Natl. Étude Agr. Congo, Bruxelles, 66 pp.

TAMURA, T., HANNA, R. M. and SHEARIN, A. E., 1959. Properties of brown podsolic soils. *Soil Sci.*, 87: 189–197.

TAVERNIER, R., 1960. La carte des sols de l'Europe. *Pédologie*, 10: 324–348.

TAVERNIER, R. and MUCKENHAUSEN, E., 1960. The soil map of western Europe (scale 1/2,500,000). *Trans. Intern. Congr. Soil Sci., 7th, 1960, Madison*, 4: 44–48.

TAVERNIER, R. and SMITH, G. D., 1957. The concept of braunerde in Europe and the United States. *Advan. Agronomy*, 9: 217–289.

TAYLOR, N. H. and POHLEN, I. J., 1962. *Soil Survey Method—Dept. Sci. Ind. Res. (New Zealand), Soil Bur. Bull.*, 25: 242 pp.

TEDROW, J. C. F. and DOUGLAS, L. A., 1964. Soil investigations on Banks Island. *Soil Sci.*, 98: 53–65.

TEDROW, J. C. F., DREW, J. V., HILL, D. E. and DOUGLAS, L. A., 1958. Major genetic soils of the arctic slope of Alaska. *J. Soil Sci.*, 9: 33–45.

THORP, J. and BALDWIN, M., 1938. New nomenclatures of the higher categories of soil classification as used in the Department of Agriculture. *Soil Sci. Soc. Am. Proc.*, 3: 260–268.

THORP, J. and SMITH, G. D., 1949. Higher categories of soil classification. Order, suborder and great soil groups. *Soil Sci.*, 67: 117–126.

TYURIN, L. V. and KONONOVA, M. M., 1963. Biology of the humus and problems of soil fertility. *Sov. Soil Sci.*, 1963: 205–213

TYURIN, L. V., ROZOV, N. N. and RUDNEVA, E. N., 1960. International soil map of the eastern part of Europe. *Trans. Intern. Congr. Soil Sci., 7th, 1960, Madison*, 4: 36–43.

UGOLINI, F. C. and TEDROW, J. C. F., 1963. Soils of the Brooks Range, Alaska, 3. Rendzina of the Arctic. *Soil Sci.*, 96: 121–127.

ULRICH, R., ARKLEY, R. J. and NELSON, R. E., 1959. Characteristics and genesis of some soils of San Mateo County, California. *Soil Sci.*, 88: 218–227.

U. S. DEPARTMENT OF AGRICULTURE, 1938. *Soils and Man— Yearbook Agr., U.S. Dept. Agr.*, 1938.

U. S. DEPARTMENT OF AGRICULTURE, SOIL SURVEY SERVICE, 1960. *Soil Classification, a Comprehensive System, 7th Approximation*. U.S. Dept. Agr., Washington, D.C., 265 pp.

VAN DER MERWE, C. R. and HEYSTEK, H., 1955. Clay minerals of South African soil groups, 2. Subtropical black clays and related soils. *Soil Sci.*, 79: 147–158.

VAN ROYEN, W., 1954. *The Agricultural Resources of the World*. Prentice-Hall, Englewood Cliffs, N. J., 258 pp.

VAN WAMBEKE, A., 1961. The soils of Rwanda–Burundi. *Pédologie*, 11: 289–353.

VINE, H., 1953. Notes on the main types of Nigerian soils. *Nigeria Dept. Agr., Spec. Bull.*, 5: 6 pp.

WHITE, E. M. and RIECKEN, F. F., 1955. Brunizem-gray brown podsolic soils biosequences. *Soil Sci. Soc. Am. Proc.*, 19: 504–509.

WHITE, M. E., 1961. Calcium-solodi or planosol genesis from solodized solonetz. *Soil Sci.*, 91: 175–177.

WILLAIME, P., 1959. *Les Sols du Confluent Tchi–Couffo*. Office Rech. Sci. Territ. Outre Mer, Cotonou, Dahomey.

WILLAIME, P., 1962. *Étude Pédologique du Secteur de Boukombe*. Office Rech. Sci. Territ. Outre Mer, Cotonou, Dahomey.

WILLIAMSON, W. T. H., 1959. The discipline of soil science. *J. Soil Sci.*, 10: 1–4.

WILLS, B. J., 1962. *Agriculture and Land Use in Ghana*. Oxford Univ. Press, London, 504 pp.

WILSON, G. V., 1958. The leetonia series in Virginia. *Soil Sci. Soc. Am. Proc.*, 22: 565–570.

WRIGHT, C. S., 1961. *Reports of the Assessor in Soils*. F.A.O., Chile.

WRIGHT, J. R. and LEWICK, R., 1956. Development of a profile in a soil column leached with a chelating agent. *Congr. Intern. Sci. Sol., 6e, Rappt.*, E: 257–262.

WURMAN, E., 1960. Pedogenic and petrogenic characteristics of soil profiles developed in silt-mantled acid shale. *Soil Sci.*, 90: 348–356.

WURMAN, E., WHITESIDE, E. P. and MORTLAND, M. M., 1959. Properties and genesis of finer textured subsoil bands in some sandy Michigan soils. *Soil Sci. Soc. Am. Proc.*, 23: 135–143.

ZONN, S. V., 1962. Some questions of the genesis of soils of the spruce forest of Tyan-Shan. *Pochvovedenie*, 5: 24–39.

ZVORYKIN, I. A., 1960. *Edaphologic Study of the Soils of the Plain of Thessaly*. Inst. N. Kanellopoulos, Piraeus, Greece (in Greek).

References Index

AGARWALL, R. R. and MUKERJI, S. K., 34, 58, 193
AHN, P. M., 49, 54, 63, 193
AKHTYRSTEV, B. P., 43, 193
ALEXANDER, L. T., see SHERMAN, G. D. and ALEXANDER, L. T.
ALLEN, J. S., see MCCLELLAND, J. E. et al.
AMERYCK, J., 52, 54, 193
ANDERSON, B., see MUIR, A. et al.
ANDREWS, E. D., see JACKMAN, R. H. et al.
ARAUJO, W. A., see FAGUNDES, A. B. et al.
ARIMURA, S., see KANO, I. et al.
ARISTARKHOV, A. N., see PANIN, P. S. and ARISTARKHOV, A. N.
ARKLEY, R. J., see ULRICH, R. et al.
ARNAUD, J. ST., see MOSS, H. C. and ARNAUD, J. ST.
ASAMOA, G. K., see OBENG, H. B. et al.
AUBERT, G., 193
AUBERT, G., see LENEUF, N. and AUBERT, G.
AVERY, B. W., STEPHEN, I., BROWN, G. and YAALON, D. H., 52, 193
AZEVEDO, A. L., see BOTELHO DA COSTA, J. V. et al.

BALDWIN, M., KELLOGG, CH. E. and THORP, J., 193
BALDWIN, M., see THORP, J. and BALDWIN, M.
BARDE, N. K., see ROY, B. B. and BARDE, N. K.
BARSHAD, B., 42, 193
BARSHAD, I., 193
BARSHAD, I. and ROJAS-CRUZ, L. A., 54, 193
BASINSKI, J. J., 193
BAUMGART, I. L., 193
BEAR, F. E., 193
BEATTY, M. T., see HOGAN, J. D. and BEATTY, M. T.
BECKETT, P. H. T., 50, 54, 193
BENGTSON, N. A. and VAN ROYEN, W., 193
BIDWELL, O. W., see JARVIS, L. L. et al.
BLOOMFIELD, C., 193
BOBKOV, V. P., 193
BONFILS, C. G., GALCAGNO, J. E., ETCHEVEHERE, H., IPUCHA AGUIRRE, J., MIACZYNSKY, C. R. O. and TALLARICO, L. A., 37, 42, 57, 193
BONNET, J. A., 193
BOTELHO DA COSTA, J. V., AZEVEDO, A. L., CARDOSO FRANCO, E. P. and PINTO RICARDO, R., 49, 58, 193
BRAMAO, D. L. and SIMONSON, R. W., 193
BRAMMER, H., 12, 166, 179, 193
BROWN, G., see AVERY, B. W. et al.
BROWN, J. and TEDROW, J. C. F., 39, 193
BRYDON, J. E., see CLARK, J. S. et al.

CALCAGNO, J. E., see PAPADAKIS, J. et al.
CARDOSO FRANCO, E. P., see BOTELHO DA COSTA, J. V. et al.
CARVALHO CARDOSO, J., see SACADURA GARCIA, J. A. and CARVALHO CARDOSO, J.
CIRIC, M., see NEJGEBAUER, V. et al.
CLARK, J. S., BRYDON, J. E. and FARSTAD, L., 193
CLINE, M. C., 193
CLINE, M. G., see FREI, E. and CLINE, M. G.
COLEMAN, N. T., see MILLER, E. V. and COLEMAN, N. T.
COSTA LIMA, J. W., 37, 49, 58, 193
CROMPTON, E., 18, 193

CRUELLAS, J., see GOLLAN, J. et al.

DABIN, B., LENEUF, N. and RIOU, G., 36, 37, 41, 50, 54, 193
DAMMAN, A. W. H., 193
DAN, J. and KOYUMDJISKY, H., 193
DA SILVA TEIXEIRA, A. J., 49, 52, 193
DEL VILLARS, E. H., 34, 193
DEWAN, M. L. and FAMOURI, J., 193
D'HOORE, J. L., 77, 193
DONALD, C. M., see STEPHENS, C. G. and DONALD, C. M.
DOUGLAS, L. A. and TEDROW, J. C. F., 54, 193
DOUGLAS, L. A., see TEDROW, J. C. F. and DOUGLAS, L. A.
DOUGLAS, L. A., see TEDROW, J. C. F. et al.
DREW, J. V., see TEDROW, J. C. F. et al.
DUCHAUFOUR, C., 33, 35, 36, 37, 38, 39, 40, 42, 47, 51, 53, 54, 55, 56, 58, 59, 60, 61, 67, 182, 193, 194
DUCHAUFOUR, C. and JACQUIN, F., 4, 25, 194
DUDAL, R. and MOORMANN, F. R., 194
DURING, C., 180, 194
DURING, C., see JACKMAN, R. H. et al.

ENDREDY, A. S., see OBENG, H. B. et al.
ENGLAND, C. B. and PERKINS, H. F., 49, 194
ETCHEVEHERE, H., see BONFILS, C. G., et al.
ETCHEVEHERE, P. H., see PAPADAKIS, J. et al.

FAGUNDES, A. B., ARAUJO, W. A., RAMOS, F. and KEHRIG, A. G., 49, 194
FAMOURI, J., see DEWAN, M. L. and FAMOURI, J.
FARSTAD, L., see CLARK, J. S. et al.
FAUCK, R. 50, 194
FAUCK, R., MOUREAUX, C. and THOMANN, C., 192, 194
FOSTER, Z. C., see SHERMAN, G. D. et al.
FREI, E. and CLINE, M. G., 194
FRIDLAND, V. M., 50, 194
FUJIMOTO, C. K., see SHERMAN, G. D. et al.

GALCAGNO, J. E., see BONFILS, C. G. et al.
GERASIMOV, I. P., 194
GILE JR., L. H., 194
GLENTWORTH, R., 194
GOLLAN, J. and LACHAGA, D., 37, 42, 55, 57, 194
GOLLAN, J., CRUELLAS, J. and NICOLLIER, V., 37, 55, 58, 194
GOUVEIA, D. H. and GOUVEIA, J., 58, 194
GOUVEIA, J., see GOUVEIA, D. H. and GOUVEIA, J.
GRAY, F., see MOLTHAM, H. D. and GRAY, F.
GUICHARD, E., see PIAS, J. and GUICHARD, E.

HALLSWORTH, E. G., 5, 8, 45, 194
HANNA, R. M., see TAMURA, T. et al.
HAPSTEAD, M. and RUST, R. H., 51, 194
HARDY, F. and RODRIGUEZ, G., 194
HARRADINE, FR., 194
HARRADINE, F., see STORIE, R. E. and HARRADINE, F.
HEYSTEK, H., see VAN DER MERWE, C. R. and HEYSTEK, H.
HILL, D. E., see TEDROW, J. C. F. et al.
HOGAN, J. D. and BEATTY, M. T., 194

HONJO, Y., *see* KANO, I. et al.
HUBERT, P., 49, 194
HUTCHESON JR., T. B., 49, 52, 194
HUTCHESON JR., T. B., LEWIS, R. J. and SEAY, W. A., 52, 194

IPUCHA AGUIRRE, J., *see* BONFILS, C. G. et al.
IVANOVA, Y. N., 194

JACKMAN, R. H., DURING, C., ANDREWS, E. D. and LYNCH, P. B., 194
JACKSON, M. L., 21, 25, 194
JACKSON, M. L. and SHERMAN, G. C., 18, 194
JACQUIN, F., *see* DUCHAUFOUR, C. and JACQUIN, F.
JANZEN, W. K., 56, 57, 194
JANZEN, W. K. and MOSS, H. C., 51, 56, 57, 194
JARVIS, L. L., ROSCOE JR., E. and BIDWELL, O. W., 36, 42, 58, 61, 194
JENNY, H., 3, 38, 54, 194
JENSEN, S. T., 194
JOHNSON, W. M., *see* McCLELLAND, J. E. et al.

KAMOCHITA, Y., 194
KANO, I., 194
KANO, I., HONJO, Y. and ARIMURA, S., 50, 194
KARIM, A. and KHAN, D. H., 57, 194
KARIM, A. and QUASEM, A., 34, 50, 58, 194
KEHRIG, A. G., *see* FAGUNDES, A. B. et al.
KELLOGG, CH. E., 194
KELLOGG, CH. E., *see* BALDWIN, M. et al.
KHAN, D. H., *see* KARIM, A. and KHAN, D. H.
KING, L. C., 16, 194
KOBO, K., 180, 194
KONONOVA, M. M., *see* TYURIN, L. V. and KONONOVA, M. M.
KOVDA, V. A., 55, 194
KOVDA, V. A., *see* LOBOVA, E. V. and KOVDA, V. A.
KOYUMDJISKY, H., *see* DAN, J. and KOYUMDJISKY, H.
KUBIENA, W. L., 11, 24, 38, 39, 194, 195

LACHAGA, D., *see* GOLLAN, J. and LACHAGA, D.
LEAHEY, A., 195
LEE, C.-K., 195
LEEPER, G. W., 187, 195
LENEUF, N., 16, 195
LENEUF, N. and AUBERT, G., 14, 16, 195
LENEUF, N., *see* DABIN, B. et al.
LEWICK, R., *see* WRIGHT, J. R. and LEWICK, R.
LEWIS, R. J., *see* HUTCHESON JR., T. B. et al.
LICHMANOVA, A. I., 195
LOBOVA, E. V. and KOVDA, V. A., 195
LYNCH, P. B., *see* JACKMAN, R. H. et al.

MACKNEY, D., 195
MANIL, G., 195
MARBUT, C. F., 51, 195
MARKS, V. Y., *see* STEVENSON, F. Y. et al.
MARTIN, D., 49, 195
MARTIN, W., *see* STEVENSON, F. Y. et al.
McALESEE, D. M., *see* McCONAGHY, S. and McALESEE, D. M.
McCALEB, S. B., *see* NYUN, M. A. and McCALEB, S. B.
McCLELLAND, J. E., 195
McCLELLAND, J. E., MOGAN, C. A., JOHNSON, W. M., SHROER, F. W. and ALLEN, J. S., 42, 56, 195
McCONAGHY, S. and McALESEE, D. M., 195
MEHLICH, A., *see* MILLER, F. T. and MEHLICH, A.
MIACZYNSKI, C. R. O., *see* BONFILS, C. G. et al.
MILLER, E. V. and COLEMAN, N. T., 195
MILLER, F. T. and MEHLICH, A., 41, 195
MOGEN, C. A., *see* McCLELLAND, J. E. et al.
MOHR, E. C. J. and VAN BAREN, F. A., 73, 195

MOLFINO, R. H., 37, 38, 54, 195
MOLTHAM, H. D. and GRAY, F., 58, 195
MOORMANN, F. R., *see* DUDAL, R. and MOORMANN, F. R.
MORTLAND, M. M., *see* WURMAN, E. et al.
MOSS, H. C. and ARNAUD, J. ST., 195
MOSS, H. C., *see* JANZEN, W. K. and MOSS, H. C.
MOUREAUX, C., *see* FAUCK, R. et al.
MUCKENHAUSEN, E., *see* TAVERNIER, R. and MUCKENHAUSEN, E.
MUIR, A., ANDERSON, B. and STEPHENS, I., 37, 49, 61, 195
MUIR, A., *see* STORRIER, R. R. and MUIR, A.
MUIR, J. W., 195
MUKERJI, S. K., *see* AGARWALL, R. R. and MUKERJI, S. K.

NEJGEBAUER, V., CIRIC, M. and ZIVKOVIC, M., 195
NELSON, R. E., *see* ULRICH, R. et al.
NEW ZEALAND SOIL BUREAU, 195
NICOLLIER, V., *see* GOLLAN, J. et al.
NYE, P. H., 195
NYUN, M. A. and McCALEB, S. B., 49, 52, 195

OAKES, H., 34, 36, 41, 56, 195
OBENG, H. B., ENDREDY, A. S., ASAMOA, G. K. and SMITH, G. K., 49, 195
OLLIER, C. D., *see* RADWANSKI, S. A. and OLLIER, C. D.

PAHAULT, P. and SOUGNEZ, N., 34, 39, 52, 195
PANABOKKE, C. R., 50, 195
PANIN, P. S. and ARISTARKHOV, A. N., 57, 195
PAPADAKIS, J., 3, 5, 7, 8, 15, 17, 24, 27, 41, 56, 63, 65, 67, 70, 71, 75, 177, 178, 179, 188, 191, 192, 195
PAPADAKIS, J., CALCAGNO, J. E. and ETCHEVEHERE, P. H., 195
PEDRO, G., 2, 11, 14, 18, 191, 196
PENDLETON, R. L., 196
PENDLETON, R. L. and SHARASUVAVA, S., 196
PERKINS, H. F., *see* ENGLAND, C. B. and PERKINS, H. F.
PIAS, J. and GUICHARD, E., 34, 36, 37, 61, 196
PICKERING, S., 177, 196
PIÑEIRO, A. R. and ZUKARDI, R., 37, 40, 42, 58, 196
PINTO RICARDO, R., *see* BOTELHO DA COSTA, J. V. et al.
POHLEN, I. J., *see* TAYLOR, N. H. and POHLEN, I. J.
POLYNOV, B. B., 17, 196

QUASEM, A., *see* KARIM, A. and QUASEM, A.

RADEKE, R. E. and WESTIN, FR. C., 52, 196
RADWANSKI, S. A. and OLLIER, C. D., 39, 49, 52, 196
RAESIDE, J., 196
RAMANN, E., 196
RAMOS, F., *see* FAGUNDES, A. B. et al.
RAYCHOUDHURI, S. P., 196
REED, W. E., 196
RETZER, J. L., 38, 196
RIECKEN, F. F., *see* WHITE, E. M. and RIECKEN, F. F.
RIOU, G., *see* DABIN, B. et al.
RIQUIER, J., 196
RODE, A. A., 10, 196
RODRIGUEZ, G., *see* HARDY, F. and RODRIGUEZ, G.
ROJAS-CRUZ, L. A., *see* BARSHAD, J. and ROJAS-CRUZ, L. A.
ROSCOE JR., E., *see* JARVIS, L. L. et al.
ROY, B. B. and BARDE, N. K., 196
ROZANOV, V. G. and ROZANOVA, J. M., 36, 41, 50, 57, 58, 196
ROZANOVA, J. M., *see* ROZANOV, V. G. and ROZANOVA, J. M.
ROZOV, N. N., *see* TYURIN, L. V. et al.
RUBLIIN, E. V., 196
RUDNEVA, E. N., *see* TYURIN, L. V. et al.
RUST, R. H., *see* HAPSTEAD, M. and RUST, R. H.

SACADURA GARCIA, J. A. and CARVALHO CARDOSO, J., 36, 37, 39, 49, 196

SALAM, M. A. A., 196
SEAY, W. A., *see* HUTCHESON JR., T. B. et al.
SHARASUVAVA, S., *see* PENDLETON, R. L. and SHARASUVAVA, S.
SHEARIN, A. E., *see* TAMURA, T. et al.
SHERMAN, G. C., *see* JACKSON, M. L. and SHERMAN, G. C.
SHERMAN, G. D. and ALEXANDER, L. T., 196
SHERMAN, G. D. and UEHARA, G., 196
SHERMAN, G. D., FOSTER, Z. C. and FUJIMOTO, C. K., 196
SHROER, F. W., *see* MCCLELLAND, J. E. et al.
SILVA CARNEIRO, L. R., 40, 49, 57, 196
SIMONSON, R. W., 29, 196
SIUTA, J., 196
SLAVNYI, YU. A., 196
SMITH, G. D. and RIECKEN, F. F., 196
SMITH, G. D., *see* TAVERNIER, R. and SMITH, G. D.
SMITH, G. D., *see* THORP, J. and SMITH, G. D.
SMITH, G. K., 49, 52, 57, 196
SMITH, G. K., *see* OBENG, H. B. et al.
STACE, H. C. T., 33, 34, 36, 38, 39, 40, 42, 48, 49, 51, 52, 54, 56, 57, 59, 196
STEFANOVITS, P., 196
STEPHEN, J., *see* AVERY, B. W. et al.
STEPHENS, C. G., 196
STEPHENS, C. G. and DONALD, C. M., 196
STEPHENS, J., *see* MUIR, A. et al.
STEVENSON, F. Y., MARKS, V. Y. and MARTIN, W., 196
STOBBE, P. C., 39, 42, 53, 56, 59, 196
STORIE, R. E., 196
STORIE, R. E. and HARRADINE, F., 196
STORRIER, R. R. and MUIR, A., 39, 196
SOUGNEZ, N., *see* PAHAULT, P. and SOUGNEZ, N.
SWANSON, C. L. W., 196
SYS, C., 33, 45, 47, 48, 196

TALLARICO, L. A., *see* BONFILS, C. G. et al.
TAMURA, T., HANNA, R. M. and SHEARIN, A. E., 39, 52, 196
TAVERNIER, R., 77, 196
TAVERNIER, R. and MUCKENHAUSEN, E., 196
TAVERNIER, R. and SMITH, G. D., 196
TAYLOR, N. H. and POHLEN, I. J., 187, 196
TEDROW, J. C. F. and DOUGLAS, L. A., 39, 196
TEDROW, J. C. F., DREW, J. V., HILL, D. E. and DOUGLAS, L. A., 38, 59, 61, 196

TEDROW, J. C. F., *see* BROWN, J. and TEDROW, J. C. F.
TEDROW, J. C. F., *see* DOUGLAS, L. A. and TEDROW, J. C. F.
TEDROW, J. C. F., *see* UGOLINI, F. C. and TEDROW, J. C. F.
THOMANN, C., *see* FAUCK, R. et al.
THORP, J. and BALDWIN, M., 196
THORP, J. and SMITH, G. D., 39, 40, 42, 51, 52, 53, 55, 56, 60, 196
THORP, J., *see* BALDWIN, M. et al.
TYURIN, L. V. and KONONOVA, M. M., 196
TYURIN, L. V., ROZOV, N. N. and RUDNEVA, E. N., 38, 197

UEHARA, G., *see* SHERMAN, G. D. and UEHARA, G.
UGOLINI, F. C. and TEDROW, J. C. F., 34, 197
ULRICH, R., ARKLEY, R. J. and NELSON, R. E., 39, 42, 57, 58, 197
U.S. DEPARTMENT OF AGRICULTURE, 197
U.S. DEPARTMENT OF AGRICULTURE, SOIL SURVEY SERVICE, 197

VAN BAREN, F. A., *see* MOHR, E. C. J. and VAN BAREN, F. A.
VAN DER MERWE, C. R. and HEYSTEK, H., 36, 39, 41, 197
VAN ROYEN, W., 197
VAN ROYEN, W., *see* BENGTSON, N. A. and VAN ROYEN, W.
VAN WAMBEKE, A., 197
VINE, H., 197

WESTIN, FR. C., *see* RADEKE, R. E. and WESTIN, FR. C.
WHITE, E. M. and RIECKEN, F. F., 42, 52, 197
WHITE, M. E., 58, 197
WHITESIDE, E. P., *see* WURMAN, E. et al.
WILLAIME, P., 36, 37, 50, 52, 197
WILLIAMSON, W. T. H., 197
WILLS, B. J., 36, 49, 54, 57, 197
WILSON, G. V., 197
WRIGHT, C. S., 197
WRIGHT, J. R. and LEWICK, R., 197
WURMAN, E., 197
WURMAN, E., WHITESIDE, E. P. and MORTLAND, M. M., 197

YAALON, D. H., *see* AVERY, B. W. et al.

ZIVKOVIC, M., *see* NEJGEBAUER, V. et al.
ZONN, S. V., 42, 197
ZUKARDI, R., *see* PIÑEIRO, A. R. and ZUKARDI, R.
ZVORYKIN, I. A., 34, 37, 41, 57, 58, 197

Geographic Index [1]

Afghanistan, *168*
Africa, *134*
Alaska, 34, 38, 39, 54, 59, 61, *80*, 87
Algeria, *134*, 135, 136
Angola, 49, 52, 58, *150*
Antarctica, 37, 38, 54
Argentina, 8, 34, 37, 40, 41, 42, 54, 55, 57, 71, *105*, 106
Asia, 42, 155
Australia, 33, 34, 36, 38, 39, 40, 42, 48, 49, 52, 54, 56, 57, 58, 59, 170, 171, *172*
Austria, 126, *127*

Basutoland, *152*, 153
Bechuanaland, *152*, 153
Belgium, 34, 39, 52, *126*, 127
Belize, 90, *91*
Bhutan, *166*
Bolivia, 98, *105*
Brazil, 33, 37, 38, 40, 47, 48, 49, 52, 54, 57, 58, 59, 110–115, *119*
British Honduras, 90, *91*
Bulgaria, 128, *129*
Burma, 36, 41, 50, 57, 58, *157*, 158
Burundi, *146*, 147

Cambodia, 158, *159*
Cameroons, *141*, 142
Canada, 7, 41, 42, 51, 52, 53, 56, 57, 59, 78, *79*
Canary Islands, 135, *136*
Central African Republic, *141*, 142, 143
Central America, 90, *91*
Ceylon, 165, *166*
Chad, 34, 36, 37, *141*, 142, 143
Chile, 99–101, *105*
China, *155*, 156
Colombia, 38, 54, 93, *94*
Congos (both), 33, 48, 49, 148, 149, *150*
Costa Rica, 90, *91*
Cuba, 90, *92*
Cyprus, 132, *134*
Czechoslovakia, *127*, 128

Dahomey, 36, 37, 50, 52, *141*, 142
Denmark, 123, *124*

Ecuador, *94*, 95
Egypt, *136*, 137
Ethiopia, *141*, 144, 145
Europe, *122*

Fernando Póo, *141*
Finland, *122*, 123
France, 33, 35, 36, 38, 39, 40, 42, 51, 58, 59, 61, *125*
French Somaliland, *141*, 144, 145, *146*

Gabon, 148, 149, *150*

Gambia, *136*, 139
Germany, 126, *127*, 128
Ghana, 36, 47, 48, 52, 53, 57, 139, *140*
Great Britain, 39, 52, *124*, 125
Greece, 34, 37, 41, 57, 58, 132, *134*
Guatemala, 90, *91*
Guiana, 90, *94*
Guinea, *136*, 138, 139

Hawaiian, Islands 81, *87*
Honduras, 90, *91*
Hungary, *128*

India, 34, 58, 164, 165, *166*
Indochina, *157*, 158
Indonesia, *159*, 160–163
Iran, *167*, 168, 169
Iraq, *167*, 168
Ireland, 124, *125*
Israel, *167*
Italy, *131*
Ivory Coast, 36, 37, 41, 49, 50, 54, 139, *140*

Jamaica, 90, *92*
Japan, 50, *157*
Jordan, *167*, 168

Kashmir, *166*, 169
Kenya, *146*, 147

Laos, *159*, 158
Lebanon, *167*, 168
Liberia, 139, *140*
Libya, *136*, 137
Luxembourg, 126, *127*

Madagascar, 151, *152*
Madeira, 135, *136*
Malawi, 148, 149, *150*
Malaya, 160, 161, *166*
Mali, 138, 139, *140*
Mauritania, *136*, 138, 139
Mexico, 88, 89, *91*
Morocco, 34, 36, *134*, 135
Mozambique, 58, *150*, 151

Nepal, 164, *166*
Netherlands, *126*
New Zealand, *172*, 174
Nicaragua, 90, *91*
Niger, *141*, 142, 143
Nigeria, *140*, 142, 143
North America, 78, *79*, 81–83, 87, 88, 89
Norway, 123, *124*

---

[1] Page numbers in italics refer to individual country and regional texts.

Pakistan, 34, 50, 58, 164, *166*, *169*
Panama, 90, *91*, 92
Paraguay, *119*
Persia, *see* Iran
Peru, *94*, 96, 97
Poland, *127*, 128
Portugal, 130, *131*
Portuguese Guinea, 49, 52, 54, *136*, 139
Puerto Rico, 81, *87*

Rhodesia, *152*, 153
Rumania, 128, *129*
Russia, 41, 43, 57, 71, 120, 121, *122*, *155*
Rwanda, *146*, 147

Salvador, 90, *91*
Santo Tomé, 36, 37, 39, 49
Sarawak, 50, 54, *159*, 161
Senegal, 136, 138, 139
Siberia, 7
Sierra Leone, *140*
Singapore, *166*
Somalia, 145, *146*
South Africa, 36, 39, 41, 49, 52, *152*, 153
South America, 90, 92, 93, *94*, 95, 96, 97, 98, 99–101, 106, 110–113
South West Africa, *152*, 153
Spain, *130*
Spanish Guinea, 148, *150*
Spanish Sahara, *136*, 138
Sudan, 142, 143, *146*

Surinam, *94*
Swaziland, *152*, 153
Sweden *122*, 123
Switzerland, 126, *127*
Syria, *166*, 168

Tanzania, 37, 41, 49, 61, *146*
Tchad, 61, *141*
Thailand, *157*, 158
Togo, 139, *140*
Trinidad, *92*
Tunisia, 135, *136*
Turkey, 34, 36, 41, 56, 133, *134*

Uganda, 39, 49, 54, *146*, 147
United Arab Republic, *136*, 137, *167*
United Kingdom, *see* Great Britain
United States, 33, 36, 38, 39, 40, 41, 42, 47, 49, 51, 52, 53, 55, 56, 57, 58, 61, 71, *79*, 81–83, 191
Upper Volta, 139, *140*
Uruguay, 104, *119*

Venezuela, 92, *94*
Vietnam, 50, *159*, 158

West Indies, 90, *92*

Yugoslavia, *129*

Zambia, 148, 149, *150*

Subject Index [1]

A horizon, 25
Absorbing capacity, 179, 181, 182
Acid clay, 35, *39*
—ferruginous, *47*, 48, 49, 50
—horizon, *19*, *23*
Acidification, *7*, 10, 17, 18
Acidity 4, 15, *19*, 23, 72, 73
Acrohumox, 192
Acrorthox, 192
Acrustox, 192
Active solonchack, *55*
Aeolian, *10*, *11*, 24, 33, 34, 36, 37, 41, 46, 47, 48, 49, 50, 53, 54
Age, soil, 1, 7, 8, 14, *15*, *16*, 18, 73
Agricultural potentialities, *177*, 192
Albic horizon, *23*
Alfisol, 12, 13, *192*
Alkaline, *56*, 57, 58, 59, 60, 182
Allitic weathering, 1, 11, 14, 45, 63, 72, 73, 185
Allophane, 1, 18, 34
Allophanic horizon, 18, *19*, *20*, 21
—weathering, 1, 3, 10, 11
Alluvial, *36*, *37*, *38*, 40, 41, 47, 48, 49, 50, 61, 69, 71, 72, 73, 180, 181, 183, 184, 192
Alpine, 37, 38, 53, 54, 59, 74
Analytical data, 17, 18
Andept, 34, 35
Andine, 53, 54
Ando, 10, 11, 34, 38, 45, 47, 67, 68, 69, 70, 73, 74, 75, 179, *180*, 181, 182, 183, 184, 185, 191
Anmoor, 61
Antibiotics of the rhizosphere, 177, *179*, 180, 181
Aquents, 192
Aquods, 192
Aquolls, 192
Arctic brown, *38*, 39, 66, 191
Arenoferral, 47, 48, 49, 50
Argialbolls, 192
Argiaquolls, 192
Argiborols, 192
Argillic horizon, *12*, 13, *191*
Argiudols, 192
Argixerols, 192
Arid cinnamonic regions, 69, *70*, 75, *184*
Aridisols, *191*
Ashy horizon, 1, 11, 12, 13, 14, *21*, 52, 53
Associations, soils, 63
Atlantic podsol, 66, *67*, 182, 183
Azonal, 37

B horizon, 25
Base reserve, *18*
—status, *19*, *23*, 46
Bed rock, 24
Biologic migration, 7, 17

Black cotton, *see* dark clays
—gley horizon, *23*
—horizon, *23*
—tropical, *see* dark clays
Bleached horizon, *23*
—humic horizon, *24*
Bleaching, *13*
Blown sands, *see* desert sands
Braunerde, *38*, 40, 41, 74, 181, 192
Braunification, *11*
Braunified horizon, *23*, *24*, 38, 39, 191
Broad-leaf forest, 65, *68*
Brown, arid, 4, *37*, 180, 181, 191
—chernozemic, *41*, 42, 43
—clays, 36, 37
—earth, 40, 41, 57, 58
—forest, 33, 34, 39
—horizon, *23*
—podsolic, 47, 48, 49, 50, *53*, 54, 56, 57, 58, 74
—recent, *36*, *37*, 38, 39, 56, 57, 180, 181
Brun acide, *38*, 39, 61, 74, 181, 192
—calcaire, 33, 34
Brunisolic, 11, *38*, *181*, 183, 184
—regions *68*, 75, 76, 192
Brunizem, *see* prairie
Brun lessivé, *51*, 52

C horizon, 25
Calcareous accumulation horizon, *22*
—chernozemic, 41, 42
—crust, *24*
—hardpan, *see* petrocalcic horizon
—horizon, 5, 9, 10, *23*, 25, 41
—, lateritic, 47, 48, 49, 50
Calcic horizon, *191*
Calcification, 5
Calcimorphe, 33
Calciorthids, 192
Calcium, 4
Cambic horizon, *191*
Camborthids, 192
Capillary fringe, 10, 12, 59
Cat clay, *35*, 183
Catena, 63
C/clay ratio, *24*
Cation exchange capacity, 1, 13, 18, 25, 45
Cation exchange capacity/clay ration, *18*, 20
Chernozem, sensu stricto, *41*
Chernozemic, 14, 34, *41*, 46, 47, 60, 74, 75, *181*, 184, 192
—braunerde, 38, 39
—ando, 34, 35, 36, 74, 181, 192
—clays, *35*, 36, 42, 43, 179, 180, 192
—ranker, *37*, *181*
—rendzina, *33*, 36, 42, 43, 56, 57

[1] The index also serves as vocabulary. Numbers in italics refer to the page in which the definition of the term is given, or the process is described.

Chernozemic *(continued)*
—regions, 63, 70, 76, *184*
—solonetz, *56*, 57
Chestnut, *see* kastonozem
Chloride saline, *55*, 56
Cinnamon(ic), 11, 38, *39*, 42, 45, 56, 57, 74, 75, 181, 192
—clays, 35, 36, *40*, 41
—horizon, *24, 40*
—para-rendzina, *33*
—regions, 63, *69*, 71, 76, *183*, 184, 185
—serozem, *33*, 40, 41
Classification of soils, 27, *187*, 191
—, artificial, 27
—, natural, 27
—, rigidity, 27, 187
Clay eluviation, 3, 4, *7, 12, 13*, 45, 56, 59
—mineralogy, *see* mineralogy
—, soil texture, 25
—skins, 8
—, synthesis, *1*
Claypan planosol, 56, 57, 58
Climate, influence on soil form, 1, 2, 3, 4, 5, 7, 9, 10, 11, 13, *14*, 41, 52, 53, 63, 70, 191
C/N ratio, 4, 7, 13
Coconut sands, 72, 180, 181
Colour, *23*
Concretionary, 40, 45, *46*, 192
—horizon, *11*, 12, *24*, 45, 72, 73
Coniferous forest, 4, 7, 13, 14, 52, 53, 65, *66*, 74, 182
Continental podsol, 66, *67*, 183
Correlation of soils, 33, 34, 35, 36, 37, 38, 39, 40, 41, 42, 43, 47, 51, 53, 55, 56, 58, 59, 60, 61, 191
Cropping, influence on soil, 4, 14, 181, 182, 183, 184, 192
Crust, 10, 33, 34
Cryochrepts, 192
Cryumbrepts, 192
Cuirassé, *46*, 47, 192

Dark clays, 34, *35*, 37, 74, 179, 180, 191
—gley horizon, *23*
—humic horizon, *24*, 39, 41, 191
Decalcification, *5*, 10, 17, 18
Definitions, 28
Degraded chernozem, *38, 39*
Density, apparent, 18, 25, 34
Depth, 14, 45
Derno-podsol, *see* sod-podsol
Description, soil, 29
Desert, 4, 33, 34, 36, 40, 41, 57, 63, 71, 72, 76, *184*
—clay plains, *71*, 72
—crust, *24*, 33, 71, 72
—pavement, *24*, 40, 41, 71, 72
—sands, *36*, 40, 41, 191
—table-lands, 71, 72
Diagnostic horizon, *19*, 191
Diagnostics, *17, see also* each soil groups and subgroups
Drainage, and soil formation, 4, 8, 10, *15*, 56, 72
Dune, 36, 37, 71, 72, 180, 181, 182, 191
Durandepts, 192
Duraquods, 14
Duripan, 9, 10, *24*, *191*
Dust, 10
Dystrandepts, 192
Dystrochrepts, 192
Dystrophic, 46, *47*, 181, 182
—horizon, 19, 22, *23*

Eluvation, *5, 21*
Eluvial horizon, *21*, 25, 45, 46, 56

Entisols, 37, 191
Erg, 71, 72
Erosion, 10, 71, 74, 179
Eucalyptus, 13, 14, 76
Eutrandepts, 192
Eutroboralfs, 192
Eutrochrepts, 192
Eutrophic, 22, *46*, 181, 192
—horizon, 19, *23*
Eutrorthox, 192
Eutrustox, 192
Experimental pedology, 2, 3, 8
Exsudational soil moisture regime, *10*, 12, 55

Ferralitic, *47*, 48, 49, 50, 181, 192
Ferralsol, 48, 49, 50
Ferrisol, 48, 49, 50, 181
Ferruginous, 2, *11*, 17, 19, 21, 34, 38, 39, 40, 41, 45, 46, 52
Fertility, soil, 177, *183*
—, actual, *183*
—, mineral, *184*
—, potential, *183*, 192
Fine clay, 24
—, loamy, 24
Flooding, 10
Fluvents, 191
Forest, 4
—, rendzina, *33*
Formation, soil, 1
Fragiaquods, 192
Fragihumods, 192
Fragipan, 24, 52, 53, 61
—, podsol(ic), *50*, 51, 52, *53*, 54, 61, 192
Free iron, 1, 11, 25
Fulvic acid, 3, 4, 7, 9, 11, 12, 13, 23, 25

Giant podsol, 53, 68, 75
Gibbsiaquox, 192
Gibbsihumox, 192
Gibbsiorthox, 192
Gleisol(ic), 14, 47, 48, 49, 50, 59, *183*, 191
Gleization, *12, 14*
Gley, *60*
—horizon, *23*, 52, 53, 191
—podsol, 53, 54, *60*, 61, 66, 67, *68*, 183
Grassland, 4, 7, 14, 41
—, ranker, *41*, 70, 71, 184
Gray-brown podsolic, 7, 13, *50*, 51, 52, 57, 58, 60, 61, *68*, 181, 182, 183
Gray calcareous 33, 34
Gray wooded, *50*, 51, 52, 75, 181, 182, 192
Ground-water laterite, 47, 48, 49, 50, *51*, 52, 60, 61, 182, 192
—podsol, 9, 14, 53, 54
Growth index, 3
Grumusol, 35
Gypsic horizon, 10, *23*
Gypsisols, *55*, 59, 60
Gypsum saline, 55

Halaquepts, 192
Hamada, 71
Haplargids, 192
Haploxeralfs, 192
Hapludults, 192
Haplumberts, 192
Haplustalfs, 192
Hardpan, 57, 58
Heath, 7, 13, 52, 53
Histic epipedon, *191*

Humaquepts, 192
Humic acids, 3, 4
—gley, *60*, 61, 182, 192
—horizon, *24*, 34
Humification, *3*
Humods, 192
Humogenic index, 3
Humolytic index, 2, 3
Humults, 192
Humus, *see* organic matter
—, eluviation, 9, 13, *13*
—, podsol, 13, 14, *53*, 54, 66, 182, 192
—, podsolic B, 9, 13, 14, 21, 52, 53
—, type, 1, 3, 7, 13, *24*, 25, 38, 40, 42, 50, 52, 53, 63
Hydrandepts, 192
Hydrol, *34*, 35, 56, 192
Hydromorphic, 47, 48, 49, 50, *55*, 56, 57, 60, 61

Illitic horizon, 19, *20*, 21, 50
Illuvial horizon, *21*, 25
Illuviation, 5, *21*
Inceptisols, *192*
Individual areas of soil mapping, 187, 188, 189, 190
Iron eluviation, 9, 11, *12*, 13
—pan, *10*, 12, *24*, 53, 54
—podsol, 13, 14, *53*, 54, 192
—, podsolic B, 9, 13, *22*, 52, 53
Irrigation, *10*

Jeltozem, 122, 155, 156, 166, 184

Kaolinitic, 2, 11, 46, 47, 56, 57, 60, 61
—horizon, *19*, 20, 21, 45, 46, 191
—, podsol(ic), 13, 46, 50, *51*, 53, 60, 182
—regions 63, *72*, *76*, *184*, 185
Kaolisol, *45*, 46, 179, 181, 191
Kastanozem, 41, 42, 43, 46, 47
Korichnevie, *see* cinnamon
Krasnozem, 4, 47, 48, 49, 50

Land type, *188*
Lateral drainage, *10*, 73, 74
Laterite, *12*, *24*, 45, 46, 50, 72, 73
Lateritic, 4, 45, 47, 48, 49, 50, 52, 60, 61
—, podsolic, 47, 48, 49, 50, *51*, 52, 60, 192
—, red earth, 47, 48, 49, 50
Latosol(ic), 34, 37, 45, 46, 47, 48, 49, 50, 52, 57, 58
Leaching, 1, 2, *5*, 13, 50, 56
—, degree, *17*, 18, 45
—, rainfall, *5*, 6, 12, 14, 15, 45
Lessivage, 3, 4, 12
Lessivé, 12, 34, 45, *50*, 51, 60
—, slightly, horizon, *21*
Light red horizon, *23*
Lithosol, *36*, 37, 38, 40, 41, 47, 48, 49, 50, 180, 181
Loamy sand, *24*
Low humic gley, 14, *60*, 61, 192

Magnesic, 8, *56*, 57
—horizon, *23*, 56
Mapping, 187, 188
Margallitic, 34
Marmorized, 60, 61
Meadow, 41, 53, 54, 56, 57
Mediterranean cinnamonic regions, *69*, *183*
Mineral deficiencies, 181, 187, 188
Mineralogy of clay, 1, 2, 13, *18*, *19*, 45, 47, 52, 53
—of non-clay, *18*, 45
Mineral reserve, 3, *18*

Moder, 4
Moisture regime, *10*, 12, 15
—tension, 18, 19, 21, 25, 34,
Mollic epipedon, *191*
Mollisol, 42, *192*
Moor, 4, 7, *13*, 13, 53, 54, 59
Mountains, 63, 73, 74, *185*
—, dry, *74*, 76, 185
—, dry mediterranean, *75*
—, high latitude humid, *74*, 76, 185
—, mediterranean, *74*, 76, 183, 184, 185
—, tropical, *75*, 185
Mull, 4, 7

Nano-podsol, *53*, 61, 66, 74
Natralbolls, 192
Natraqualfs, 192
Natraquolls, 192
Natrargids, 192
Natriboralfs, 192
Natriborols, 192
Natric, 35
—horizon, *23*, 56, 191
Natrixeralfs, 192
Natrixerols, 192
Natrudalfs, 192
Natrustalfs, 192
Natrustolls, 192
Neutral horizon, *19*, 23
Nitrogen, 4, 183
Nodules, 9, 14
Nomenclature, 25, *29*, 30, 31, 77
Non-calcic brown, 40, 41, 47, 48, 49, *50*, 51, 52, 181, 182

Ochric epipedon, 191
Ochrosol, 47, 48, 49, 50
Ocre podsolic, 53
Organic, 33, 34, 55, *58*, 60, 61, *182*
—horizon, 19, *24*, 191
—matter, *3*, *24*, 181, 183, 184
—pan, *10*, *24*, 52, 53
Orthods, 192
Oxic horizon, *191*
Oxisols, *192*
Oxysol, 47, 48, 49, 50, 52

Paddy, 59, 180
Palaeo-kaolisol, 45, 46, *47*, 48, 49, 50
Paleo-kaolinitic, regions *76*, *185*
Paleo-lateritic, 45, *51*
Pale cinnamon, 40, 41
Paleargids, 192
Pallid zone, 12
Pans, 7, 10, 14
Para-chernozems, *41*, 42, 43, 56, 57, 58, 182
Para-rendzina, *33*, 36, 179
Para-serozem, *37*, 180, 181
Parent material, 1, 2, 11, 13, 14, *15*, 52, 184
Peat(y), 4, 13, 52, 53, 55, 59, 66
Peaty horizon, 19, *24*, 191
Permafrost, 9, 14, 37, 53, 66, *68*, 183
Petrocalcic, 5, 9, *24*, 41, 52, 53, *191*
Phase, soil, 187
Planosol(ic), 8, 9, 11, 12, 23, 34, 41, 42, 45, *56*, 60, 61, 68, 71, 72, *182*, 192
Plinthaquepts, 192
Plinthaquoxs, 192
Plinthaquults, 192
Plinthudults, 192

Plinthustalfs, 192
Plinthustults, 192
Plinthoxeralfs, 192
Podsol, 1, 7, 10, 12, 51, *52*, 58, *182*, 192
—, humo-cendreux, 13, 52, 53
Podsolic, 12, 13, 45, *50*, 57, 58, *181*, 192
—regions, 63, *63*, 76, *183*, 185
—weathering, *1*, *2*, 3, 4, 12, *13*, 52, 53, 182, 185
Podsolization, 8, 10, *12*, 14, 17, 38, 52, 53, 63, 64, 74
Podsolized chernozem, 42, 43
Podsol ranker, 13, 52, *53*, 181
Polygenesis, 15, 45
Polygonal, 36, 37, 66, 74
Positive, *34*
—horizon, *21*
Postpodsol, 53, 54
Prairie, 37, 38, 39, 41, 42, 43, *51*, 52, 57, 58, 74, 181, 182
Prepodsol, 51, 52
Proto-rendzina, *33*, 34, 38, 66, 74
Psamments, 191
Pseudogley, 51, 52, 53, *60*, 61
—horizon, *23*

Ranker clays, 35, 192
Rankers, 37, 74, *181*, 192
Raw soils, *36*, *180*, *181*, 191
Red-brown earth, 57
Red cinnamon, 40, 41
Reddish brown, 33, 34, 36, 39, *40*, 41, 47, 48, 49, 50, 51, 52, 57, 58, 70, 74, 75
—para-chernozem, 40, *41*
—desert, *40*, 56, 57, 58, 192
Red(dish) prairie, *40*, 42, 54
Red earth, 52
—horizon, *23*
—, podsolic, 40, 41, 47, 48, 49, 50, 52
Red-yellow mediterranean, 47, 48, 49, 50
—podsolic, 7, 47, 48, 49, 50, *51*, 52, 73, 192
Regosolic, 37, 38, 39, 47
Regur, 35, 36
Rendoll, 33, 192
Rendzina, *32*, 36, 37, 39, 42, 43, 60, 61, *179*, 192
—clays, 33, 34, *35*, 179, 192
—rankers, *33*, 34
—solonetz, 33, 34, *56*
Rhodoxeralfs, 192
Roof clays, *35*
Rubification, *11*, 38, 39, 63, 69, 71, 72, 74, 75
Rubified horizon, *24*, 191
Rubrozem, 39, 40, 47, 48, 49, 50, 75, 185

Salic horizon, 10, *23*, 25
Saline, *55*, 59, 60, *182*, 192
—, alkaline, *55*
—, chernozemic, *55*
—clays, 35, 36, *55*, 56
—rendzina, *55*, 56, 57
—solonetz, *55*, 61
Salorthids, 192
S/clay ratio, *19*, 22, 45
Sedimentation volume, 21, 25
Self mulching, 35
Semi-gley horizon, *23*, 52, 53
Series, soil, *187*, *189*, *190*
Serozem, *33*, 34, 56, 57, 192
Seventh American Approximation, 12, 14, 17, 19, 29, 33, 35, 36, 38, 39, 40, 41, 42, 47, 51, 52, 53, 54, 55, 56, 57, 59, 60, 61, 191
Siallitic weathering, 1, 11, 14

Silty loamy, *24*
SiO_2/Fe_2O_3 ratio, 8
Skeletal, *24*, 33, 34, 36, 37, 39
Slightly lessivé horizon, *21*, 191
Slope change, 12
Soda solonchack, 55
Sod podsol(ic), 14, *50*, *53*, 66, 67, *68*, 182, 183
Soil distribution, 63
—regions, *63*, 64, 65, 66
—relations, of plants, *177*
Solod, 7, 12, 56, 57, 58
Solodized, 57, 58
—solonetz, 56, 57
Solonchack, *see* saline
Solonetz, 7, 8, 9, 10, 34, 51, 52, *56*, 59, 60, 61, *182*, 192
Solonization, 7, 8, 14, 59
Spodic horizon, 1, 12, *20*, 52, 53, 191
Spodosol, *53*, 192
Stagnogley, 60, 61
Stone line, 11
Structure, 25
Subglacial desert, 65, *66*
Subtropical kaolinitic regions, *73*
Sulphate saline, *55*
Super kaolinitic horizon, *19*, 45, 46, 20, 191
Survey, soil, *187*
Swamp, acid, 59
Symbols, *19*, 30, 31, *32*, 66

Takyr, 71, 72 .
Tca/clay ratio, *18*, *19*, 20
Temperature, soil, 192
Terra rossa, 39, *40*, 41, 47, 48, 49, 50, 55, 56
—roxa, 34, *47*, 48, 49, 50, 181, 184
Terre de barre, *47*, 48, 49, 50, 51, 181, 182
Texture, *24*
Textural horizon, *21*, 50, 52, 53, 56, 191
Time, *see* age
Tir, 33, 34, 36, 69, 70
Tropical, 7, 11, 12, 14, *45*
—, ferruginous, 40, 41, 46, 47, 48, 49, 50, 51, 181, 192
Tropudalfs, 192
Tundra, 9, 37, 38, 52, 53, 54, 60, 65, *66*, 74
Turf podsol(ic), *see* sod podsol(ic)
Type, soil, 187

Ultisol, 47, *192*
Umbric epipedon, *191*
Ustalfs, 192

Vegetation, 2, 4, 7, 13, 50, 55, 63, 64, 65, 74
Vertisol, *35*, *191*
Vertisolic horizon, *20*, 21
Very fine clay, 24, 25
Volcanic, 1, 7, 8, 10, 11, 34, *see* ando

Water, soil as reservoir of, 177
Waterlogging, 4, 9, 10, 12, 13, 14, 52, 53
Water movement, 5
—table, *10*, 12, 14, 59
Weatherability, 3, 7, 9, 11, 12, 15, *18*
Weathering, *1*, 11, 191
—degree, *18*, 45
—types, *1*

Xeralfs, 192
Xero-ferrisol, 47, 48, 49, 50

Yellow horizon, *23*
—podsol(ic), 52, 53, 66, 67, 68
Young kaolinitic regions, *73*